C000071826

VEER ECOLOGY

VEER ECOLOGY

A COMPANION FOR ENVIRONMENTAL THINKING

JEFFREY JEROME COHEN AND
LOWELL DUCKERT *Editors*

Foreword by Cheryll Glotfelty
Afterword by Nicholas Royle

UNIVERSITY OF MINNESOTA PRESS
MINNEAPOLIS • LONDON

Poetry in "Obsolesce" was originally published as Brenda Hillman, "Phone Booth," in *Practical Water* (Middletown, Conn.: Wesleyan University Press, 2009), 17. Copyright 2011 by Brenda Hillman. Reprinted with permission of Wesleyan University Press.

Poetry in "Sediment" was originally published as Mark Nowak, "Procedure," in *Coal Mountain Elementary* (Minneapolis: Coffee House Press, 2009), 150; copyright Mark Nowak 2009; and as Rita Wong, from *undercurrent*, blewointment, 2015; copyright Rita Wong 2015.

Poetry in "Try" was originally published as Norman Jordan, "A Go Green Flash Dream," in *Where Do People in Dreams Come From? And Other Poems* (Ansted, W.V.: Museum Press, 2009). Reprinted with permission.

Poetry in "Rain" was originally published as Edward Thomas, "Rain," in *Collected Poems* (Highgreen: Bloodaxe Books, 2011).

Copyright 2017 by the Regents of the University of Minnesota

All rights reserved. No part of this publication may be reproduced, stored in a retrieval system, or transmitted, in any form or by any means, electronic, mechanical, photocopying, recording, or otherwise, without the prior written permission of the publisher.

Published by the University of Minnesota Press
111 Third Avenue South, Suite 290
Minneapolis, MN 55401-2520
http://www.upress.umn.edu

Printed in the United States of America on acid-free paper

The University of Minnesota is an equal-opportunity educator and employer.

22 21 20 19 18 17   10 9 8 7 6 5 4 3 2 1

Library of Congress Cataloging-in-Publication Data
Names: Cohen, Jeffrey Jerome, editor. | Duckert, Lowell, editor.
Title: Veer ecology : a companion for environmental thinking / Jeffrey Jerome Cohen and Lowell Duckert, editors ; foreword by Cheryll Glotfelty ; afterword by Nicholas Royle.
Description: Minneapolis : University of Minnesota Press, 2017. |
Includes bibliographical references and index. |
Identifiers: LCCN 2017009420 | ISBN 978-1-5179-0076-2 (hc) | ISBN 978-1-5179-0077-9 (pb)
Subjects: LCSH: Nature—Effect of human beings on. | Environmental protection. | Environmental degradation. | Global environmental change.
Classification: LCC GF75 .V45 2017 |
DDC 304.2—dc23
LC record available at https://lccn.loc.gov/2017009420

Contents

Foreword

CHERYLL GLOTFELTY

If I ask you to brainstorm verbs that we commonly associate with the environmental movement, you might come up with *reduce, recycle, reuse, conserve, preserve, protect, save, clean up, bike, garden, regulate, legislate,* and *restore.* I would argue that these actions are still necessary but no longer sufficient. Most of these words describe work we can do to help the environment, but few of them tell us how to work on ourselves in a time of environmental upheaval. Taking a cue from Nicholas Royle's recent book *Veering: A Theory of Literature* (2011), let us unpack the noun *environment* to discover the embedded French root, *virer,* an action verb that means "to turn." As Royle proposes, life and literature abound in sudden veers, unexpected turns that point us in new directions. Indeed, the environment itself veers from time to time—meteor strikes, volcanic eruptions, earthquakes, ice ages, mutations, and extinctions. In the age of the Anthropocene, climate veers, icebergs calve, sea level rises, alien species invade, and habitats transform into novel ecosystems. Given that the pace of environmental change appears to be accelerating, we need a new set of verbs that will help us think—and perhaps act—outside the box. What are the words that will help us to conceptually veer along with the environmental flux of which we are a part? What verb might *you* suggest?[1]

Jeffrey Cohen and Lowell Duckert asked nearly thirty brilliant thinkers to each propose one "vital term to think with for ecological and environmental theory." As they explained to their contributors, "Ecotheory is an emergent field that makes use of recent advances in philosophy, psychology, sociology, literature, sustainability studies and cultural studies to deepen our understanding of how we can better frame our ethical, historical and cognitive relations to the world, especially at a time of anthropogenic climate and resource crises." Cohen and Duckert requested verbs, because they want the collection to veer—that is, to send us in new directions and propel change. In polite but pointed email messages to contributors during the composition process, the editors warned their authors that "your verb will want to become a noun, so that it freezes into a concept rather than transports unexpectedly: watch out for that peril, and veer rather than stabilize. Other than that, feel free to be as creative and provocative as you wish with your essay."

The authors stepped up to the challenge. The verbs featured in this collection are like the springboard that gymnasts jump on to launch themselves into the air to perform amazing feats on the vault. Some verbs are closely tied to natural processes—*compost, saturate, seep, rain, shade, sediment, vegetate, environ.* Others pertain to human affect and action—*love, represent, behold, wait, try, attune, play, remember, decorate, tend, hope.* Many are vaguely unsettling—*drown, unmoor, obsolesce, power down, haunt.* And still others are enigmatic or counterintuitive—*curl, globalize, commodify, ape, whirl.* These writers exhibit dazzling acrobatics of the mind, performing intellectual leaps, philosophical flips, epistemological twists, ontological somersaults, ethical reaches, and stylistic flourishes that ultimately return and connect them—and the reader—to Earth. The performance ends with a heightened sense of possibility, our imaginations awakened by "an invitation to think the world anew" (Oppermann). Even though you will be made freshly aware of the daunting problems plaguing our biosphere, I think you'll find this book exhilarating.

While I was reading these essays I became hyperaware of verbs and began obsessively circling them. One slightly unexpected verb that threads its way through the collection is *entangle,* along with the noun *entanglement.* Although the essays are strikingly diverse in subject and style, there does appear to be a paradigm emerging like a blurry bird seen through binoculars, sharpening into focus to become a pinyon jay (to give a nod to a raucous bird where I live). *Entangle* describes human relations with our feathered, furry, finny, scaly, vegetal, and microbial kin. *Veer Ecology* decenters the human, trouncing anthropocentrism and exploring the possibilities of an ecocentric perspective. *Veer* reimagines people as materially embodied, ecologically embedded beings with the capacity to enter into reciprocal relationships with nonhuman persons. If neoliberal capitalism is the context, and change is the condition, the ethic espoused by these essays entails "respect . . . relationship . . . responsibility" toward our ecological family, to cite LeMenager quoting First Nations scholar Leanne Simpson's idea of resurgence.

Veer Ecology completes a trilogy of recent ecotheory anthologies pioneered by Jeffrey Cohen, joined by Lowell Duckert, and published by the University of Minnesota Press. *Prismatic Ecology: Ecotheory beyond Green* (2013), edited by Cohen, observes how the color green dominates our thinking about ecology, connoting balance, sustainability, and the natural, all notions that we are now questioning in light of serial catastrophes and "discordant harmonies," as Daniel Botkin's important book on ecological discontinuities is entitled. Breaking the green grip, *Prismatic Ecology* is organized by a full spectrum of colors, inviting thought experiments on what associations hues such as pink, orange, chartreuse, beige, ultraviolet, and gray might have for ecological thinking. *Elemental Ecocriticism: Thinking with Earth, Air, Water, and Fire* (2015), coedited by Cohen and Duckert, adds momentum to the material turn in ecotheory, regarding the elements as fellow actors rather than props or backdrop in the unfolding drama of life on Earth—or, perhaps better put, life *in* and *with* Earth. While *Veer Ecology* explicitly signals the jolt of veering out of timeworn cognitive pathways, *Prismatic Ecology* and *Elemental Ecocriticism* likewise wrench the reader out of mental ruts. The constellation of writers and thinkers brought together in these three volumes includes recognized luminaries and rising stars. If nothing else, like rainmakers who disperse silver iodide crystals to seed the clouds, Cohen and Duckert, by providing nuclei for these brilliant minds to formulate thoughts around and by then disseminating their essays in widely available form, precipitate fruitful natureculture conversations. Their work affirms the ecocultural value of the verb *anthologize*.

Despite *Veer Ecology*'s focus on verbs, reading this book is less likely to propel one into environmental action than to stimulate ecological reflection, with ecology here understood to include the human and the *noosphere*, Teilhard de Chardin's word for the sphere of human thought. In a sense, this collection demands, "Don't just do something, sit there," to quote the title of a recent book on meditation and mindfulness. Of course, it could be argued that *to contemplate* is an active verb: Thinking, it's what we *do*. In any case, I don't regard *Veer*'s spur to sit as a shortcoming but, rather, a strength. Just as we should do good work in the world, we also need to take care of ourselves. To be well—and who does not want to be well?—we cannot be daily assaulted by bad news without developing cognitive tools for coping. *Veer Ecology* is a mental toolkit that can help us gain perspective and become wise. This ecotheory companion prompts awareness, insight, play, and contemplation. And these verb-ignited essays stand to enrich our experience of living with an alarmingly dynamic planet.

Note

1. For what it's worth, my verb would be *reinhabit,* as Peter Berg defines it. See Cheryll Glotfelty and Eve Quesnel, eds., *The Biosphere and the Bioregion: Essential Writings of Peter Berg* (London: Routledge, 2015).

Introduction

Welcome to the Whirled

JEFFREY JEROME COHEN AND
LOWELL DUCKERT

We call this book a *companion* in the hope of offering to its readers a ready partner and congenial fellow traveler, a vade mecum for fostering ecological attentiveness and encouraging further wandering. Through the transports of environmentally inclined verbs familiar and unexpected, this collaborative project aims not to provide encyclopedic overviews or definitive accounts of critical concepts (*all* concepts are critical) but to forge a welcoming and heterogeneous fellowship, a colloquy for pondering possibilities for environmental thinking, ecological theory, and engaged humanities practice during a time of widespread crisis. Imagining futures by rethinking possibilities present and past, *Veer Ecology* extends to its readers an invitation to shared endeavor. Across historical periods, geographies, archives, and fields of expertise, its authors attempt disanthropocentric modes of apprehending agency and urgency. We strive to multiply points of view, to harness the ability of language to convey cognition and affect beyond the small orbit of the human.

Not every contributor to this volume self-identifies as an expert in the environmental humanities. Our hope is that *Veer Ecology* might make evident that we (a first-person pronoun meant to include you, the reader) are always thinking environmentally, but our modes of engagement could be intensified through better recognition of collective precarity

1

and unlooked-for but wide companionship—even within the modest and fugitive shelters that projects like book building offer.

Ecocriticism is a lively confluence of ecology, philosophy, anthropology, sociology, literature, feminism, sustainability studies, environmental justice (especially within indigenous and postcolonial studies), queer theory, and numerous adjacent fields that seek to deepen our understanding of the intimacy of humans and nonhumans. Striving to better frame ethical, historical, and cognitive relations to the world, especially at a time of anthropogenic climate change, ecotheory ranges across the environmental humanities, green studies, social activism, and the new materialisms (including material feminism, object studies, and vibrant materialism).[1] Literature, history, and the arts bring to environmental science a long and spirited conversation about the relation of human activity (intellectual and industrial) to a world that exceeds anthropomorphic capture. Working against the concretizing tendency of a research guide or definitive overview, this book traces environment in motion, as an arcing verb, as *veer*. Thinking ecologically is after all a ceaseless spur and a doing, a way of apprehending from the thick of things, not the cementing of an extant body of knowledge into perduring form or a sedate collation of facts to be glimpsed from some exterior point of view.[2] Our title's relation to queer ecological studies is more than homophonic, since queer ecology so well articulates the transportive power of desire, the challenges of overlapping intimacies, and nonnormative trajectories of thriving.[3] An actively contemplative response to contemporary and historical states of emergency, *Veer Ecology* attempts to complicate understandings of human entanglement within a never-separable nature, emphasizing a material enmeshment perceived long before the Anthropocene arrived.[4] This collaboration therefore aims for catalysis rather than mastery, incitement rather than codification. Because of its relation to environmental activism, ecocriticism cannot be divorced from multimodal forms of protest and attempts at social change (including writing), fostering unforeseen alliances as a form of challenge.

Veer Ecology emphasizes through its title the etymology of the noun *environment* in deviation and spiral. The French verb *virer* means "to turn." This book responds to an intensified interest within the environmental humanities in directionality. The "animal turn," "material turn," "geologic turn," and "hydrological turn" designate an array of incisive investigations into how the ecological works: it spins. Our endeavor therefore takes the ecological turn quite literally. Far from merely environing the human in anthropocentric ways—Michel Serres's worry about the term *environment*—*Veer Ecology* acknowledges a world of inhuman forces, dynamic matter, and story-filled life that inevitably go off course.[5] They act, they drift, they swerve, and resist. In diverging from human domination, they disrupt secure dwelling in ways that are catastrophic, pleasurable, orbit changing. Besides a swift change of subject or direction, *veer* describes wind's swirling motion. Though not the world's only sudden element, air well conveys the dynamism embedded in *veer*, the propensity it designates to circle back, to whirl as a vortex.[6] As the intensity and frequency of superstorms have made evident, climate does not conform to a bounded system. Affect and atmosphere at once, meteorological and bodily, a shifter of scale and breaker of partition, climate is not to be encompassed or controlled. *Veer Ecology* stresses the forceful potential of inquiry, weather, biomes, apprehensions, and desires to swerve and sheer—with unevenly distributed and insistently material impacts. This companionate project is therefore not a compass, not a closed system of neatly arranged points to orient readers. Each word is a spur to more turns. Veering enables ontological, epistemological, and ethical positions to curl, converge, converse.

We invite you to accompany us along some spiraling trajectories, topographies in motion. We enumerate five kinetic possibilities that inhere within this volume's essays, but our list is incomplete and alternative tracks manifold. They await your deviations. *Welcome to the whirled.*

Verb (to spur motion)

Welcome is a passionate imperative: an opening, not a capture; an energetic interruption, seldom a habitual state. We address its injunction toward ourselves as a reminder of this project's aspirations, but we hope that you will accompany us, for a while, beneath its shelter. Bring whatever gear you suspect will assist in creating harbors and havens, in launching shared adventure. The ecology that *welcome* opens is a house awhirl, a moving castle, a domain in disarray. Dwelling in such a mess is often uncomfortable. We seldom seem to finish constructing these spaces of refuge, these hearths for warmth and story, before we find ourselves in company too boisterous for even the most capacious limit.[7] It is as if the unfinished Tower of Babel were being inhabited as spiraled encyclopedia or library, a perpetual gathering and emission device for soundings and new languages, loquacious reverberation, polyglot and heterogeneous collectivity. Its future is ruin, of course, but also the generation of new languages that will forever strive to connect. Art and activism are works of translation.

House is a humane verb. Although at their secret interiors nouns are words in motion, they have a habit of obscuring the eventuation of the world, its ongoingness. Ecology is a doing, emergence more than structure, housemaking more than household.[8] The cleft of definition has a way of too securely stabilizing the dichotomies it founds: noun/verb, language/world, stasis/mobility, book/stone, home/wild.[9] Segregations are imposed with lasting and unevenly felt costs.[10] Motion only appears to cease. If a dictionary is a house of letters, then its *oikos* must be a restless one, never perfectible. What might more open books welcome? Thick with ecological possibility and narrative drift, words move, speak the world, convey whirl.[11] No alphabet can still a vibrant lexicon for long. Ecology and every word it houses attunes us to verbose multidirectionality, the unmooring of terms: a veercabulary, never monoglot or merely reiterative, a tongue-twisting surge of disanthropocentric energy to challenge human soliloquizing. *Tend, attend, tender*: word-life thrums with wildlife, with world-life.[12] To verb is to find the motion in the noun, the play in the preposition, the transport of the metaphor, the intensification of the adverb, the escalation of the adjective, the doing of the word.

Keywords unlock doors. This companion would not have been possible without foundational projects like Raymond Williams's *Keywords: A Vocabulary of Culture and Society,* a work that had the forethought to offer some blank final pages as part of its arrangement, a signal that lexical inquiry "remains open" and that "the author will welcome all amendments, corrections, and additions."[13] The capacious *Keywords for Environmental Studies* offers a collaborative terminological inventory for building a bracingly cross-disciplinary future for the environmental humanities.[14] The volume is arranged alphabetically, and its companion website offers pedagogical tips for reshuffling its contents in the classroom. Greg Garrard's indispensable *Ecocriticism* organizes the field into thematic strands, articulating parameters through key terms *(apocalypse, animals, pollution, wilderness, pastoral).*[15] The *Oxford Handbook of Ecocriticism,* which Garrard edited, offers a gregarious survey of the field's contours.[16] Every such project demands a foundational principle of order: arrangement by time period, discipline, alphabet, genre, topic. Yet attempting to gather the field into comprehensiveness risks obscuring its turbulence, a multivectored proliferativeness made clear by its burgeoning number of readers, handbooks, and research guides.[17] Rather than explicate terms so that we can ensure "that we get our lexical and conceptual bearings straight," rather than attempt to articulate lasting boundaries for ecocritical significations, this collection follows a rather different path, accompanying some ecologically rich verbs, companioning their trajectories, seeing where they lead as they perturb disciplines, boundaries, domains.[18] Against a humanities that too often becomes a war of the words, we hope for a shared ethics of veering, a turning *toward* and *with* that entails deep attunement to human and nonhuman thriving.

Companion (to accompany, even when difficult)

Keys are essential tools. They divulge. They explain. A companion does not necessarily unbolt anything and will not likely provide quick access to a storehouse of provisions or knowledge. Yet a companion may hold open some hospitable doors, invite conversation, wander with you along unexpected paths. *Companion* is a reliable verb.

You seem like you are having a lot of fun. So stated an audience member at a conference panel we arranged for the project that opened a door for this volume, *Elemental Ecocriticism: Thinking with Earth, Air, Water, and Fire.*[19] We were pleased that the critical conviviality propelling that collaboration was palpable—and we hope that a similar joy in working together is evident in this anthology. We also trust that happiness in shared venture does not obscure the seriousness or urgency of the themes contemplated here. Collaboration involves challenge, and the contributors to this book pushed us, repeatedly, to do better: refine our terms, embrace new ambits or disrupt old ones, contemplate possibilities and limits. The twenty-nine essays, foreword, and afterword arose from sustained dialogue among thirty-one contributors. Our invitation to *Veer Ecology* arrived in the mailbox of each participant with a list of suggested terms, only some of which were welcomed. Many chose their own verbs, or realized that a verb had already chosen them. Others changed their word in the process of composing an essay, so far had trajectories veered from origins. We cautioned our collaborators that their verbs would want to become nouns, freezing into concepts rather than transporting toward the unexpected. "Watch for that peril," we wrote, "and veer rather than stabilize—but other than that, feel free to be as creative and provocative as you wish." We never policed, but we did push, wonder with, and companion. We found that those who wrote with us brought *Veer Ecology* along paths we could not have predicted. Lines whirled into spirals, coils became rhizomes. The ethos we attempted to cultivate was one of intensification, a building together of fugitive havens for thoughts that might not thrive in solitude. In these days of narcissistic nationalisms, closed borders, gated communities, human-engineered ecological disaster, neoliberal resourcism, and proliferating hatreds, we are attempting to place a little more motion into concepts like home and haven. The essays offer a series of capacious hearths around which communities of humans and nonhumans might cluster to shade themselves or find a roof against the weather when it rains, maybe even to remember some sustaining stories. *Shelter* is a necessary verb.

The welcome we extend opens a door through which unexpected things will pass, including monsters. In an ancient but weirdly contemporary poem about fire, ecology, refuge, and entanglement, Grendel invades the hall of Heorot because its music—its foundational story and divine place setting—excludes him, leaves his family to roam a distant moor, to inhabit a sunken home. Example and warning, a creature not so different from the community that built its timber walls against him, Grendel smashes doors, benches, and tables. His havoc is unsettling, making a mess of what had been hierarchy and order. But he also brings the outside within, an unwanted change of climate to spur a community to think more deeply about imposed limits, to contemplate sustained violence, the drowning of those declared off a nation's maps. In retelling the medieval tale it is difficult not to behold the shape of economies and ecologies to come. So let us overturn the epistemological tables—or at least allow unexpected guests their seats. Admit that any shelter is likely to prove temporary. Enable the place by the fire to be capacious, community difficult, admittance and sustenance just.

The varied contributions to this book offer somewhat spontaneous definitions of ecology: not a perfected house that walls the project but a transhistorical hostel, a lively commons taking shape to gather humans, animals, plants, elements. A companion, Donna Haraway observes, is someone to break bread with *(cum panis),* a messmate.[20] We love tables that welcome and homes that invite. We also love the making that happens around the fire, the leavening and enlivening. This ecology or open house or mobile hearth is a space in which we experiment rather than merely consume, where we share story and song rather than arrange ourselves into a hierarchy of prearranged seating. Interspecies and eco-material, parasitical and hospitable, a commons as shared refuge includes the ingredients and the debris, airborne yeast and insects, the bread and its crumbs, the ants and the rats, a perturbed ecology in which to dwell. Intellectual, physical, material, and social energy propels these endeavors, threatens to exhaust, potentially disempowers. Welcome to the whirled, a crowded place where we power down, eat together, speak together, story together in gyred conviviality.

Spiral (to move forward by curving back)

Rotation occurs around an axis, a center that may wander. From inside the whirl it is difficult to know if motion is inward or outward, a loosening or a tightening. Things move apart and thereby touch. Ends become beginnings while contiguities proliferate. Roland Barthes once declared of spirals that inside their trajectories *"nothing is first yet everything is new,"* by which he also meant nothing is last and everything is already ancient.[21] *Time* is a complicated verb.

What a whirligig: forward and backward, here and elsewhere at once, a topography for perspective shift. To see the world from multiple viewpoints curves senses into motion. Propulsion can offer a falling behind, a sudden touching of history thought long surpassed.

Recycle, Repurpose, Restory (to rework a mantra)

Recycle. Three arrows fold on each other, their trajectory an eternal loop. Designed by college student Gary Anderson and intended to represent a Möbius strip, the universal recycling symbol is a corporately sponsored, public domain figure for how to dwell sustainably within a closed system. *Re-cycle.* In the Pacific Ocean a gyre of trash spins, aping that rotation a little too literally while challenging the assumption that environmental cycles can remain closed. The apposition of these two ecological circlings suggests that sustaining our current modes of existence is neither possible nor desirable.[22] The universal recycling symbol conveys the spin of a system in which everything supposedly remains inside. Recycling means using obsolescing things over and over with no waste, no exterior, like a nation that imagines itself behind a secure wall. A turbulent whirl of debris, the Pacific gyre spins with the actual vectors of waste and profligacy that propel contemporary capitalism, a whirlpool global in scale, open and lethal.[23] Love child of petroleum culture run amok, plastic is outsourced for recycling but keeps coming back, churned through border and bodily crossings.[24] As pellets and microtoxins, this waste inhabits the bodies of fish, seeps into human and animal bodies, litters shores. The refuse of industrial nations clings to lands only imagined as distant, the "away" to which unwanted things are "thrown."[25] Rocks, hills, even islands sediment from unwanted landfill, because some spaces are too full of discarded objects and substances, sent out of sight and attention, globalizing the local. The Pacific garbage patch is a spiral of slow-churning violence, the bending together of a series of transfer stations that aim to obscure the transits of waste, a "patch" that does not mend its harms but exposes its wounds, a "matterphor" for unwanted ecological intimacies.[26] *Recycle* too easily greenwashes, commodifies, obscures its motivating imperatives *consume* and *forget*. A mantra of enmeshment becomes a motto for malfeasance. We convince ourselves that our trash must surely be a treasure for others. So we send it to them: corporate outsourcing, the offloading of a heavy burden in

Figure 1. The universal recycling symbol.

the guise of a virtuous circle. Behind the closed circle of the recycling symbol is a maelstrom of accident and intention, a violent gathering of things, porous zones, forceful global currents, the agency of matter and elemental forces in drift. Matter is storied.[27] No system is closed. *Environ* is a troubling verb.

Repurpose. We are not against recycling. We cannot dismiss any practice that reduces ecological harm, that decreases environmental injustice, that assists the arrival of less toxic futures. Nor are we against being green. But the spectrum holds a diversity of hues, many of which human eyes cannot perceive. Every color is ecological, even the ones we cannot see.[28] The grounds from the coffee that fueled these sentences will soon be atop a compost pile, the paper bag that held the beans deposited in a bin for transport to a facility that will render its fibers into something else. But we do not suppose that because the bag and the "fair trade" grounds have been recycled that we are off any hooks. Because "we shape the world through living," we always want to know better the intimacy of small choices to larger networks of possibility and harm and to make better collective and individual choices.[29] Although well intentioned, the universal recycling symbol is a bounded system, a gated community. Too often we aspire to enact some version of its call to plenty within closure, striving toward an exclusive totality that to sustain its endless cycling imposes a high a price on "offsite" humans and nonhumans alike.[30] We want to open re-cycling systems to the gyres that underlay their motion. Matter may be transferred, transported, thrown to some unthought away, but matter does not disappear.

Neither does story. *Restore* might be repurposed to mean "reactivate and intensify story." Linear histories and crystalline origin myths anchor the world that we know in a world that we believe has always been. As shelters they are always too small. They delimit and justify exclusive community. Counternarratives make such stories spin: books of beginnings over a single book of Genesis, vegetating chaos over walled and perfect gardens. *Reverse* could mean decorate, magnify lyricism, unmoor aesthetic force from the merely human. Re-versing counters narrative forces of containment, opens meager and ungenerous homes to widened refuge. Skew story. Companion the plot twists. *Veer Ecology* gathers a historically diverse archive because we have a hunch that the re-story-ation of the world will be aided by a return to narratives and modes of thought

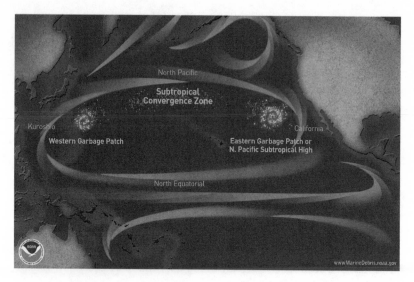

Figure 2. Marine debris accumulation locations in the North Pacific Ocean.
Courtesy of the National Oceanic and Atmospheric Administration's Marine
Debris Program. https://marinedebris.noaa.gov/.

carelessly relegated to the dust heap of the past.[31] This "heap" is actually
a teeming site for activating new possibilities, for reframing more capacious futures in a time of austerity, catastrophe, and the widespread
inflicting of harm. *Turn back to forge ahead.* We are not speaking about
retrieving all things lost, but renewing how we story, all together. *Haunt*
is distantly related to *home.*[32] Resilience is not a stiffening against but a
bending toward, a winding up with others. Rather than forget or abandon, we might try to slow down, engage, attend heavy weights and long
waits, contemplate more to act better. Carving out a space of communal
thought that does not have a predetermined outcome can in these times
of relentless productivity and assessment offer an act of ecotheoretical
resistance.

Veer (to anthologize unexpectedly)

Nothing in good order, everything in motion, weird, ardent, curving.
"Desire," Nicholas Royle writes, "is a veering thing."[33] Veering things are
in turn saturated with revolutionary desire. *Veer* is therefore the difficult
verb we have chosen to place not only in our book's title but here in lieu

of an ending, hoping it will convey an unsettling motion that inheres within all ecological thinking. To veer is to gather *(anthologize)* unlikely but passionate companions, and in that sudden community to hope.[34] To veer is to enlarge, to break closed circles into spirals, to collect for a while, to dwell in revolution. This collection of essays is meant to affirm an ongoing project.

Widened belonging is seldom comfortable. Our contributors render snug habitations strange, opening them to a world agentic and wide. The curving trajectories of *veer* do not abandon the past. *Reduce, reuse, recycle*: these are words for matter, words that matter. As imperatives to a less oppressive mode of dwelling we take them seriously, even as our collaborators find their kin in some less conventionally ecological verbs. We do not aspire to complete, transcend, or otherwise leave behind the academic and activist work that has laid the wide foundations for ecocriticsm. This companion would not have been possible without the challenges of queer theory, environmental justice, ecofeminism, indigenous studies, or any other interrogator of how limited the *human* in the humanities has too often proven. *Veer Ecology* collects beneath the fleeting refuge of its covers some shared labor, some provisional attempts to follow the ecological trajectories of linguistic organisms, especially as vectors of disanthropocentric story. To behold the summons and provocations of the worldly and nonhuman agencies that thrum within narrative requires the estranging of what has become too familiar, the widening of our house, its opening up or repurposing, sometimes its abandonment, always the building of wider sanctuaries, ecologies in wandering motion.

The contributions that follow offer an anthology of verbs that spiral and gather. Companion us. *Welcome to the whirled.*

Notes

1. Inspirational to us in this endeavor has been the work and support of Jane Bennett, whose sustained attention to matter's agency as both political and ethical inspired our first collaborations. See *The Enchantment of Modern Life: Attachments, Crossings, and Ethics* (Princeton: Princeton University Press, 2001); *Vibrant Matter: A Political Ecology of Things* (Durham: Duke University Press, 2010).

2. Donna Haraway famously describes this impossibly disembodied perspective as the "god trick" in "Situated Knowledges: The Science Question in

Feminism and the Privilege of Partial Perspective," *Feminist Studies* 14, no. 3 (1988): 575–99.

3. We are deeply indebted in the framing of this book to queer ecocritical work like Catriona Sandilands and Bruce Erickson, eds., *Queer Ecologies: Sex, Nature, Politics* (Bloomington: Indiana University Press, 2010); Stacy Alaimo, *Bodily Natures: Science, Environment, and the Material Self* (Bloomington: Indiana University Press, 2010); Mel Y. Chen, *Animacies: Biopolitics, Racial Mattering, and Queer Affect* (Durham: Duke University Press, 2012).

4. Influenced by the fondness of Isabelle Stengers, Bruno Latour, and Donna Haraway for variations on the word *cosmos,* Jamie Lorimer writes hopefully of the advent of a Cosmoscene that "would begin when modern humans became aware of the impossibility of extricating themselves from the earth and started to take responsibility for the world in which they lived." See *Wildlife in the Anthropocene: Conservation after Nature* (Minneapolis: University of Minnesota Press, 2015), 4. While we agree with Lorimer in principle, we see this Cosmoscene as something that has long existed within human perception, so that future-making involves a project of renewal, return, and re-story-ation. For such entanglement in action, see Laura A. Ogden, *Swamplife: People, Gators, and Mangroves Entangled in the Everglades* (Minneapolis: University of Minnesota Press, 2011).

5. "So forget the word *environment.* . . . It assumes that we humans are at the center of a system of nature." Michel Serres, *The Natural Contract,* trans. Elizabeth MacArthur and William Paulson (Ann Arbor: University of Michigan Press, 1995), 33.

6. We have explored the shape of elemental ecocriticism as a vortex in "Eleven Principles of the Elements," the introduction to *Elemental Ecocriticism* (Minneapolis: University of Minnesota Press), 1–26; we are attempting an even wider topography of ecological reading through the form here.

7. On the "mess" as fecund gathering, see J. Allan Mitchell, *Becoming Human: The Matter of the Medieval Child* (Minneapolis: University of Minnesota Press, 2014). On the "messmate" and companioning, see Donna Haraway, *When Species Meet* (Minneapolis: University of Minnesota Press, 2008).

8. On emergence, "unexpected detours and happy accidents" as a careering vector inherent to nature, see Eben Kirksey, *Emergent Ecologies* (Durham: Duke University Press, 2015).

9. See, for example, Jean E. Feerick and Vin Nardizzi's introduction to *The Indistinct Human in Renaissance Literature* (New York: Palgrave Macmillan, 2012), "Swervings: On Human Indistinction": "We see it as our burden to create a useful roadmap for these essays while encouraging and facilitating a reading practice that bends—or swerves across—our own categories, parts, and pairings" (6). As Steve Mentz puts it in *Shipwreck Modernity: Ecologies of Globalization, 1550–1719* (Minneapolis: University of Minnesota Press, 2015), "No island is an island" (51).

10. On the enduring effects of the American creation of "wild" space on the people of color long excluded from them, for example, see Carolyn Finney, *Black Faces, White Spaces: Reimagining the Relationship of African Americans to the Great Outdoors* (Chapel Hill: University of North Carolina Press, 2014). These lasting divisions also tend to be far more exclusive than they seem, with the "riotous presence" of those who are not Christian and male being foundationally excluded by big ecological terms like *human*. Obscured stories of entanglement are essential to thinking beyond some of the impasses such bifurcations have established. See especially Anna Lowenhaupt Tsing, *The Mushroom at the End of the World: On the Possibility of Life in Capitalist Ruins* (Princeton: Princeton University Press, 2015).

11. Tim Ingold's thoughts on the "weather-world" underlay our words here. See chapters 9 and 10 in *Being Alive: Essays on Movement, Knowledge and Description* (New York: Routledge, 2011).

12. Nicholas Royle's term "wordlife" is from a book fundamental to this project, *Veering: A Theory of Literature* (Edinburgh: Edinburgh University Press, 2011).

13. Raymond Williams, *Keywords: A Vocabulary of Culture and Society* (Oxford: Oxford University Press, 2015), xxxvii.

14. Joni Adamson, William A. Gleason, and David N. Pellow, eds., *Keywords for Environmental Studies* (New York: New York University Press, 2016).

15. Greg Garrard, *Ecocriticism*, 2nd ed. (New York: Routledge, 2012).

16. Greg Garrard, ed., *The Oxford Handbook of Ecocriticism* (Oxford: Oxford University Press, 2014).

17. The sheer number of such collections makes an exhaustive list impossible, but some that have been essential to our framing of this introduction include Cheryll Glotfelty and Harold Fromm, eds., *The Ecocriticism Reader: Landmarks in Literary Ecology* (Athens: University of Georgia Press, 1996); Laurence Coupe, ed., *The Green Studies Reader: From Romanticism to Ecocriticism* (London: Routledge, 2000); Michael P. Branch and Scott Slovic, eds., *The ISLE Reader: Ecocriticism, 1993–2003* (Athens: University of Georgia Press, 2003); Timothy Clark, *The Cambridge Introduction to Literature and the Environment* (Cambridge: Cambridge University Press, 2010); Louise Westling, ed., *The Cambridge Companion to Literature and the Environment* (Cambridge: Cambridge University Press, 2014); Stacy Alaimo and Susan Hekman, eds., *Material Feminisms* (Bloomington: Indiana University Press, 2008); Ken Hiltner, ed., *Ecocriticism: The Essential Reader* (New York: Routledge, 2015).

18. The quotation is from the foreword by Lawrence Buell to Adamson, Gleason, and Pellow, *Keywords*, viii. See also Buell's prescient *The Future of Environmental Criticism: Environmental Crisis and Literary Imagination* (Oxford: Blackwell, 2005).

19. The comment was made by Jesse Oak Taylor at the MLA convention in Austin (2016) and was meant to capture the general sense of pleasure in

shared thinking, even during a time of crisis, that the presenters (and audience) evinced.

20. And as a verb, Haraway writes in her rich exploration of what inheres in *companion,* the word means "'to consort, to keep company,' with sexual and generative connotations always ready to erupt." *When Species Meet,* 17.

21. Quoted in Nico Israel, *Spirals: The Whirled Image in Twentieth-Century Literature and Art* (New York: Columbia University Press, 2015), 22. It is also Israel's point about a spiral's ambiguous—both centrifugal and centripedal— turn: "Does the spiral travel outward from the fixed point, thereby increasing its distance from that point, or curve inward, diminishing that distance?" (23). He has helped us frame what a spiral does *in* and *beyond* the twentieth century.

22. See especially the cluster of essays on "sustainability" in *PMLA* 127, no. 3 (2012): 558–606.

23. For Israel, spirals "assert their relation to the geopolitical . . . turning . . . both in toward itself to observe its own torsions *and* out to the 'globe.'" *Spirals,* 41. Local-global "eco-cosmopolitanism" is the topic of Ursula K. Heise, *Sense of Place and Sense of Planet: The Environmental Imagination of the Global* (Oxford: Oxford University Press, 2008).

24. For a smart reading of petroleum ardor and energy's deep costs, see Stephanie LeMenager, *Living Oil: Petroleum Culture in the American Century* (Oxford: Oxford University Press, 2014).

25. Timothy Morton's point in *The Ecological Thought* (Cambridge, Mass.: Harvard University Press, 2010): "We can't throw empty cans into the ocean anymore and just pretend they have gone 'away.' Likewise, we can't kick the ecological can into the future and pretend it's gone 'away'" (119).

26. We have in mind here the work of Rob Nixon, *Slow Violence and the Environmentalism of the Poor* (Cambridge, Mass.: Harvard University Press, 2011).

27. See Serenella Iovino and Serpil Oppermann's "Introduction: Stories Come to Matter" to their edited volume *Material Ecocriticism* (Bloomington: Indiana University Press, 2014), 1–17. As collaborators and fellow travelers, Iovino and Oppermann are constant inspirations to our own projects.

28. On this topic, see Jeffrey Jerome Cohen, ed., *Prismatic Ecology: Ecotheory beyond Green* (Minneapolis: University of Minnesota Press, 2013).

29. Quotation from Jedediah Purdy's *After Nature: A Politics for the Anthropocene* (Cambridge, Mass.: Harvard University Press, 2015), 22. Purdy argues for a critical engagement with the long human histories of making nature "real" that have disastrously shaped contemporary landscapes.

30. Serenella Iovino writes compellingly of how contemporary ecotheory must think place as entanglement, drawing on Ursula Heise's notion of "eco-cosmopolitanism" and Stacy Alaimo's "trans-corporeality" (among others) to argue that "we are at once here and elsewhere; vice versa, what affects the life of other places and beings has unsuspected reverberations in space and time,

eventually touching our bodies and backyards, too." *Ecocriticism and Italy: Ecology, Resistance and Liberation* (London: Bloomsbury, 2016), 2.

31. David Macauley discusses "re-story-ation" in *Elemental Philosophy: Earth, Air, Fire, and Water as Environmental Ideas* (Albany: State University of New York Press, 2010), 5.

32. Cf. Nicholas Royle's notion of "visiting" in "Veerer: Where Ghosts Live" (chapter 8 of *Veering*).

33. "To desire, to fear, to desire to fear, to fear to desire: veering" (*Veering,* 197).

34. *Anthology,* etymologically a "flower collection" in Greek (from *anthos-,* "flower," and *-logia,* "collection") later denoted a gathering of verses by various authors. To "anthologize," we offer, could also mean to gather ecopoetics with political potential. Through *passionate* we hope to convey what Tobias Menely describes as cross-species "passion as an opening to the world and an openness to the passion of others." *The Animal Claim: Sensibility and the Creaturely Voice* (Chicago: University of Chicago Press, 2015), 31.

Vegetate

CATRIONA SANDILANDS

Vegeō, vegēre, veguī, vegitum

In popular parlance, when people vegetate, they spend the weekend on the couch in flannel pajamas in a monster binge of multiple seasons of *Scott and Bailey* (or, at least, I might). When people vegetate *persistently*, their brainstems keep basic functions going (circulation, respiration, digestion) but they do not display what we generally like to think of as consciousness: they are considered alive and cannot be killed/let die without a lot of legal wrangling, but they do not demonstrate critical kinds of awareness or independent capability. Even more: as Mark Twain wrote famously in *The Innocents Abroad*, vegetating is directly opposed to activity, to citizenship, and to cosmopolitanism: "Broad, wholesome, charitable views of men and things cannot be acquired by vegetating in one little corner of the earth all one's lifetime."[1]

In contrast, when plants or fungi or viruses vegetate, they are understood to grow prolifically, even abnormally quickly (although the usage is a bit anachronistic, tumors are also said to vegetate when they metastasize). According to the *OED*, plants vegetate both intransitively and transitively: they vegetate by growing (as in the annual rhythms of vegetation and senescence or death that are common among temperate angiosperms) and they are also vegetated by cultivators who create landscapes by establishing particular plants to grow in particular locations with particular results. Plants also vegetate by propagating asexually: witness the rhizomatic growth of trembling aspens in the middle of North America that spread on a pretty cosmopolitan spatial and temporal scale, creating (arguably) the largest and oldest single organism on the planet.

Vegetal *veering* is thus a form of movement that is conceivably both passive and active.[2] When people and animals vegetate, they are considered largely inert: alive but not quite fully. When plants and other nonanimal organisms do likewise, they are considered abundantly alive, perhaps even excessively so. Etymology makes things even more complicated. The English word *vegetate* dates from the early seventeenth century and means to grow as plants do; the sense of leading an inert, passive life employed by Twain emerged later, in the mid-eighteenth century (not accidentally at about the same time as the beginning of the modern sciences). The word originates, however, in the Latin *vegēre*, "to be active," and applies (intransitively and transitively, actively and passively, singularly and plurally) to a range of persons, plants, and other beings, including the active first-person human singular: *vegeō*, *I* am lively, *I* am active, *I* excite, *I* arouse.

To understand this paradox, prolific vegetal (vegetating?) philosopher Michael Marder might direct us to the Greeks, whose thinking on questions of life and living has exerted, and continues to exert, a powerful influence on Western metaphysics. Very briefly: for Aristotle, there are three kinds of living force (or soul, or *psukhe*): growth, nutrition, and reproduction (nutritive or vegetative soul); perception, sensation, and locomotion (sensitive soul); and thought and intellect (rational soul). Hierarchically arranged, plants are at the lowest level and display the activities suited to vegetable beings, those of the nutritive soul; up the ladder, then, animals display sensitive soul, and only human beings rational soul. However, these soul activities are *cumulative*, meaning that beings higher up the scale demonstrate both their own unique forms of activity and those of the beings below them: animals display nutritive as well as sensitive activities, and humans have sensitive and nutritive desires as well as rational ones.[3] In this sense, when people vegetate, they are not so much being *passive* as demonstrating those *activities* that are consistent with the vegetal undergrowth of their *psukhe*: growth, nutrition, reproduction, decay. Reflecting on vegetating, then (which would, of course, no longer be vegetating: plant soul is not self-representing in this way even if we do access something of vegetal desire when we are thirsty), could indicate a practice through which people might come to feel the pulsing vibrations of our plant-selves, our kinship with plants, our common enactments of liveliness, something almost completely foreign to our habitual instrumentalization of the entire plant kingdom as wholly other, largely inert, and unquestionably consumable.

I rather like this idea, but I am getting ahead of myself. The fact that vegetating is considered an active part of the human psyche in Aristotle and in many other places in the history of Western philosophy does not, despite my desire to the contrary, mean that vegetating is *valued* in this schema any more highly than it is in modern, more taxonomic separations. Plants remain at the bottom of the ladder of living for Aristotle, and scholars as diverse as Dante Alighieri, G. W. F. Hegel, and Hannah Arendt confirm his anthropocentric order in which thinking—and doing/acting thoughtfully—is always already understood as a higher form of living than vegetating because it enacts the *noblest* elements of the human's tripartite soul, because it demonstrates an orientation of growth to a *higher* teleological purpose in the dialectical upward-progression of Spirit and/ or because it enables the individual person to conduct a reflective dialogue with herself in order to move *beyond* the biological exigencies of bodily survival. Don't get me wrong: I am a big fan of thinking (with Arendt, I firmly believe that allowing the mind to withdraw from the phenomenal, processual world of vegetating and animating for a while in order to reflect on it is a necessary—if not, unfortunately, sufficient— preventative of tyrannies both totalitarian and banal). But is it really necessary to demean the form of living that is vegetating in the process of defending reasoning, reflecting, and understanding? Is a recognition of our own vegetative life *as such*—an understanding, however tentative and fleeting, of our own constitutional plantiness—necessarily a capitulation to couch potato-dom?

Marder does not think so: his book *Plant-Thinking* includes the important observation that plants encourage us to imagine a form of living that is not always predicated on the central assumption of an individual self in encounter with discrete others: the plant is "indifferent to the distinction between the inner and the outer, it is literally locked in itself, but in such a way that it merges with the external environment, to which it is completely beholden."[4] Elaine Miller makes a similar point in *The Vegetative Soul*, which (brilliantly) follows vegetation through nineteenth-century German Idealism and Romanticism: here, "the vegetative soul, in contrast to the animated soul, emphasizes rootedness, vulnerability, interdependence, and transformative possibility rather than a separation of soul from body, actualization, and a stance of aggressiveness and self-preservation."[5] And Theresa Kelley, following similar philosophical shoots, spends a delightful chapter of her book *Clandestine Marriage* pitting Hegel against J. W. von Goethe in an extended fencing match between vegetative desire and teleological orientation: for every subordination of nutritive growth to the pointedly individuating and entelechial dictates of Spirit, she parries, via Goethe, with a defense of "the individuality, particularity, and metamorphosis of the plant form, and the contingent, unsystematic energy of nature in general."[6]

Vegetation flourishes, in these forays, because the process of *thinking like a plant* reveals to us what we have chosen to forget in dominant Western philosophical and scientific imaginations of our human selves as primarily rational, self-organizing, and independent beings over and above all others: a sense of our profound dependence on and location in the conditions of our growth and decay, including the other beings with whom we share these elements of liveliness. To vegetate, then, suggests a thinking response to our plantiness. As Marder argues, this kind of work involves not only thinking *about* plants as objects of attention and reflection (given how many people fail to notice plants at all, that is still not a bad place to start) but also thinking *with* plants "and, consequently, *with* and *in* the environment, from which they [and we] are not really separate."[7] I vegetate, you vegetate, we vegetate: despite the inevitable difficulties of representation, and debates about ethical action, that will no doubt adhere to any project of thinking like/with/as plants, it seems to me that it is ecologically important to reflect on the ways in which we are constituted by vegetal desires that both connect us with and remind us of our shared aliveness with/as plants. Although vegetating does not presume a specific right course of action toward particular plants, such as demanding the inclusion of plants in privileged liberal discourses of rights (in fact, it suggests the opposite: opening the question of ethical relationship to forms of life and relationship that are not premised on the masculine, singular, sovereign agent and may instead align more closely with feminine, plural modes of subjectivity),[8] it most certainly suggests an attentive and lively practice in which possibilities for ecological kinship are able to germinate, proliferate, and even effloresce.

Vegeta(ria)t(e)

Despite important criticisms of his anthropocentrism, Michel Foucault remains a key interlocutor in contemporary thinking about multispecies biopolitics. As he writes in *The History of Sexuality, Volume 1*, "For millennia, man remained what he was for Aristotle: a living animal with the additional capacity for a political existence; modern man is an animal whose politics places his existence as a living being in question."[9] What is important to note here is that, for Foucault, it is precisely the embodied *animality* of humans—our shared biological capacities for living and dying that can be harnessed directly by forces of power primarily oriented to *bodies* rather than, say, to consciences—that renders us biopolitical subjects. Although as Nicole Shukin and others have pointed out, this human-animalization does not mean that animals are generally well treated in biopolitical relations (some animals are highly regarded and appear to become humanlike in law, policy, and popular culture even as others are exploited, enslaved, and killed en masse),[10] it does mean that matters of life and organismic kinship—including ecological understanding—potentially come to the fore in new ways: humans become biological beings, and other biological beings begin to look more like us as a result. Animal suffering, for example, emerges as an important ethical/political concern in the same context as laboratory rats are considered useful models for human reactions to medication: we are understood to share elements of physiology, response, and affect, and to function optimally or wither (for Foucault, to make live or let die) in response to similar kinds of variables.

As Jeffrey Nealon writes in his provocative intervention into biopolitical understanding, it is thus not *the animal* that is abjected in and excluded from modern understandings of life, but rather *the plant*: animals do not function in modernity as our Others because we are so powerfully rendered animal ourselves. Instead, what remains in the sphere of absolute Otherness is the vegetal. "Following Foucault's reading," Nealon writes:

> One might suggest that [the] role of abjected other as having been played throughout the biopolitical era [is] not by the animal but by the plant—which was indeed forgotten as the privileged form of life at the dawn of

biopower. In this context it is probably worth recalling that the biomass of plant life on Earth's terra firma does remain approximately one thousand times greater than the combined zoomass of all humans and other animals.[11]

In other words, he argues, it is important not just to trace the ways in which Western thinkers have historically subdivided life by contrasting humans with animals (as is Giorgio Agamben's project in *The Open*),[12] but also to examine how we currently and institutionally organize our planetary preeminence by equating meaningful life *(bios)* with animality and expendable "mere" life *(zoë)* with vegetality.

More precisely, continental philosophy—the main subject of Nealon's inquiry—remains largely opposed to a consideration of the ways in which plants *as plants* enact, complicate, and model life and living in a biopolitical era: despite (for example) Gilles Deleuze and Félix Guattari's multiple invocations of the rhizome as a mode of nonteleological subjectivity and liberation, the metaphor does not get at the actual conditions of vegetation (and thus life) in neoliberal, biopolitical capitalism. In this moment, he argues, "the vegetable *psukhe* of life is a concept or image of thought that far better characterizes our biopolitical present than does the human-animal image of life."[13] To vegeta(ria)t(e), here, is then to consider more fully how we are biopolitical subjects in an era that intervenes not only in our human and animal souls (rationality, subjectivation, perception, discipline) but also in our vegetal ones, in the realms of nutrition, growth, and decay: in other words, in our growing, eating, thirsting, reproducing, senescing, decaying, and composting bodies.

Plants are clearly treated badly in neoliberal biopolitics. Think, for example, of the ways in which so-called terminator technologies are employed by multinational agricorporations specifically to deny plants the ability to reproduce of their own accord by producing fertile seeds to spawn a new generation. In addition to the fact that such technologies deprive farmers of the ability to collect and save seeds for future planting (which is the point of the technology), they also intervene directly into plants' bodies in order to render their properties of growth and reproduction as sites of profit and accumulation as fully as possible. Plants are intensively hybridized, genetically altered, vegetated/germinated, and controlled on a massive scale, and their specific capacities are patented in order to serve particular corporate ends: from attempts to hybridize and/or regulate against open pollination to direct genetic and chemical manipulation of specific cultivars in order to create vegetal forms that serve specific consumer desires (strawberries in Canada in February that taste somewhat like strawberries), precisely the quality of *vegetation* is more and more harnessed to capitalist accumulation. Hence, I propose the *vegetariat*:[14] capitalist accumulation is not possible without the ever-intensifying exploitation of the surplus labor of plants. Intensive (ab)use of plants is the rule, even in a universe that has begun to question the widespread exploitation of animals, because we still do not largely consider plant-lives as meaningful.

Again, this is not to say that I think that we should pursue anything like plant rights in response to this exploitation (a proposition that many animal rights advocates find patently absurd if not outright destructive to animal rights and welfare agendas).[15] In fact, rather than imagine that plants should be granted ethical status on the basis of their resemblance to humans (e.g., their potential capacities for pain and suffering), we should instead consider the ways in which people and animals are increasingly organized and controlled *like and even as plants* in a neoliberal biopolitical universe. Our capacities for nutrition, growth, and reproduction are the precise vectors of intervention in current economic and policy debates about "proper" life and living: people are not only animalized, but people and animals are also vegetated, treated as beings whose most plant-like capacities are the stuff of concern. To hell with questions of perception, sensation, and rationality in this era: what is at the forefront of current political debate is where and when and how we are to live as reproducing, productive bodies who serve the polis by way of being, simply, alive. Growing. Populating. Spreading. Invading. Vegetating. Vegetariating.

In this context, recent literature on "plant intelligence" gives me pause. On the one hand, I am very pleased to see plants get recognition for the important work they do to keep life going (including, but obviously not only, under capitalism) and also for the ways in which they participate as complex, sensate, and interactive beings in the process (in fact, so much so that the line drawn between "vegetal" and "animal" forms of liveliness no longer holds firm). According to Stefano Mancuso and Alessandra Viola, for example, "A compelling body of research shows that higher-order plants really are 'intelligent': able to receive signals from their environment, process the information, and devise solutions adaptive to their own survival."[16] Further, their intelligence is both singular and collective: "They manifest a kind of 'swarm intelligence' that enables them to behave not as an individual but as a multitude."[17] On the other hand, then, this research is all too easily parlayed into new modes of control: understanding *and harnessing* collective vegetal intelligence, here, becomes another mode of biopolitical intervention into life as, for example, plant signaling chemistry comes to be used in new agricultural, communicative, and even robotic technologies. As Mancuso and Viola themselves enthuse: "For some time now, there's been talk of plant-inspired robots, a real generation of planetoids. . . . Plans are also under way for the construction of plant-based networks, with the capacity to use plants as ecological switchboards and make available on the Internet in real time the parameters that are continuously monitored by the roots and leaves. . . . Soon the plant Internet may become part of everyday life for all of us."[18] The point is that the exploitation of plant intelligence is not, here, only about plants: in this context, because we all vegetate, we all *vegetariate*.

That plants have intelligence is not really a new understanding. That plants are understood to have intelligence in much the same biological manner as animals and humans have intelligence is, however, a relatively recent incorporation of vegetation into popular discourses of animation/perception and even cogitation/thinking that were once considered the sole realm of *Homo sapiens*. I do not mean to suggest that it is a bad idea to extend understandings of intelligence to include plants. I do mean to suggest that it is a mistake to equate the consideration of plants as intelligent, responsive, thinking beings with the idea that this understanding means that plants will necessarily benefit from this new understanding or that an understanding of shared vegetative intelligence is necessarily liberatory. Human beings have become animals in the biopolitical age: we are members of a species, population, race, sex, group, vector, heredity. This incorporation has not spelled better treatment for people. Animals have become "people" in the manner that people are now animals: beings that can suffer, emote, relate. This treatment spawns new ethical responses to animals but only within a very limited range. And now: plants. It's an open question.

Vege(bili)tate

I am walking through a grove of old Douglas-fir trees on the southeastern tip of Vancouver Island, in the place that was ancestrally, and is now again, known as PKOLS. In SENĆOŦEN, PKOLS means "White Head," possibly referring to the fact that the place was the last from which the glaciers receded from Vancouver Island (for about 160 years, the place was also known by white colonists as "Mount Douglas," but after a different Douglas).[19] The trees are large, brown, wet, thick, textured, and imposing. There are enormous sword ferns everywhere in the understory, green, dense, reaching, enclosing. It is December, and everything drips with the winter rain. The trees, ferns, salal, and Oregon-grape shine slick green against the pewter sky.

Coast Douglas-firs (properly hyphenated because they are not true firs) typically live to be over 750 years old and can reach ninety meters in height. The W̱SÁNEĆ people (who call them JSÁY) have used them extensively for thousands of years: their thick bark is an excellent, hot-burning fuel; their durable wood can be crafted into all manner of useful implements, like poles for salmon weirs; their prolific and sticky pitch is "used as a cement to patch canoes and water containers . . . [and also] as a salve for wounds."[20] Starting in earnest in the mid-nineteenth century, others of the firs' properties made them the most important industrial tree species on North America's west coast, Vancouver Island included: their immense size and straight, strong, tightly grained wood made them ideal as building material, and in places like PKOLS, their proximity to the ocean made them easily accessible commodities. There are almost no old growth Coast Douglas-firs left as a result: one study "estimated that only one-half of one percent (about 1100 hectares) of the low coastal plain is covered by relatively undisturbed old forests."[21] Environmental organizations such as the Ancient Forest Alliance (AFA) are thus determined to preserve what few are left, such as in the Edinburgh Mountain Ancient Forest, part of which has already been industrially clear-cut (leaving behind the second largest Douglas-fir on record, nicknamed Big Lonely Doug, a fine figure of charismatically tragic megaflora). There is some protected second growth (as at PKOLS, which was logged but set aside as a reserve in 1858), but large areas of coastal forest have been converted into tree farms (more accurately, fiber farms), in which a clear-cut forest area is slash and burned to get rid of snags and stumps, replanted, and harvested again once the trees have reached a marketable size.

The massive commodification of Douglas-firs is a stunning example of the ways in which plants are the vegetal foundations of capitalism and colonialism. Treated as individual units whose value is estimated almost entirely in terms of board-feet of timber, they cannot be anything other than resource, standing reserve, bare life: the vegetariat, writ especially large. It is interesting, then (if not all that surprising), that environmental organizations like the AFA and ecoluminaries such as David Suzuki and Wayne Grady have sought to develop greater public respect for the trees by portraying their mode of living up the vitality hierarchy, emphasizing their singularity, individuality, and even quasi-personhood. So there is Doug, there is Luna of Julia Butterfly Hill fame, and there is also the rather anthropomorphized individual portrayed "arbobiographically" in *Tree: A Life Story*, which is simultaneously a knowledgeable foray into the ecological interconnectivity of coastal forest life and a striking example of the ongoing tendency to think of certain trees as heroic exceptions to the multiplicity, contextuality, and lack of self-boundedness associated with vegetation.[22] (For example, Aristotle allowed that trees have a telos, Deleuze and Guattari specifically contrast the rhizomatic with the arborescent, and Suzuki and Grady's is neither the first nor the last work in the arbobiographical genre.)

Recent research, however, suggests that Douglas-firs are not at all heroically singular; it also emphasizes that thinking about the trees' value in terms of the particular commodity they are understood to contain misses almost all of what is going on in their lives and communities (quite literally a matter of not seeing the forest for the trees). What is going on, of course, is vegetating. As Suzanne Simard, for example, has documented extensively, Douglas-firs are active participants in a complex, subterranean network of mycorrhizae in which roots and fungi engage in an elaborate process of chemical symbiosis: the tree gives photosynthesized carbon to the fungi and the fungi transmit inaccessible soil nutrients and moisture back to the tree.[23] As the fungus spreads, it also links tree to plant to tree (not just Douglas-firs), creating a vast, interconnected forest network in which trees also communicate carbon to each other.[24] In this context, thinking of a tree as a singular and person-like being grossly misrepresents the fact that, even though Douglas-firs germinate from seed and grow progressively larger from that origin (in the manner of individuals), they are also inextricably linked to rhizomatic soil fungi, so much so that it would be impossible to have the tree without the fungi. In this respect, it is not the trees that represent the forest at all: "Mycorrhizal fungi are considered to be *the* keystone of coastal Douglas-fir forests,"[25] meaning that what tends to be *valued* in the forest—old trees as singular lives, board-feet of timber as sources of profit—is not at all related to what is most *lively* in the forest (and also that what is most lively is something that is less easily anthropomorphized).

The point, then, is to not imagine *for a second* that we give rightful value to plants by making them appear "like us." Walking through PKOLS, I try instead to *vege(bili)tate*: to restore my connection to the vegetal liveliness of the forest by connecting into the network of mycorrhizal relationships that define and sustain this place; by becoming *plural*, attending to the decentered vegetality of the forest as it resonates with my own multiple plant capacities; by paying attention to the ways in which insects, birds, and mammals also plurally interact with the trees and fungi (and other plants, mosses, and lichens) in an extraordinary dance of sustenance and relationship (which, to their credit, Suzuki and Grady depict well); and by imagining what it means for me, a white settler-colonist used to treating this place as a "park," to be *part* of these relationships rather than just an admiring observer of their exuberant green-ness. Eduardo Kohn might allow that I am trying to think *with* the forest: "Forests are good to think with," he writes, "because they themselves think. Forests think. I want to take this seriously."[26] Or perhaps I am trying to think *as* the forest, as part of the "we" that is our collective *psukhe*. Conceiving of complex multispecies forms of biosemiotic relationality as *thought,* for Kohn, does not mean that we all think in similar ways or that I can ever remotely apprehend a mychorrizal *umwelt*. What it does mean is that, by understanding myself as participating in the lively, thoughtful interactivity of the forest as a self among selves—or as an element in a plural, distributed "forest" selfhood—I might be able to see my own relationships to plants and others as ecologically embedded, and myself as something other than a fiber-user (or even mychorrizae user, if the Wood Wide Web pans out): perhaps even as mindfully and multiply vegetating among the many others who are doing the same.

Robin Wall Kimmerer sums it up neatly in one chapter of her book *Braiding Sweetgrass,* in which she describes replanting sweetgrass on a property in the Mohawk Valley in Pennsylvania. The place, Kanatsiohareke, is a loving and thoughtful reinhabitation of a site that was an ancient Haudenosaunee Bear Clan village, an attempt to wrest the land from white botanical settlement (it is now largely populated with timothy, clover, and daisies) and white settler racism (it is, very intentionally, an antidote to the cultural genocide perpetrated at the nearby Carlisle Indian Industrial School). Kimmerer describes the science of restoring sweetgrass to the place:

> The most vigorous stands [of sweetgrass] are the ones tended by basket makers. Reciprocity is a key to success. When the sweetgrass is cared for and treated with respect, it will flourish, but if the relationship fails, so does the plant. . . . What we contemplate here is more than ecological restoration; it is the restoration of relationship between plants and people."[27]

For Kimmerer (and for the many Haudenosaunee people before her in this place), planting sweetgrass is an act of vege(bili)tation: participation in an ancient, ongoing ritual of planting, harvesting, and respectful use that draws on the precise vegetal properties of sweetgrass (which mostly spreads rhizomatically) and the agricultural proclivities of humans (who know collectively how and where to plant and harvest the grass in order to make the best use of these properties) in order to achieve flourishing for all concerned. To plant sweetgrass, here, is to engage in restorative ecological relationship, in a process of attentive intertwining of the capacities of people and plants in concert in a mode that, at least potentially, defies the compulsion to capitalist accumulation. Together, then, perhaps *we* can vegetate, even in the complicated, capitalist-mycorrhizal landscapes of Douglas-fir forests.[28]

Vegeō. Vegeta(ria)t(e). Vege(bili)tate. You pick: think like the plants that we are, think as we are rendered plant, think with the plants with whom we are (or should be) in communicative and productive relation. But don't forget the many ways in which our lives are constituted vegetally: for the love of life in this biopolitical era, vegetate.

Notes

1. Mark Twain, *The Innocents Abroad: or; The New Pilgrim's Progress* (Hertfordshire: Wordsworth Classics, 2010), 427.

2. *Vegetate* and *veer* do not share an etymological origin, even though one might easily imagine the genesis of *veer* in the particularly vegetal movements of climbing plants toward the objects of their attachment.

3. Michael Marder, *Plant-Thinking: A Philosophy of Vegetal Life* (New York: Columbia University Press, 2013) and "Plant-Soul: The Elusive Meanings of Vegetative Life," *Environmental Philosophy* no. 1 (2011): 83–99.

4. Marder, *Plant-Thinking*, 32.

5. Elaine Miller, *The Vegetative Soul: From Philosophy of Nature to Subjectivity in the Feminine* (New York: State University of New York Press, 2002), 18.

6. Theresa Kelley, *Clandestine Marriage: Botany and Romantic Culture* (Baltimore: Johns Hopkins University Press, 2012), 240.

7. Marder, *Plant-Thinking*, 181, emphases in original.

8. See, for example, Miller's discussion of Luce Irigaray's vegetal feminism in *The Vegetative Soul*, 189–200.

9. Michel Foucault, *The History of Sexuality, Volume1: The Will to Knowledge* (New York: Pantheon, 1978), 143.

10. Nicole Shukin, *Animal Capital: Rendering Life in Biopolitical Times* (Minneapolis: University of Minnesota Press, 2009).

11. Jeffrey T. Nealon, *Plant Theory: Biopower and Vegetable Life* (Stanford: Stanford University Press, 2016), 11.

12. Giorgio Agamben, *The Open: Man and Animal* (Stanford: Stanford University Press, 2004).

13. Nealon, *Plant Theory*, 106.

14. I thank Hannes Bergthaller for the term *phyto-Marxism*, which inevitably gave rise to the *vegetariat*.

15. For a conversation about plant versus animal ethics, see "Michael Marder and Gary Francione Debate Plant Ethics," http://www.cupblog.org/?p=6604; see also my essay "Floral Sensations: Plant Biopolitics," in *The Oxford Companion to Environmental Political Theory*, ed. Teena Gabrielson et al. (Oxford: Oxford University Press, 2016), 226–37.

16. Stefano Mancuso and Alessandra Viola, *Brilliant Green: The Surprising History and Science of Plant Intelligence* (Washington, D.C.: Island Press, 2015), 5.

17. Ibid.

18. Ibid., 157.

19. See http://crdcommunitygreenmap.ca/story/history-pkols-mount-douglas.

20. Nancy Turner and Richard J. Hebda, *Saanich Ethnobotany: Culturally Important Plants of the W̱SÁNEĆ People* (Victoria: Royal BC Museum, 2012), 56.

21. Samantha Flynn, *Coastal Douglas-fir Ecosystems* (Victoria: BC Ministry of Environment, Lands and Parks, 1999), 2.

22. David Suzuki and Wayne Grady, *Tree: A Life Story* (Vancouver: Greystone Books, 2004).

23. See, for example, Suzanne Simard and Daniel M. Durrall, "Mycorrhizal Networks: A Review of Their Extent, Function, and Importance," *Canadian Journal of Botany* 82 (2004): 1140–65.

24. Not surprisingly, this network is now popularly called the "Wood Wide Web," as in, for example, Manuela Giovannetti et al., "At the Root of the Wood Wide Web: Self Recognition and Non-Self Incompatibility in Mycorrhizal Networks," *Plant Signaling and Behavior* 1, no. 1 (2006): 1–5. It's a provocative metaphor in many ways but, as above, the slope between the recognition and the abuse of plant capacities is a slippery one.

25. Flynn, *Coastal Douglas-fir Ecosystems,* 4, emphasis in original.

26. Eduardo Kohn, *How Forests Think: Toward an Anthropology beyond the Human* (Berkeley: University of California Press, 2013), 21. Needless to say, his understanding of "thought" is not Aristotelian: his beautiful book is based on research with the Ávila Runa people in the Upper Amazon region of Ecuador, in conversation with a posthumanist Peircean semiotics in which "we all live with and through signs" (9).

27. Robin Wall Kimmerer, *Braiding Sweetgrass: Indigenous Wisdom, Scientific Knowledge, and the Teachings of Plants* (Minneapolis: Milkweed Editions, 2013), 262–63.

28. Although she is less restoratively inclined than I, this practice is related to Anna Tsing's in *The Mushroom at the End of the World: On the Possibility of Life in Capitalist Ruins* (Princeton: Princeton University Press, 2015).

Globalize

JESSE OAK TAYLOR

No circumstance characterizes the cartographical art of the Modern Age—and *eo ipso* its way of thinking—more profoundly than the fact that no globe we have ever seen shows the earth's atmosphere.

—PETER SLOTERDIJK, *In the World Interior of Capital*

The globe is in our computers. No one lives there.

—GAYATRI CHAKRAVORTY SPIVAK, *Death of a Discipline*

We don't live *on* Earth. We live *in* it. If there is a single perspectival shift demanded by the Anthropocene, it lies in acknowledging our own planetary internality. Earth's atmosphere envelops all of history, a vaporous archive in which molecules exhaled by the dead remain suspended along with an ever-increasing quantity of industrial effluent. History bubbles up from the depths of an alien planet whose interior remains as mysterious as the heavens. Modernity runs on the fossilized remains of our prehistoric ancestors sucked and blasted from subterranean fissures. But the world we know has always been shaken, stirred, and, occasionally, shattered by what lies beneath.

Because it marks a phase transition at which human influence scales up to affect the Earth System as a whole, the Anthropocene can only be approached at planetary scale. The injunction to "think globally" has never been truer. But what does it really mean? Images of the global from Archimedes's lever to Apollo 8's *Earthrise* have depended on a fiction, the fantasy of an external perspective. But we can only think the Anthropocene from the inside out. Viewed from the outside, it is easy to think of globalization as a condition, but it isn't. It is a process. As humans, as organisms, as earthlings, we are globalized and ever globalizing, expanding our worlds from within using the resources at hand.

Globalization may be the least popular idea in environmental thinking since the Great Chain of Being. It smacks of elitism, capitalism, profit; of abstraction and mediation; of technocrats and the silver bullets they ride in on. It isn't local. It isn't organic. It isn't green. It is human without being humanitarian, corporate rather than cooperative, hybrid without being hip, robotic instead of posthuman. The Anthropocene injunction to globalize seems apt to translate into geoengineering schemes and planetary surveillance systems, a Promethean gesture akin to Stewart Brand's oft-quoted quip, "We are as Gods, and we have to get good at it."[1] And yet we need it all the same.

Living in an advanced capitalist society means being brought into physical contact with distant reaches of the planet—from its elemental innards to the biota of other continents—by the time you finish breakfast. Your drive to work leaves a vaporous trail that will soon be dispersed across the atmosphere. If you stay home and watch TV or surf the Internet you will sift through images from around the world, bouncing through space. In other words, *all* of our thoughts and *all* of our actions are always already global. Even our most intimate encounters are mediated by a derivative of the rubber tree—an Amazonian native now cultivated mostly in Asia.

Globalizing thought, action, and awareness thus begins by acknowledging the global networks within which thought and action are already embedded and then seeks to repurpose that web of interconnection into the fabric of a more hospitable world. It also means engaging in political innovations and experimentations capable of translating global thought into global action. As Jedediah Purdy points out, "All serious responses to global climate change—and to inequality in global capitalism—face the same basic problem: there is no political body that could adopt and enforce them."[2] Globalizing democracy, not in the sense of extending "free" markets at gunpoint, but rather in conjuring genuinely new polities predicated on ecological citizenship in the global commons, may well be the greatest challenge of all. It means not only extending suffrage in what

Bruno Latour calls the "parliament of things" made up of nonhuman actants, some of which (methane, carbon dioxide, the Western Antarctic Ice Sheet) are already among the most powerful entities in the Anthropocene, but also rethinking the bounds of human polities.[3] As democratically elected governments seal their borders against climate refugees fleeing the catastrophes driven by economic growth and consumption in democratic societies, it becomes ever clearer that democracy itself needs to change if it is to be worthy of the name. Globalizing democracy means extending the principle of participatory governance beyond the nation, beyond the state, and beyond the human.

Globalizing from the inside out refigures both local and global as processes rather than geographical positions. On a long enough timescale, there is no local; animals, plants, rocks, and even continents move, shift, and meld into one another. On the other hand, without locality there is no form, no identity, no agency. Evolution depends on location. Species exist because of where they are (or were) at a certain point in time; ditto for cultures; ditto for persons. Many of the Anthropocene's extinctions arise out of the obliteration of locality. Habitats are overrun by corporate monocultures and invasive species, local traditions are swamped by a homogenizing culture industry, languages and species vanish forever—lost experiments in the organization of life. All organisms inhabit a life-world, an *Umwelt* consisting of those features of their environment that have signification or meaning for them and to which they must attend to ensure their own survival.[4] Globalizing our modes of dwelling (in the sense of *oikos*, the root of ecology) means self-consciously extending our *Umwelten*, in both geographical and temporal terms, to planetary scale without losing the contours of the local and immediate, finding ways of increasing biological, cultural, and linguistic diversity, living in and as open wholes.

Earth is a single system, encompassing all past and future history. Everything every human has ever made has been a thing of this planet. Every thought that every human has ever had has been a snippet of planetary self-consciousness. But they are not the only ones. The emergence of the Anthropocene marks the threshold at which human action—however unequally distributed among humans—must be acknowledged as a force operating at the level of the Earth System as a whole. The Anthropocene marks the globalization of the Human, but in such a way that seemingly renders us lowly humans powerless to avert oncoming catastrophe. We have crossed a threshold, emerging as a "geophysical force" akin to climate or plate tectonics, and thus demanding consideration in decidedly inhuman terms.[5] Despite the fact that the Anthropocene is by definition unprecedented, dwelling within it means confronting an old truth that it is no longer possible to forget: the planet is bigger than we are. We are not gods, we are Earthlings—and we must get good at it.

The stratigraphic debate over the Anthropocene presents this challenge in an acute form. The effort to date the Anthropocene hinges on the search for an incontrovertible marker that will remain legible as a profound shift in the geologic state of the planet even in the absence of all correlating evidence. To justify affixing a Global Boundary Stratotype Section and Point (GSSP), or "golden spike," the marker must be sufficiently clear—and sufficiently durable—to remain legible even in the event of human extinction or at any rate the obliteration of all known languages and schemes of signification.[6] The trace of the Anthropocene must also be globally synchronous, which is to say it must occur at the same time all around the world—a criterion that is possible in part due to the sheer scale of geologic time such that thousands or even millions of years can be compressed into a single trace, the contemporaneity that arises out of deep time. The Anthropocene marks a moment of global synchronization, a truly globalized event. As such, it forces us to globalize our thinking, introducing a scale on which many of our conceptual and methodological practices, usually based in discrete languages and national or religious traditions, are inadequate. In this sense, it is telling that the rise of the term Anthropocene dovetails with the renewed interest in "world literature" as a geographically and temporally expansive category around the turn of the millennium.

The emergence of the Anthropocene is paradoxical in that its global synchrony is precisely the thing that our established models of the global are inadequate to achieve. Whereas humans can only "act locally," the Anthropos *only* acts globally because it only comes into being across the threshold at which the aggregation of human action affects the operation of the Earth System as a whole. If we are to apprehend the Anthropos, we must find the means to globalize—to globalize our response, to globalize our dissent, and to globalize our conception of space, time, and species being. By "globalizing" I mean not only "expansion to global scale" but also the utilization of globes, models that depict, represent, or simulate the planet as a whole. A globe in the familiar sense is a four-dimensional map—it not only folds space around a globular object but also spins, simulating the gravitational motion without which there would be no time.[7] Time occurs because the earth turns. The Atlassian finger swipe that spins a globe creates a blurred image of simultaneity as nations, regions, and oceans pass more quickly than the eye can apprehend. On a moving globe the world becomes one before our eyes.

A globe is also a world-making model. It presents the world in miniature, simplified form. To "globalize," then, is not merely to expand to planetary scale or to model planetary processes but to deploy model worlds *as a constitutive element of that expansion to planetary scale*. With the emergence of the Anthropocene, globes and other planetary simulations cease to be merely models of an external world and become *internal* models of the Anthropocene as a distinct phase within the operation of Earth systems. In this sense, those models cease to be merely human tools and become instead modes of planetary self-knowing. If humans are part of the Earth System then human knowledge is a part of that system as well. The Anthropocene is thus not only the moment at which humans become aware of our own agency within the Earth System but also the moment that the Earth System becomes aware of itself. If examining the Anthropocene as a kind of planetary singularity seems entirely too sanguine, this trajectory nonetheless suggests that the models we build to imagine the world *matter* in profound and literal ways. If we are to *veer* from the present catastrophe, we must create models for planetary democracy that can help bring a new order into being.

The emergence of the Anthropocene has occurred through a series of phase transitions in the ecology of human being. Understanding these shifts as *globalizing* transitions highlights not only the way they each draw a greater network of organisms, energy sources, waste dumps, signs, social organizations, and technologies into complex interrelation, but also the fact that each has also remade the human image of the world in its own image. Below, I briefly sketch out this process at three signal moments in Anthropocene history: 1610, 1784, and 1945. Each has been proposed as a candidate for the "golden spike" defining the origin of the Anthropocene, and each subtly redefines both the Anthropocene concept and its eponymous agent. While the Anthropocene Working Group has indicated that a mid-twentieth-century boundary event is "stratigraphically optimal," my goal is not to fix a particular point of origin but to highlight a connection between the globe-as-model and the world remade by that model.[8]

~

1610: The 'Orbis' Hypothesis. The geographers Simon Lewis and Mark Maslin suggest that the Anthropocene be dated to 1610 and the "Columbian Exchange" of biota brought about by European conquest of the Americas, "a swift, ongoing, radical, reorganization of life on Earth without geological precedent."[9] Lewis and Maslin hypothesize a drop in atmospheric CO_2 (legible in ice cores) is linked to the deaths of more than *fifty million* Native Americans from slaughter, famine, enslavement, and disease, resulting in widespread reforestation and thus an uptick in carbon sequestration. This dating makes the Anthropocene an inherently imperial condition, an epoch wrought in death.[10] It also makes it a condition defined by ecological globalization. The spread of plants, animals, people, money, technology, and microbes becomes the operative force of the epoch, as reflected in their term of choice, the "Orbis Hypothesis."

Orbis is Latin for *world* and etymologically linked to *globe*. And globes, in the familiar sense of spherical maps, were in fact constitutive technologies in this process. The spherality of Earth had been known for centuries—hefted by Atlas, measured by Eratosthenes, and depicted on medieval maps that placed the earthly orb at the center of the cosmos. Despite this knowledge, the rotundity of the planet remained beyond reach, an object of pure science, or cosmology, rather than navigation. In the early modern era, the globe became the world around which one might sail. Coupled with celestial globes of the heavens, terrestrial globes helped sailors navigate their way across the aqueous surface of the planet by making it possible to calculate how the stars would look from different coordinates. The earliest extant terrestrial globe was created in 1492, and, while it had no direct bearing on Columbus's voyage, the coincidence could not be more telling.[11] The cartographic developments that went into modeling the planet as a single unit of navigable space were as essential as the caravels on which the conquistadors sailed to the voyages of "discovery" that brought the Old and New Worlds into contact. Globes must thus be reimagined within the historical ecology of conquest that Lewis and Maslin describe as the origin of the Anthropocene.

The history of globes entwines scientific exploration and imperial power. The first English globe was dedicated to Elizabeth I, and depicted not only the continents and seas but also the routes traveled by explorers like Francis Drake and Walter Raleigh, enabling the queen to "see at a glance how much of the seas her naval forces could control."[12] Innovations in globe making were entangled with a number of other developments, including perspective in painting, the rationalization of space in geometry, the invention of longitude, and the rise of experimental science, all of which brought new frontiers into capitalist exploitation and helped facilitate "the swift, ongoing, radical reorganization of life on Earth" that we have come to call the Anthropocene.[13] Globes thus not only depicted the world as a sphere, they actively participated in the reintegration of life on Earth into a single ecosystem in a manner unseen since the breakup of Pangaea 175 million years ago. Globes helped bring the Anthropocene Orbis into being.

～

1784: James Watt and the Steam Engine. When Paul Crutzen and Eugene Stoermer initially proposed the idea of the Anthropocene, they dated it to 1784, with James Watt's patent of the double-acting steam engine, an event legible in rising CO_2 levels in Antarctic ice cores beginning at the end of the eighteenth century.[14] However, the steam engine was not only a technology capable of performing physical work, it also modeled principles of heat and energy that both drew on and helped inspire James Hutton's conception of geology. As Jussi Parikka argues, "For Hutton, the planet *is* a machine . . . modeled on the steam engines of his age, primarily the Newcomen steam engine [i.e., the engine that preceded Watt's]; its principles of expansion of steam inspired Hutton with the idea of elevation of the crust. This machine also assumes organic unity and cyclical renewal, and feeds off the heat at its core."[15] The steam engine thus becomes a model for the conception of the earth that it is in the process of transforming. Furthermore, those transformations themselves arise out of the same principles of heat and convection that enable steam engines to function. As Manuel De Landa explains, "When we say that 'a hurricane is a steam motor' we are *not* simply making a linguistic analogy; rather, we are saying that hurricanes embody the same diagram used by engineers to *build* steam motors—that is, we are saying that a hurricane, like a steam engine, contains a reservoir of heat, operates via thermal differences, and circulates energy and materials through a (so-called) Carnot cycle."[16] Geology and the steam engine are linked not by a random fluke of history but by what Stuart Kaufman calls "adjacent possibility," in which the operation of a system affects the conditions of possibility in its vicinity. Coincidence becomes synchrony. The science of stratigraphy and the steam engine did not just happen to emerge at the same time and place. Rather, both emerged in part *because* that shared geohistorical space enabled each to serve as a model for the other. Simon Schaffer has argued that Hutton, in turn, helped inspire Adam Smith's economic theory, especially the "invisible hand," an image that I read as an attempt to characterize an emergent phenomenon—the "second-order" operation of the market as a separate agent that looks a lot like that other "higher power," God.[17] Karl Marx would, in turn, conceptualize the transformative power of capitalism in terms of evaporation: "All that is solid melts into air," an all-too-appropriate, all-too-literal image for an industrial modernity driven by combustion.[18]

To recapitulate: a steam engine works because it mimics, re-creates, intensifies, and accelerates processes of heat transfer that are also at work in the earth's physical processes. The deployment of steam engines, in turn, ultimately fed into ever more far-reaching systemic shifts that are now altering planetary processes through the vaporization of solid carbon into the atmosphere, in turn accelerating and intensifying Earth's internalization of solar energy. At the same time, the steam engine as a simulation was transferred laterally into Hutton's geology and Smith's economics—both of which are crucial to extending the systemic conditions that enabled the scaling up of industry now known as the Industrial Revolution and thus, ultimately, the Anthropocene. Each internal model brings certain features of its immediate environment into view, making them available for the system in which it is embedded, while at the same time affecting the conditions of possibility in its vicinity in ways that ultimately enable that system to undergo a phase transition or scalar shift, often by coupling the operations of one system to another. Such phase transitions are the points at which a system *veers* into a new state, displaying unexpected attributes, strange behaviors, and morphing forms. Even though such veering can neither be forced nor forecast, internal models enable systems to couple, uncouple, and intermix, generating unruly offspring and unintended consequences, planting the seeds of veerings to come. Unlike seventeenth-century globes, the steam engine did not present itself as a model. Nonetheless, its pervasiveness as a metaphor in the industrial imagination makes clear that it was, among other things, a tool to think with, a model that remade the world both materially and conceptually. The Anthropos globalized under a head of steam.

⌒

1945: Cybernetics and the Great Acceleration. While the geological agency of the human may have begun with the Industrial Revolution, Jan Zalasiewicz and his colleagues in the Anthropocene Working Group have recently argued that it was only "from the mid-20th century that the worldwide impact of the accelerating Industrial Revolution became both global and near-synchronous."[19] Unlike the invention of the steam engine, this dating does not focus on an "event" in the traditional sense. While 1945 aligns with the first nuclear tests that left a legible, globally synchronous trace in the stratigraphic record, the real force behind the choice is the scaling up of population growth, fossil fuel combustion, and urbanization known as the Great Acceleration.[20] Dating the Anthropocene to the Great Acceleration doesn't locate it at a point of invention where something "new" happens but rather traces the phenomenon to a scalar shift in both size and intensity, a moment at which the vectors of carbon-based capitalism veered skyward. All of the dynamics in question—industrialization, fossil fuel use, population growth—had long been underway. They simply pass a threshold of both acceleration and scope in the mid-twentieth century, beyond which they emerge as forces at planetary scale. Whereas dating the Anthropocene to the Columbian Exchange makes it a phenomenon driven by globalization and empire and dating it to industrialization makes it about technology and the shift to fossil fuels, dating it to the Great Acceleration suggests that *it is all about scale.*

Dating the Anthropocene to the Great Acceleration also makes it coincident with the emergence of cybernetics, systems theory, and, most importantly, global climate modeling. As Anna Tsing explains, "The idea of global climate change articulated the new realization that places far apart from each other were still connected for their basic survival, especially through the circulation of air and water."[21] Climate modeling gave rise to what Tsing describes as "a specific—even a peculiar—kind of globe," one in which "the global scale is privileged above all others."[22] Climate is an abstraction that can only ever be witnessed via models, instruments, and simulations. If the development of climate modeling in the mid-twentieth century coincides with the emergence of anthropogenic climate change, then the abstraction that emerges from the model is not simply a newly quantifiable phenomenon but a phenomenon newly entwined with human impacts—a "new kind of globe" indeed. Just as the alignment between geology and the steam engine suggests that there may never have been a geologic record without human beings operating as agents within that record, dating the Anthropocene to the mid-twentieth century suggests that there has never been a model of the "global climate"—that is climate as single, planet-spanning, interconnected phenomenon—without human agency operating within it on a similarly globalized scale. Furthermore, in yet another example of the adjacent possible in action, those global climate models were themselves adapted from technologies developed to trace the fallout from nuclear weapons testing. Hence, the globalized "signature" now poised to be adopted as the GSSP formalizing the Anthropocene's emergence was itself integral to the rise of Earth systems science. As Joseph Masco succinctly puts it: "The early Cold War nuclear program thus enabled a changing understanding of the planet."[23]

Once is strange, twice is coincidence, three times is enemy action. The above trajectory suggests a correlational (if not necessarily causal) relationship between modeling and world-altering technologies, akin to what climatologists call a "teleconnection," an association between phenomena too distant and complex to be mapped but that nonetheless appears to hold true. The global synchrony of the Anthropocene needs to be approached as a question not merely of geophysical processes and stratigraphic traces but of the technological, historical, and aesthetic processes (and prostheses) of cognition, the technics of thinking globally. In contrast to the division between global thought and local action, these models for thinking globally seem to translate into *global* action in the sense that they remake the world in their own image using technologies that purport to remake the world in its own image. These examples highlight the recurrence of globalizing models at potential dates for the Anthropocene's emergence. Each redefines not only the Anthropocene itself but also its definitive agent. The Anthropos is a conquistador in 1610, an industrialist in 1784, and a posthuman cyborg in 1945. However, in no case is the Anthropos an individual, except as a kind of synecdoche, a part that conjures an image of the whole. The imperial Anthropos might be personified in Sir Francis Drake, who was often depicted with a terrestrial globe following his circumnavigation—a fitting emblem of the world-making ambitions of Renaissance humanism that also echoed from Shakespeare's "Globe."[24] However, this image is accurate only if we view it as a stand-in for an entire system of ships, instruments, humans, animals, plants, microbes, financial arrangements, ocean currents, wind patterns, and millions and millions of dead bodies. James Watt's steam engine is a fitting image of the Anthropos only if imagined not as a local act of invention but as a node bound up with coal seams, Corn Laws, the Acts of Enclosure that privatized the commons, Huttonian geology, the outer atmosphere, and the Antarctic icepack. Dating the Anthropos to 1945 converges with the formation of the United Nations, the Universal Declaration of Human Rights, the Bretton Woods institutions and the idea of a single global economy—all designed to both further and manage a newly globalized humanity at the precise point that the Anthropos was emerging as a force within the Earth System. In each case, the Anthropos can only refer to the global agent, not its local manifestation. As Anthropos, we are globalized and ever globalizing, not simply in the sense that our actions have globalized impacts but also in our deployment of internal models to apprehend the planet of which we are a part. Taken together, these moments can *all* be read as points at which the entwinement of Earth and humanity veered into a new state of complexity, scope, acceleration, and intensity. The slow process of tectonic shifts and species drift was accelerated by caravels, the gradual compression of carbon into stone was vaporized with a vengeance, the absorption of solar energy by Earth's atmosphere was intensified by the hothouse of modernity—and through it all, the Anthropos expanded and adapted, ate, mined, fucked and extracted, killed, made merry, and laid waste to all it could survey.

The Anthropocene marks a shift in the Earth System as a whole, which is to say Earth as understood as a single system. Even the now familiar images of Earth from space do not capture this sense of unified global simultaneity. After all, they only show half of the earth at a time while rendering invisible the intermediate zone, sandwiched between atmosphere and molten stone, that makes up the encrusted world on which we actually live. More troublingly, the vantage point they offer is the very one we cannot afford to imagine if it allows us to indulge in a fantasy of planetary escape: "Please stop this planet, I'd like to get off." Earth looks very different when imagined from the inside out, in part because the medium in which the projected image is produced is itself part of the larger totality it helps bring into focus. Ursula Heise's "eco-cosmopolitanism" offers one model for globalizing from the inside out. Heise notes that technologies like Google Earth allow us to tack back and forth between our situated, on-the-ground perspectives and the view from space, replacing the singular "Blue Marble" image with a form of modernist montage.[25]

Any time we use the GPS function in our smartphones we are tracking our movements from space by way of orbiting satellites. While this may seem like an extraplanetary perspective, those satellites remain tethered by Earth's gravitational pull like planetary selfie-sticks. Pausing in an awareness of such technological encounters by way of Jussi Parikka's work on the "geology of media" highlights the fact that such extraplanetary encounters are not only routed beyond Earth's atmosphere but also bits of geological history—the rare earth minerals on which media technologies depend.[26] The extraction of those minerals is, in turn, remaking large swathes of Earth's surface and riddling its subterranean regions with fissures of absence, while the discarded technologies themselves accumulate into what will eventually constitute new strata in the ongoing mineral encrustation of geologic history.[27] The prosthetic vantage point that allows us to see globally, whether in visions of the cloudy blue bubble glimpsed from space or the more banal perspective afforded whenever we pull out our smartphones to find the nearest coffee shop, entails a traffic between a planetary and an extraplanetary point of view that is also channeled through deep time and mediated by rare earth minerals—it extends from beneath the ground to above the sky. Globalizing entails not only a spatial but also a temporal doubling, between the vastness of the planetary and the concentration of techno-modernity's propensity to cram ever-faster processors into ever-smaller spaces (whether those be computer chips or cubicle-bound employees).[28]

Globalization is a trans-scalar endeavor. It doesn't simply mean expansion but rather translation from one scale to another and depends on concomitant recognition that different forces, energies, and concerns operate at different scales. Systems are nested within systems, and while each is entangled with those within which it is nested, its operations also remain self-enclosed and self-enclosing such that the operative dynamics that must be considered within it may well cease to matter at larger (or smaller) scales. Globalizing our thinking, our acting, and our seeing is not simply a matter of privileging one scale over another (say the planetary over the local, national, or bioregional) but rather of re-imagining our modes of dwelling—cognitively, physically, spiritually, imaginatively, sensually—around negotiating scalar shifts and veering with grace and humility, if not with ease. Doing so depends on models, simulations, and fictions. Globes may be abstractions where no one lives, but we cannot dwell in Earth without them.

The entirety of human thought is a mode of planetary self-knowledge. We live *in* Earth, are made *of* Earth, and to Earth we will return, along with all our inventions. Technological innovations and planetary explorations are encoded deep within the Antarctic icepack even as the banalities of daily existence entwine the earth with their invisible contrails. "Thinking globally" is not a project of extending our ideas to the bounds of the planet—they are already there. Rather, it is a project of modeling those planetary entanglements on a scale at which they can become present to us, from within. The extraplanetary vantage that would be adequate to the Anthropocene is, by definition, beyond us. We have to make do with what we have, here, today, inside this swirling ball of vapor, stone, and plastic we call home.

Notes

1. Stewart Brand, *Whole Earth Discipline: An Ecopragmatist Manifesto* (New York: Viking, 2009). Brand notes that photographs of Earth from space first suggested this "God's eye view."

2. Jedediah Purdy, *After Nature: A Politics for the Anthropocene* (Cambridge, Mass.: Harvard University Press, 2015), 20.

3. Bruno Latour, *We Have Never Been Modern,* trans. Catherine Porter (Cambridge, Mass.: Harvard University Press, 1993).

4. Jakob von Uexküll, *A Foray into the Worlds of Animals and Humans, with a Theory of Meaning,* trans. Joseph D. O'Neil (Minneapolis: University of Minnesota Press, 2010).

5. Dipesh Chakrabarty, "Postcolonial Studies and the Challenge of Climate Change," *New Literary History* 43, no. 1 (Winter 2012): 1–18 (13).

6. Simon L. Lewis and Mark A. Maslin, "Defining the Anthropocene," *Nature*, March 23, 2015, 171–80.

7. Jeffrey Cohen has reminded me that the tides (which may have been integral to the origins of life) are affected by the moon—the Earth System is always already extraterrestrial.

8. Colin N. Waters et al., "The Anthropocene Is Functionally and Stratigraphically Distinct from the Holocene," *Science*, January 9, 2016, aad2622.

9. Lewis and Maslin, "Defining," 174.

10. See Dana Luciano, "Inhuman Anthropocene," *Avidly*, March 22, 2015, http://avidly.lareviewofbooks.org/2015/03/22/the-inhuman-anthropocene/; Steve Mentz, "Enter Anthropocene, c. 1610," *Glasgow Review of Books*, September 27, 2015, http://glasgowreviewofbooks.com/2015/09/27/enter-anthropocene-c-1610/.

11. Sylvia Sumira, *Globes: 400 Years of Exploration, Navigation, and Power* (Chicago: University of Chicago Press, 2014), 14.

12. Ibid., 65.

13. Lewis and Maslin, "Defining," 174.

14. Paul J. Crutzen and Eugene F. Stoermer, "The Anthropocene," *Global Change Newsletter* 41 (2000): 17–18.

15. Jussi Parikka, *A Geology of Media* (Minneapolis: University of Minnesota Press, 2015), 40.

16. Manuel De Landa, *A Thousand Years of Nonlinear History* (New York: Swerve/MIT Press, 2000), 58.

17. Quoted in Parikka, *Geology*, 40.

18. See Tobias Menely, "Anthropocene Air," *minnesota review* 83 (2014): 93–101.

19. Jan Zalasiewicz et al., "When Did the Anthropocene Begin? A Mid-twentieth-century Boundary Level Is Stratigraphically Optimal," *Quaternary International* 383 (2015): 196–203 (201).

20. Ibid., 198–99.

21. Anna Tsing, *Friction: An Ethnography of Global Connection* (Princeton: Princeton University Press, 2005), 102.

22. Ibid.

23. Joseph Masco, "Bad Weather: On Planetary Crisis," *Social Studies of Science* 40, no. 1 (Feb. 2000): 7–40 (18).

24. Sumira, *Globes*, 18.

25. Ursula Heise, *Sense of Place and Sense of Planet: The Environmental Imagination of the Global* (Oxford: Oxford University Press, 2008), 67.

26. Parikka, *Geology*, 50–52.

27. Ibid., 113–15.

28. See Rob Nixon, *Slow Violence and the Environmentalism of the Poor* (Cambridge, Mass.: Harvard University Press, 2011), 11–13.

Commodify

━━━━━━▶   ◀━━━━━━

TOBIAS MENELY

We start with what is immediately at hand, what we see and touch, what supports and sustains us. In societies where capitalism "prevails," Karl Marx writes, "wealth" *(Reichtum)* "appears" as a prodigious heap of commodities, *"eine 'ungeheure Warensammlung.'"*[1] Marx begins *Capital* by reminding his readers of the historical distinctiveness of such a society, in which the existential richness of life comes to be measured in the profuse variety of goods available for purchase. Folklore describes the *Ungeheure* as an immense man-eating ogre. The joke, of course, is that human beings—what Marx calls living-labor, "the expenditure of human brain, nerves, muscles and sense organs" (164)—are actually inside the heap of things, which had initially appeared to be "external" (126). This human labor, Marx hints, will turn out to be the elementary commodity and the real, rather than apparent, source of capital's distinct measure of value. The ever-growing pile of commodities is monstrous because it consumes human bodies, *congealing* labor—Marx elsewhere describes capital as a vampire, a werewolf—but also because its ontological status is ambiguous, "a species," as Jacques Derrida defines the monster, "for which we do not yet have a name."[2] It *appears* as an ungainly composite of inert things yet is defined by some principle of motion, an almost organic unity of purpose as it grows and mutates.

Marx immediately turns, in the first chapter of *Capital,* from this shape-shifting jumble to the single commodity, a *"triviales Ding,"* if only to discover in the sensuous immediacy of boot polish, a jacket, or a table the spectral trace of exchange. *Capital,* as many readers have noticed,

44

infuses a technical critique of liberal political economy with evocations of supernatural horror, a persistent figuring of capital as "dead labour" feeding on "living labour" (342). The commodity is haunted by its passage through the immateriality of exchange but also by its origin in the exploitation of the labor of those who lend their "human flesh" to production (379). The unsurpassed insight of *Capital* is to link expanding value, the concentrated surplus of which the heap of commodities is the monstrous signifier, with an understanding of labor as a "definite social relation between men" (165). Marx insists, however, that human labor—as "vital force" (341), as commodity and agent of commodification—cannot be conceptualized outside of a broader planetary metabolism, relations to lives and forces beyond the human. There is, for instance, in *Capital* an insistent tropology of animal life and death. Dinesh Wadiwel points to the metaphors of tanning and skinning, "the flaying of the skin from a living entity," which Marx employs to describe the laborer's "experience of commodification," the "transition—from subject to commodity."[3] Nicole Shukin notes Marx's use of "mere jelly" to describe "homogeneous labor time," which, evoking "animal fats and gelatins," implies "that human labor and (animal) nature are cosubstantial matters of real subsumption."[4] The etymological link between *capital, chattel,* and *cattle* (livestock) is widely noted. Moreover, as Marx observes, human labor power *(Arbeitskraft)*— the "activity" by which "man changes the forms of the materials of nature in such a way as to make them useful to him" (163)—itself can be understood only in its correspondence with planetary energy exchange.[5] To produce commodities, humans not only organize and appropriate "natural forces" *(Naturkraft)*; in doing so, they "proceed as nature does herself" (133).

A century and a half before Marx wrote *Capital,* Alexander Pope captured, in an arresting couplet, this haunted apprehension of commodities, "trivial Things," as corpses, a materialized remainder of life force.[6] *The Rape of the Lock* (1712) is concerned primarily with the psychosocial consequences of commodification, for Pope the process by which people and things, the sacred and the profane, are brought into the same domain of value. But it is also a poem that gestures at absent scenes of appropriation: invisible "Labours" (1.148), mercantilist war, and the original input of nature, "all that Land and Sea afford" (5.11). Pope's heroine, Belinda, is awakened from a portentous dream by her lapdog. Seeking comfort in the "mystic Order" of her dressing-table—"here / The various Off'rings of the World appear"—she gazes at her reflection in the mirror, her self-regard amplified by a halo of luxury goods: Arabian perfumes, Indian gemstones (1.120–34).[7] The poet tracks the provenance of Belinda's "Treasures" further still (1.129): "The Tortoise here and Elephant unite, / Transform'd to *Combs*, the speckled and the white" (1.135–36). Pope alludes to the oriental fable of the elephant and the tortoise, which he would have found referenced in John Locke's account of our imprecise language of "substance." An "Indian," writes Locke, "who, saying that the world was supported by a great elephant, was asked, what the elephant rested on? To which his answer was, a great tortoise: but being again pressed to know what gave support to the broad-backed tortoise, replied, something, he knew not what."[8] The upshot is not entirely clear, since Locke remains uncertain about the need to posit a "substance"—matter, nature—that precedes, and unifies, our diverse sensory perceptions. Pope's joke, not unlike Marx's in the first line of *Capital*, is that to commodify—to *transform* one thing into another more useful thing—is to produce a chimeric, monstrous *unification* of the diverse orders of reality. Pope, in other words, literalizes the allegory: living organisms do actually supply the substratum, the material in Belinda's combs, an origin faintly apprehensible in the difference between marbled tortoiseshell and white ivory. The point, of course, is that the tortoise and elephant are not really "here." There is a lacuna between the living beings and the inanimate things on a dressing table, the difference between a *commodity*, a substantial object, and *commodification*, a metamorphic process within a web of relations. In this gap transpires not only transport, manufacture, and exchange but also a capture and a killing.[9] The difficulty of representing this gap can be perceived in the clunky catachresis of *commodify* as a verb. To commodify is to destabilize the identities of subjects and objects.

In *History and Class Consciousness*, György Lukács invokes the same oriental fable in his discussion of reification, the conceptual expression of generalized commodity exchange. Reification is a kind of misrecognition whereby the distinct historical forms taken by capitalist society, its patterns of crisis and conflict, are concealed in the apparent objectivity of commodities, the (delusive) immediacy of things Lukács calls "second nature." "Modern bourgeois thought," he writes, finds itself in the "situation of that legendary 'critic' in India who was confronted with the ancient story according to which the world rests upon an elephant. He unleashed the 'critical' question: upon what does the elephant rest? On receiving the answer that the elephant stands on a tortoise 'criticism' declared itself satisfied."[10] The elephant and tortoise allegorize the hypostatized categories and specialized, calculative cognitive habits adopted by "crude empiricists" (77). The fable's particular resonance, for Lukács, surely has to do with the status of animals as exemplary fetishes. As Shukin observes, animals remain ubiquitous features of advertising because they endow "the historical products of social labor . . . with an appearance of innate, spontaneous being," seeming to "speak from the . . . place of nature" (3–5). For Lukács nature is a "social category" (130), the appearance of givenness that prevents us from recognizing commodification—"the central, structural problem of capitalist society in all its aspects"—as a sociohistorical "relation between people" (83). Lukács is known to amplify a particular tendency in Marxist thought: he "dissolves nature, both in form and in content, into the social forms of its appropriation."[11] Yet even Lukács subtly reintroduces a biological language, intimations of "life and death" that exceed the human (110), when he asks to what extent "commodity exchange together with its structural consequences [is] able to influence the *total outer and inner life [ganze äußere wie innere Leben]* of society" (84, emphasis added).

Here I define *commodification* in its difficult double movement, as a social relation and a relation between human beings and the biosphere, "the envelope of life where the planet meets the cosmic milieu."[12] This relation to *life,* as productive force, has less to do with politics, the "bare" or "precarious" life central to theorizations of biopolitical sovereignty, than with institutions of value and infrastructures of production. Not just a relation between people concretized in things, a commodity, as Noel Castree writes, embodies "socio-natural relations."[13] Commodification occurs in a "general economy" that begins in planetary surplus, the superabundance of solar radiation.[14] The biosphere, the other Earth systems, and no less human labor, are manifestations of this free gift, the sun's expenditure without return. Life is a distinctive, though variegated, process for storing and converting energy flows, one outcome of which is the production of living matter: tusks and shells, fat and muscle, cellulose and woody tissue (the material from which, etymologically, we get "matter"). Locating commodification within this general economy of constantly renewed surplus and universal productivity allows us to better define capitalism's adaptability, the ways in which it overcomes ecological "limits" by restructuring its metabolism. It also allows us to delineate forms of social and biospheric violence, as certain lives, invested with value, are incorporated in the circuits of industrial production while others—less useful, less valuable—are altogether expropriated. While this essay draws on recent research in Marxist ecotheory, my emphasis is less on the social asymmetries generated in the production of surplus value than on its secondary effect: the *turning* of the world toward the needs of (certain) human beings, the reshaping of the earth's environs for the *accommodation* of a single species.

∼

Etymologically, a commodity is something *commodious*: convenient, useful, fit to human need or desire. In this sense, to commodify is always to fit to a measure defined by the human. Locke introduces the distinction between a commodiousness given in nature and a commodiousness added by labor. As he writes in the *Second Treatise* (1690):

> whatever *Bread* is more worth than Acorns, *Wine* than Water, and *Cloth* or *Silk* than Leaves, Skins, or Moss, that is wholly *owing to labour* and industry. The one of these being the Food and Rayment which unassisted Nature furnishes us with; the other provisions which our industry and pains prepare for us, which how much they exceed the other in value, when any one hath computed, he will then see, how much *labour makes the far greatest part of the value* of things, we enjoy in this World.[15]

He estimates that human labor, "our industry and pain," introduces ten, even a hundred or a thousand, times the value supplied by nature. Notably, Locke does not define this increase only in terms of additional "Conveniencies" (296). Already, for Locke, commodification involves an abstract form of value that derives specifically from exchange and the accumulation of surplus, two facets of commodification facilitated by money. "Commodities," he writes, are "movables, valuable by money, the common measure." A commodity is a useful object that can be relocated, exchanged, and made commensurate with other commodities. It is not the landed estate, for Locke, but the abstract measure and temporal stability of money, "some lasting thing" (300), that enables "larger possessions" (293) and creates the psychic imperative toward accumulation, "the desire of having more than Men needed, [which] altered the intrinsic value of things" (294).

Locke's labor theory of value is less an analytical than a normative principle, a revolutionary justification for political authority premised on the need to safeguard property. Commodification is, for Locke, privatization, a sociolegal relation between persons and things that begins in humankind's embodied self-possession and capacity for labor: "Every Man has a Property in his own Person" (287). From this initial "right" derives an entitlement to appropriate what is outside of the body: "Thus the Grass my Horse has bit; the Turfs my Servant has cut, and the Ore I have digg'd in any place where I have a right to them in common with others, become my *Property*" (289). The authority of the state, endowed with the "*Right* of making Laws with Penalties of Death, and consequently all less Penalties, for the Regulating and Preserving of Property," is thus justified as an auxiliary to commodification (268).

Even as he attributes the creation of "value" in the commodification process to human labor, Locke requires a contribution from nature. Nature is figured in terms of a living generativity that precedes, and augments, human labor: the "spontaneous hand of Nature," "the Fruits it naturally produces, and Beasts it feeds" (286), the creatures that may be "killed, caught, or tamed" (294). This productivity is spatialized as territorial exteriority, the *terra nullius* of imperial fantasy: the "vast Wilderness" and the "vacant places of *America*" (293), the ocean, "that great and still remaining Common of Mankind" (289).[16] Alongside the unique form of value produced by labor, Locke posits as necessary the input of a generative and expansive nature, not yet owned or appropriated. Accumulation requires this reserve, an unlimited surplus of productive potential that precedes commodification.

This contradiction—value accrues as surplus only from human labor yet depends on what Locke calls the "spontaneous Products of Nature" (295)—will inform subsequent theories of commodification. For Marx, nature's role in commodification, under capitalism, is as immaterial to "value," which derives solely from labor productivity, as are the qualities that constitute a commodity's usefulness. Yet commodification requires the input of natural forces and materials just as it requires that commodities fulfill human need. Commodification, under capitalism, conjoins two independent imperatives, the turning of nature to human use and the asymmetrical distribution of a surplus: "Use-values are produced by capitalists only in so far as they form the material substratum of exchange-value, are the bearers of exchange-value" (*Capital*, 293). The challenge for contemporary thinkers drawing on Marx to explore the crucial ecology/economy nexus has been to integrate this human-centric principle of value, an accumulative imperative premised on an internal species relation, with attention to the place of planetary life—as productive force, congealed energy, embodied being, provider of services, and extraneous surplus—in the commodification process.

In *Capitalism in the Web of Life*, Jason Moore, extending the pioneering work of Marxist feminists, fine-tunes the categories of political economy to reconcile the specificity of the value-form with what he terms the "value-relation." "Every act of exploitation (of commodified labor-power)," he writes, "depends on an even greater act of appropriation (of unpaid work/energy)."[17] "Value does not work," as he puts it pithily, "unless most work is not valued." Generalized human labor is the unique engine of value-in-motion, yet the appropriation of uncompensated work, not least biospheric productivity, determines "socially necessary labor time" (100). Though his ecohistorical examples range widely—from silver mines to sugar plantations, fisheries to forests—Moore conceptualizes the role of nature, including "human nature," in terms of "work," equivalent to waged human labor except in that it may be appropriated without compensation. Drawing on the hard-won axiom that "the value form and the value relation are non-identical," Moore reinterprets environmental history, under capitalism, as a series of crises—underproduction and overproduction, exhaustion and expansion—that begin in the fifteenth century (65). "Commodity frontiers"—"bundles of uncapitalized work/energy" accessed through geographical expansion and technoscientific innovation (144)—offer temporary means of increasing labor productivity, solutions that amplify the scale of capitalization. In his recent work on "fossil capitalism," Andreas Malm has identified a similar dialectic, a value-form that depends on the appropriation of fossil fuels as "the material substratum of exchange-value."[18] Malm's focus on the role of fossilized energy in facilitating circulation and restructuring labor relations does much to explain carbon modernity's scale-shift, our accelerative veering into the Anthropocene.

In describing a "new phase of the production of nature," Neil Smith invokes the cosmic tortoise as a figure for planetary totality, describing the "commodification and financialization of nature 'all the way down.'"[19] Morgan Robertson similarly observes that "we now confront an environment that can be defined as potential commodities in nearly every aspect of its material existence, and at every scale from the atmospheric to the biochemical."[20] Recent scholarship has investigated the manner in which capital, in a "neoliberal" period of overproduction and diminished frontiers, deepens its penetration of the biosphere, redefining commodification through biotechnological and financial innovation. This has involved examining phenomena—wetlands and the atmosphere, genes and microorganisms, bioengineered lifeforms—that are newly subject to commodification and often lack the tangible thingness of conventional commodities. For example, in a study that compares ecosystem services in carbon markets with the trade in exotic pets, Rosemary-Claire Collard and Jessica Dempsey identify a category of "lively commodities" for which commodification involves not a taking of life but a sustaining of life. Lively commodities "remain alive . . . for the duration of their inclusion in the commodity circuit."[21] While wildlife becomes commodifiable only insofar as an animal is separated from its natural habitat, carbon markets, the trade in credits for the carbon sequestration services provided by intact ecosystems, depend on the "reproductive ability of aggregated nonhumans" (2690). Though a miniscule proportion of the global economy, ecosystem services have been subject to much scholarly scrutiny.[22] Nature is valued as a provider of services—pollination, water purification, carbon sequestration, biodiversity preservation—that can be classified, calculated, compared, and traded. An infamous article in *Nature* estimated the total economic value of global biospheric services in the range of $16–53 trillion per year.[23] Several actual markets have emerged, such as the system of wetland credit exchanges that followed the Clean Water Act.

The commodification of the biosphere, as present condition and future prospect, is generally less a matter of innovative markets and bioengineered lifeforms and more a continuation of the long transition from extraction to agroindustrial monoculture (indeed, the sugar plantation is Moore's paradigmatic example of early capitalism as world ecology). Boyd, Prudham, and Schurman see a shift from industries that treat nature as an "exogenous" stock of resources to the "*real subsumption of nature*," whereby "biophysical systems are industrialized," augmented, or rerouted to serve as "productive forces."[24] This productivity, of course, remains dependent on metabolic, biological, and climatological processes, supplemented and intensified rather than supplanted by industrial infrastructure. The zone of biospheric appropriation—the life energy that we claim for our accommodation—is, increasingly, not a geographical periphery but the built environments of high-yield, high-density agribusiness: pine and oil palm plantations, shrimp and salmon ponds, confined feeding operations, soybean and wheat fields. It is this generalization of monoculture that leads Donna Haraway to propose the "Plantationocene"—defined by the "relocations of the substances of living and dying around the Earth"—as an appellation for the current epoch of anthropogenic influence.[25]

There is no more acute example of the process by which living organisms are made to act as productive forces within industrial infrastructures than the massive system of intensive animal farming, which, over the course of the twentieth century, replaced free-range poultry and livestock husbandry (which itself replaced hunting and scavenging). Asking how "biological systems" are incorporated into "the circuits of industrial capital," Boyd offers the example of the broiler chicken, which, through selective breeding and industrial efficiencies, came, in the course of the twentieth century, to produce twice as much meat in half the time with half the feed.[26] However modified by breeding and genetic engineering, animals—each a center of consciousness—remain the productive force, capable of converting overproduced grain crops into profitable proteins.[27] For this reason, I think James Stanescu is incorrect to suggest that the "deading life" of the factory farm is "life completely denaturalized, life as completely produced and constructed."[28] The suffering experienced by factory-farm animals reflects the fact that their nervous systems, sensory receptors, and behavioral patterns, by and large, evolved under natural selection. Beef cattle and broiler chickens, for example, are grown so quickly that their muscle tissue overburdens their under-formed skeletons, making embodiment a source of unceasing pain. Unable to dust-bathe, perch, or exercise, hens must be debeaked to prevent cannibalism in cramped battery cages. Naturally impelled to build nests, pregnant sows compulsively paw the concrete floors. With such alienated existence in mind, Ted Benton extends Marx's analysis of the factory worker's self-estrangement to the experience of factory-farm animals. "Sustained solely to serve purposes external to them," these living beings are "systematically denied" opportunities for the "exercise of their species-powers."[29]

Biospheric processes and forms of life are not equally commodifiable, equally available to the generation of value and use. While certain biological systems and organisms are integrated into productive infrastructures, others are dispossessed. Native ecosystems are replaced by monocultures, such as the oil palm plantations that are rapidly supplanting tropical rainforests throughout Southeast Asia, and resource extraction industries, such as the coltan mines—which may well have supplied the tantalum capacitors in my laptop—that are fragmenting gorilla habitats in Central Africa. The orangutans and gorillas, along with the multitudes of less charismatic flora and fauna that share their habitat, enter the category of life Collard and Dempsey deem the "outcast surplus": "natures . . . wholly and permanently superfluous" to value creation.[30] Orders of magnitude now separate those forms of life that are commodifiable from those that are pushed into ever-smaller peripheries due to expropriated habitat. In 2013 the United Nations estimated the global populations of domesticated animals: twenty-two billion chickens, one and a half billion cattle, more than a billion sheep, nearly a billion pigs. By contrast, there are an estimated hundred thousand gorillas and fifty thousand orangutans living in the wild, with small populations in captivity. Or, to return to the oriental fable: there are around a half million elephants and twenty thousand nesting female hawksbill turtles living in the wild.

At least since Jevons's *The Coal Question* (1865), the extraction model has been associated with inevitable exhaustion. These days, limits-talk establishes a material catalyst for the systemic reorganization of socio-ecological relations, an internal constraint on endless commodification. Moore, for example, maintains that we have reached the end of "cheap nature": "There are limits to how much new work capitalism can squeeze out of new working classes, forests, aquifers, oilfields, coal seams, and everything else. Nature is finite" (87). What is not finite, historically speaking, are the flows of solar radiation that energize all planetary production and circulation. The commodification of the biosphere now depends less on exhaustible resource frontiers—wild fisheries and old growth forests, tortoiseshell and ivory—than on the incorporation of an infinitely reproducible life into industrial infrastructures and agro-monocultures.[31] Over five centuries, capitalism has shown a miraculous adaptability, as it restructures its metabolic relation with the biosphere and other Earth systems; today, the only foreseeable absolute limit is runaway climate change. Perhaps the emphasis on capital's socioecological contradictions, and the predictions of its imminent collapse, symptomatize how difficult it is to envisage effective political resistance to comprehensive commodification. The catastrophe of this century, however, may turn out to be not capitalism's end but its persistence on a hotter, geoengineered, ecologically simplified, and more toxic Earth.

Rather than awaiting a collapse premised on biospheric finitude, we need to better account for the fissures, the incommensurate values, and the politically imposed limits within biospheric commodification. There is, after all, no single law of value. The fate of wildlife, like that of factory-farm animals, is determined by competing forms of legal regulation, social concern, and economic valuation. Thirty-five thousand species of flora and fauna are protected under the 1973 Convention on International Trade in Endangered Species, which regulates, but does not necessarily ban, their trade. Along with national laws, such as the Endangered Species Act, and the animal welfare statutes that govern the treatment of food animals, such treaties do provide barriers to commodification, real limits on intervention, exchange, and value creation. To altogether dismiss legal protection, to simply conflate capitalism and the liberal state, is to miss one key inflection point whereby animal suffering is diminished and biodiversity preserved. There are, moreover, ways within the commodification process of vesting value in biospheric flourishing: in the ecosystemic integrity enjoyed by an ecotourist, in the ecosystem services traded in new markets, and in the animal welfare that adds value to meat.[32] Precisely because commodification appears so totalizing, so inescapable, Anna Tsing asks us to track, within capital's environs, instances of "contaminated diversity," such as what she finds in the errant ecology and irreducible locality of matsutake mushrooms and their foragers.[33] While it is tempting to read neoliberal conservation or markets for ecosystem services within the dominant logic of capital accumulation,[34] Tsing asks us to develop analytics capable of distinguishing degrees of biospheric diversity and flourishing within different regimes of value.

~

My worry, though, is that such veering ecologies—the unpredictability and persistence of life in "capitalist ruins," as Tsing puts it—become ever less recognizable to us who benefit from the planet's commodification. In Marx's account, after all, commodity fetishism begins not in false consciousness but in our direct experience of the commodity's palpable usefulness. Our failure to retrace the commodity's history, to learn about the *lives* that gave it shape, reflects less a willed blindness than our perceptual capture in what is, for Marx, most *natural* about the commodity: its capacity to actualize human need and desire.[35] A communist society eschews exchange value, and the exploitation of human labor, but not the need to appropriate useful things from the biosphere.[36] In other words, even if we imagine a world in which commodification occurred entirely without the imperative to accumulate value through the appropriation of surplus labor time, the problem of human accommodation, of the human as end, would remain. It is, moreover, not easy to distinguish the need to use things from the compulsion to amass them, a universal principle of utility from the socialization of consumerist desire. The Yanomami shaman Davi Kopenawa, for instance, describes first contact with the "white people" who "accumulate merchandise in vast quantities and always keep it close to them, lined up on wood boards in the back of their houses."[37] When these objects were first introduced to his ancestors, he observes, "they were euphoric and did not yet suspect that these trade goods carried *xawara* smoke epidemics and death with them" (the word *xawara* refers to disease and also to gold).[38] There is certainly evidence that consumerism—the experience of desiring, acquiring, and using commodities—entails its own forms of psychic disturbance and self-estrangement. Sianne Ngai, for instance, suggests that the fetishization of cute things—little, personable objects—expresses a compulsion to resist, while we recapitulate, the logic of equivalence instantiated by mass commodification.[39] This said, the critique of consumerism as a form of alienated existence is often overstated. The vast Amazon of merchandise offers real satisfactions; factory-farmed meat may be delicious. Use value is an unrelentingly difficult problem for ecotheory.

This is not to dismiss the ideological dimensions of biospheric commodification, which requires the generalization of reified patterns of thought that make the living world recognizable as property, possessable and sellable; as resource, inert things separable from living ecosystems; and as valuable and monetizable. Neoliberal ideology frames economic growth as the prerequisite of other social goods and defines markets as the solution to the socioecological problems caused by commodification. However, while universities, law courts, and the culture industry play a role in the production of what Moore calls "abstract social nature"— *nature available to commodify*—the most powerful *thinking* about the biosphere occurs in what Alfred Sohn-Rethel terms the "real abstraction" of the marketplace: the acts of exchange that commensurate heterogeneous biospheric phenomena within a homogeneous law of value, an "abstraction not in mind, but in fact," an "exchange" that precedes and enables the "knowledge of nature."[40]

To commodify is to materialize equivalence. It is also to materialize human primacy, reengineering the Earth's environs around the needs and desires of certain human beings. In societies not organized around commodities, there is an imperative to justify, in ritual or myth, the exploitation or taking of life. Consider that God's first communicative act, in Genesis, is to grant dominion to the newly created humans. Such sanction may be understood as a compensatory response to an apprehension that the Earth is not designed to fulfill human need, that the more-than-human "world [is] composed of a multiplicity of points of view," each a "center of intentionality."[41] In a society structured around commodification—the turning of the world to human use, as it comes to be conjoined with the imperative to accrue surplus—anthropocentrism need not be justified, especially insofar as the violence of commodification is occluded by global commodity chains, industrial processes, and legal regulations such as ag-gag laws. The centrality of the human— however deeply our species is itself internally riven by the commodification process—is less a matter of norms or beliefs than an order of things built into the fabric of a commodified world.

Notes

1. Karl Marx, *Capital,* vol. 1 (New York: Penguin, 1976), 125. Hereafter cited parenthetically in text.

2. Jacques Derrida, "Passages—from Traumatism to Promise," in *Points . . . Interviews, 1974–1994*, ed. Elisabeth Weber, trans. Peggy Kamuf and others (Stanford: Stanford University Press, 1995), 372–95 (386).

3. Dinesh Wadiwel, "'Like One Who Is Bringing His Own Hide to Market': Marx, Irigaray, Derrida and Animal Commodification," *Angelaki* 21, no. 2 (2016): 65–82 (68).

4. Nicole Shukin, *Animal Capital: Rendering Life in Biopolitical Times* (Minneapolis: University of Minnesota Press, 2009), 76. Hereafter cited parenthetically in text.

5. See Anson Rabinbach, *The Human Motor: Energy, Fatigue, and the Origins of Modernity* (Berkeley: University of California Press, 1992).

6. Alexander Pope, *The Rape of the Lock,* in *The Poems of Alexander Pope* (New Haven: Yale University Press, 1973), 217–42 (1.2). Hereafter cited in text by canto and line number.

7. That Pope names the distant origins of the "glitt'ring spoil" reflects a mercantilist understanding of value as derived from the circulation of rare things (132). See Wolfram Schmidgen, *Eighteenth-Century Fiction and the Law of Property* (Cambridge: Cambridge University Press, 2002), chapter 4.

8. John Locke, *An Essay Concerning Human Understanding* (Penguin: London, 1997), 268.

9. In *An Essay on Man,* Pope admires the "half-reas'ning elephant," in *The Poems,* 501–47 (1.222).

10. György Lukács, *History and Class Consciousness* (Cambridge, Mass.: MIT Press, 1971), 110. Hereafter cited parenthetically in text.

11. Alfred Schmidt, *The Concept of Nature in Marx* (London: Verso, 2014), 96.

12. Vladimir Vernadky, *The Biosphere* (New York: Copernicus, 1997), 39. "Activated by radiation, the matter of the biosphere collects and redistributes solar energy, and converts it ultimately into free energy capable of doing work on Earth" (44).

13. Noel Castree, "Commodifying What Nature?" *Progress in Human Geography* 27, no. 3: 273–97 (282).

14. See Georges Bataille, *The Accursed Share,* vol. 1 (New York: Zone Books, 1991).

15. John Locke, *Two Treatises of Government,* ed. Peter Laslett (Cambridge: Cambridge University Press, 1988), 297, emphasis in original. Hereafter cited parenthetically in text.

16. See Robert Markley, "'Land Enough in the World': Locke's Golden Age and the Infinite Extension of 'Use,'" *SAQ* 98, no. 4 (1999): 817–37.

17. Jason Moore, *Capitalism in the Web of Life: Ecology and the Accumulation of Capital* (London: Verso, 2015), 54. Hereafter cited parenthetically in text.

18. Andreas Malm, *Fossil Capital: The Rise of Steam Power and the Origin of Global Warming* (London: Verso, 2016).

19. Neil Smith, "Nature as Accumulation Strategy," *Socialist Register* 43 (2007): 19–41 (30).

20. Morgan Robertson, "Measurement and Alienation: Making a World of Ecosystem Services," *Transactions of the Institute of British Geographers* 37 (2011): 384–401 (387).

21. Rosemary-Claire Collard and Jessica Dempsey, "Life for Sale? The Politics of Lively Commodities," *Environment and Planning A* 45 (2013): 2682–99 (2684). Hereafter cited parenthetically in text.

22. See, for example, Larry Lohmann, "Performative Equations and Neoliberal Commodification: The Case of Climate," in *Nature™ Inc.: Environmental Conservation in the Neoliberal Age,* ed. Bram Büscher, Wolfram Dressler, and Robert Fletcher (Tucson: University of Arizona Press, 2014), 158–80.

23. Robert Costanza et al., "The Value of the World's Ecosystem Services and Natural Capital," *Nature,* May 15, 1997, 253–60.

24. William Boyd, W. Scott Prudham, and Rachel A. Schurman, "Industrial Dynamics and the Problem of Nature," *Society and Natural Resources* 14 (2001): 555–70 (557).

25. Donna Haraway et al., "Anthropologists Are Talking—About the Anthropocene," *Ethnos* 81, no. 4 (2015): 1–30 (23).

26. William Boyd, "Making Meat: Science, Technology, and American Poultry Production," *Technology and Culture* 43, no. 4 (October 2001): 631–64 (633).

27. See Bill Winders and David Nibert, "Consuming the Surplus: Expanding 'Meat' Consumption and Animal Oppression," *International Journal of Sociology and Social Policy* 24, no. 9 (2004): 76–96.

28. James Stanescu, "Beyond Biopolitics: Animal Studies, Factory Farms, and the Advent of Deading Life," *PhaenEx* 8, no. 2 (Fall/Winter 2013): 135–60 (148).

29. Ted Benton, *Natural Relations: Ecology, Animal Rights, and Social Justice* (London: Verso, 1993), 59.

30. Rosemary-Claire Collard and Jessica Dempsey, "Capitalist Natures in Five Orientations," *Capitalism Nature Socialism,* July 2016, http://www.tandfonline.com/doi/abs/10.1080/10455752.2016.1202294.

31. And yet even these species, protected by global treaties and subject to intense conservation efforts, remain vulnerable to intensive poaching pressures. Tens of thousands of elephants are killed each year to supply the illegal ivory trade. Lucinda Cole notes that by the mid-seventeenth century, European trade had exhausted the ivory stocks, and elephant herds, on Africa's west coast. "Guns, Ivory, and Elephant Graveyards: The Biopolitics of Elephants' Teeth," in *Animals and Humans: Sensibility and Representation, 1650-1820,* ed. Katherine Quinsey (Oxford: Oxford University Press, 2017), 35–55.

32. Sami Torssonen studies "the commodification of livestock welfare," wherein "public concern for livestock thriving now enables the sale of scientifically-certified welfare products." "Sellfare: A History of Livestock Welfare Commodification as Governance," *Humanimalia* 7, no. 1 (Fall 2015): n.p.

33. Anna Tsing, *The Mushroom at the End of the World: On the Possibility of Life in Capitalist Ruins* (Princeton: Princeton University Press, 2015), 33.

34. Bram Büscher et al., "Toward a Synthesized Critique of Neoliberal Biodiversity Conservation," *Capitalism Nature Socialism* 23, no. 2: 4–30.

35. As many have observed, Marx's fetishizes *use*. Jean Baudrillard, for example, observes in *The Mirror of Production* (St. Louis: Telos Press, 1975) that Marx failed to question the "naturalist propositions" he inherited from classical political economy: "the useful finality of products as a function of needs" and "the useful finality of nature as a function of its transformation by labor" (56).

36. Schmidt notes that in a "classless society," the "problem of nature"—as object of labor and source of appropriated use values—"continues to exist for men in their new-found solidarity." *The Concept of Nature in Marx*, 136.

37. Davi Kopenawa and Bruce Albert, *The Falling Sky: Words of a Yanomami Shaman* (Cambridge, Mass.: Harvard University Press, 2013), 331. See also Regenia Gagnier, *The Insatiability of Human Wants: Economics and Aesthetics in Market Society* (Chicago: University of Chicago Press, 2000), chapter 1.

38. Kopenawa and Albert, *Falling Sky*, 336.

39. Sianne Ngai, *Our Aesthetic Categories: Zany, Cute, Interesting* (Cambridge, Mass.: Harvard University Press, 2015), 13.

40. Alfred Sohn-Rethel, "Historical Materialist Theory of Knowledge," *Marxism Today*, April 1965, 119.

41. Eduardo Viveiros De Castro, *Cannibal Metaphysics* (Minneapolis: Univocal, 2014), 55, 58.

Power Down

JOSEPH CAMPANA

Power down—why not just power down? Everything seems to scream it: the rise of CO_2 levels, the rise of sea levels, superstorms, temperature swings, and polar vortices, not to mention the exponentially accelerating pace of extinctions, displacements, and deforestation, or the melting ice sheets, ice shelves, and glaciers, all of which make up a short list (you surely have your own) of only a few of the many symptoms of the already irrevocably altered ecosystems in which the lives of humans and other creatures will be increasingly trying and, perhaps eventually, impossible. If only we could surrender our addiction to cheap, easy fuel, change might be possible. Or so proponents of sustainability maintain as when Richard Heinberg sounds that call in *The Energy Reader: Overdevelopment and the Delusion of Endless Growth*: "As we power down, we will find new ways to use the technologies and scientific understanding developed during our brief, unsustainable, and probably unrepeatable period of high energy use in order to make the inevitable energy decline survivable and perhaps even salutary. But power down we must."[1] It seems so simple when stated bluntly: power down to prevent disaster. And yet persistent inattention to the increasingly visible problem of energy generation and consumption—from destructive extraction to wasteful expenditure to toxic remainders—leaves many to marvel at the glacial pace of change.

The more extreme forms of climate change denial represent merely the most inflammatory of the counterimpulses to fundamentally restructuring our relationship to energy. Attachment to convenience and privilege, the inhibiting expense associated with the transformation of energy generation and transmission, and the rapidly decreasing social capacity to imagine a future beyond one's own—or any future at all—represent only a few of the major points of resistance that keep us locked in our current loop statement. Indeed, more pernicious and perhaps less apparent conflicts and paradoxes make powering down both less likely to occur and, even when it does, ambivalent at best. Power fluidly refers to energy systems and sociopolitical systems alike, which constitute nonidentical but overlapping circuits of efficacy and resistance. While a great deal of writing considers the politics of energy, the entangled nature of *power* and *energy* constitutes a conceptual and political knot Imre Szeman and Dominic Boyer recently address through the term *energopolitics,* which Boyer defines as "power over (and through) energy," arguing that "the staggering significance of energy as *the* undercurrent and integrating force for all other modes and institutions of modern power has remained remarkably silent even in this era of so much talk about climate change, energy crisis and energy transition."[2] Energopolitics is thus prior to (if modeled on) biopolitics, since "power over energy has been the companion and collaborator of modern power over life and population from the beginning."

It is a struggle to find language, then, to capture both the distinctions and the collusions between power and energy. But if there is a locus, a ground zero, for that entanglement it may not be in the institutions and infrastructures of power and energy but rather in bodies. However dwarfed by infrastructures humans may be and however significantly fossil fuels have catalyzed explosive, civilization-altering growth by exceeding the energy potentialities of human or animal bodies, flesh still matters inasmuch as it can clarify not only the relationship between power and energy but also the intimate and often disastrous intertwining of affect, energy, and economics.[3] Even if we suspend any skepticism about the efficacy of sustainability,[4] the potentially species-saving benefits of powering down exist in exquisitely torturous tension with the life-draining operations of late capitalism. In spite of the pronounced emergence of off-the-grid and alternative-energy enthusiasms[5] (which has already been captured by capitalist enterprise), perhaps most perniciously we struggle to articulate what it is to power down without also avoiding the painful tedium of change, which might explain a noticeable cultural

preference for media spectacles of self-annihilation and species destruction over patient attention to changing habits and minds. For the most
part, we don't want to power down. But it isn't just that we don't want to
power down. Perhaps we can't power down because we don't know how
to think about what powering down might look like and, in fact, what it
might feel like.[6]

~

When did the world start running out of energy? Two answers to this
improbable question share the ring of truth that the certainties of myth
and science respectively imply. *The world began running out of energy
when the world began.* What is an origin myth if not a tale of inevitable
decline? Thus so many tales of the beginning of the world in so many
mythic traditions tell a story of gradual degradation. Biblical tradition
offers a fall. The Greeks were not alone in their calendar of descent:
Gold, Silver, Bronze, and Iron Ages trace increasing spiritual and environmental degradation. Thus one way to understand the world is that
which gradually exhausts itself, depletion occurring inevitably over millennia. While myth exhibits our particularly human attachment to tales
of decline, the first law of thermodynamics suggests quite the opposite
in insisting that energy can be neither created nor destroyed. A system
conserves energy, which transforms but does not disappear. Thus one
might say, *The world will never run out of energy.* One might even say
that to pose the question as such is to take a particularly and peculiarly
human point of view.

Barring miraculous technological advances, humans may run out of
easy and affordable energy sources to power the massive machinery of
twenty-first-century global capitalism.[7] However improbable our fantasies of such miracles of science are, the hangover of this era of fossil fuels
will likely last for millennia. From that point of view, other answers arise
to the question *when did the world start running out of energy?* For
instance, *March 7, 1956*, when Marion King Hubbert delivered "Nuclear
Energy and the Fossil Fuels" at the spring meeting of the Southern District of the American Petroleum Institute in San Antonio, Texas. Hubbert
argued that "the evolution of our knowledge of petroleum since Colonel
Drake's discovery of oil in Titusville, Pennsylvania, nearly a century ago,
resembles in many striking respects the evolution of knowledge of world

geography which occurred during the century following Columbus' discovery of America."[8] In the framing of what was to become Hubbert's iconic if controversial contribution to our thinking about energy futures, tropes of discovery and navigation necessitate "an approximate idea of where we are now, and where we are going."[9]

And yet how quickly the optimism and zeal of discovery are tempered by the anticipation of decline. The exponential growth in rates of production belies the seeming longevity of fossil fuel use. Hubbert notes that "coal has been mined continuously for about 800 years, and by the end of 1955 the cumulative production for all this time was 95 billion metric tons. It is somewhat surprising, however, to discover that the entire period of coal mining up until 1925 was required to produce the first half, while only the last 30 years has been required for the second half." As with coal, so too with oil but more so. Hubbert insists that of the 53 billion barrels of oil produced by 1955, "the first half of this required from 1859 to 1939, or 80 years to be produced; whereas, the second half has been produced during the last 16 years."[10] The consequence was an approaching date of "culmination of world production," which Hubbert assumed would happen, with respect to "petroleum and natural gas in both the United States and the state of Texas, . . . within the next few decades." Thus was born Hubbert's Peak, or the theory of peak oil: that given the finite nature of fossil fuel reserves, the point after which production would decline could be calculated. Perhaps it is not surprising that Hubbert uses the term *exhaustible resources,* which seems more honest and direct than *nonrenewable.*[11] Indeed, perhaps we might think more clearly about energy realities and futures if we focused our attention not on the fantasy of an infinitely renewable earth but on acts of expenditure and depletion.

Still, Hubbert by no means imagined himself as a horseman of the apocalypse. "This does not necessarily imply," he insisted, "that the United States or other parts of the industrial world will soon be destitute of liquid and gaseous fuels. . . . But it does pose as a national problem of primary importance, the necessity . . . of gradually having to compensate for an increasing disparity between the nation's demands for these fuels and its ability to produce them from naturally occurring accumulations of petroleum and natural gas."[12] Of course Hubbert had already sounded a particular alarm we still struggle to hear. On September 15,

1949, Hubbert made a presentation to the American Association for the Advancement of Science, which was later published as "Energy from Fossil Fuels." There he similarly examined the exponential and indeed aberrant growth of energy use and the human population. Already, for Hubbert, the boom times of fossil fuel were over. "In fact," he argued, "we have no choice but to proceed into a future which we may be assured will differ markedly from anything we have experienced thus far. Among the inevitable characteristics of this future will be the progressive exhaustion of the mineral fuels."[13]

Some have found fault with Hubbert's calculations and assumptions. Most controversial has been not if, but when the moment peak oil occurred, not to mention what to do as a result. Decline has become an atmosphere, a horizon of expectation. And yet it crystallizes only one part of a particularly powerful half-century arc that left in its wake a pattern of affective and energic vicissitude. Hubbert's anticipation of decline was only possible because of the exuberant emergence of the age of oil some half century earlier. Another answer to the question *When did the world start running out of energy?* might then be *January 10, 1901*, when the salt dome oil fields near Beaumont, Texas, erupted after months of failure:

> On January 10 mud began bubbling from the hole. The startled rough-necks fled as six tons of four-inch drilling pipe came shooting up out of the ground. . . . The Lucas geyser, found at a depth of 1,139 feet, blew a stream of oil over 100 feet high until it was capped nine days later and flowed an estimated 100,000 barrels a day. . . . A new age was born. The world had never seen such a gusher before.[14]

The transformations attendant upon the arrival of what Ross Barrett and Daniel Worden call *oil culture* were profound. That culture, they argue, "encompasses the fundamental semiotic process by which oil is imbued with value within petrocapitalism, the promotional discourses that circulate through the material networks of the oil economy, the symbolic forms that rearrange daily experience around oil-bound ways of life, and the many creative expressions of ambivalence about, and resistance to, oil that have greeted the expansion of oil capitalism."[15] One such rearranging of the patterns of "daily experience around oil-bound ways of life" was the interlacing of energic and affective cycles constituted by the

oscillation between boom and busts, or what Heidi Scott calls "the petroleum era paradox of exuberance and catastrophe,"[16] and which Frederick
Buell describes more broadly as part of a larger "fossil-fuel culture" constituted by an "age of exuberance—an age that is also, given the dwindling finitude of the resources it increasingly makes social life dependent
on, haunted by catastrophe."[17] Henceforth alternating rhythms of under-
and overproduction, skyrocketing and plummeting oil prices would send
shivers not merely through economic systems but through bodies and
the bodies politic throughout the twentieth century, including multiple
waves of instability: the oil crises of the 1970s and 1980s. Those not
already too cynical to find jubilation in the landmark environmental
accord recently reached in Paris may have found their celebrations cut
short as anticipated (and perhaps motivating) forms of scarcity blossom
into unexpected and potentially catastrophic abundance in energy markets, as blazoned recently on the front page of the *New York Times*: "Barely
a month after world leaders signed a sweeping agreement to reduce carbon emissions, the global commitment to renewable energy sources faces
its first big test as the price of oil collapses."[18] Radically unstable markets
are not merely one more obstacle to radical change, though they certainly obstruct change. Nor are they merely symptomatic of a world out
of balance. Boom–bust cycles are lived in the body and in the cultural
sphere. Thus powering down cannot be imagined solely in the language
of virtuous activism—turning off lights, turning off taps, turning to bikes.
Powering down must occur precisely where economy and ecology intersect, in the susceptible and interlocking circuits of feeling and flesh.

～

Capitalism exhausts. It alienates and expropriates. It compensates pain,
which is one way of describing labor. For Karl Marx, "capital is dead
labour, that, vampire-like, only lives by sucking living labour, and lives
the more, the more labour it sucks. The time during which the labourer
works, is the time during which the capitalist consumes the labour-
power he has purchased of him."[19] By the force of its strange alchemy,
capitalism drains the flesh, aspiring to make the human body available for an infinite process of extraction of both labor and energy. Thus
Marx devoted part of *Capital* to historical variations in the length of
the work day. "The prolongation of the working-day," he argues, "beyond

the limits of the natural day, into the night, only acts as a palliative. It quenches only in a slight degree the vampire thirst for the living blood of labour. To appropriate labour during all the 24 hours of the day is, therefore, the inherent tendency of capitalist production."[20] As Francesco "Bifo" Berardi puts it, in such a state, "the worker appears overwhelmed, reduced to a passive appendage producing empty time, to a lifeless carcass."[21] More recently David McNally argues that the zombie and the vampire constitute paradigmatic monsters that best capture "the monstrous forms of everyday life in a capitalist world-system."[22]

That capitalism exhausts is ever-more documented in recent decades. Bifo has lucidly explained the vampiric nature of capital: "Our desiring energy is trapped in the trick of self-enterprise, our libidinal investments are regulated according to economic rules, our attention is captured in the precariousness of virtual networks: every fragment of mental activity must be turned into capital."[23] And yet what accompanies late capitalism's (or, as Bifo puts it, *semiocapitalism's*) capture of desiring energy are manic upsurges and depressive downturns. Thus while the twenty-first century begins with "a psychopathic phenomenon of over-excitation and panic," it follows hard upon the heels of "the financial euphoria of the 1990s," which itself "gave way to spectacular depression."[24] In essence, what accompanies the drive to empty out human flesh is a poisonous intertwining of affect and energy that crystalizes in the form of a boom–bust economy mimicking the vicissitudes not only of financial capital but of the energy transformations of the twentieth century.

In *24/7: Late Capitalism and the Ends of Sleep,* Jonathan Crary too explores the advancing reality of a pernicious fantasy, as technological and pharmaceutical developments make increasingly possible, if decreasingly desirable, a state of perpetual work, which would therefore also represent a state of perpetual excitations and extraction of energies from laboring flesh. As he argues, "24/7 markets and a global infrastructure for continuous work and consumption have been in place for some time, but now a human subject is in the making to coincide with these more intensely."[25] No wonder popular culture witnesses another wave of night-dwelling vampires, Marx's figure par excellence of life-draining capitalism. No wonder we are simultaneously mobbed by yet another wave of slow-moving, brain-delectating zombies, figures par excellence of late-capitalist enervation and lobotomization.

No wonder, too, alarms have rung out from the watchtowers of Silicon Valley. In a 2011 op-ed for the *New York Times,* Tony Schwartz describes the parameters of this "personal energy crisis" impacting "corporate executives, government employees, small-business owners, hospital administrators and physicians, school administrators and teachers."[26] "As demand has increased in recent years," he argues, "fueled by advances in technology, we've relied on an external resource to get more done: time. When there's more to do, we put in more hours." Thus the task of the new worker is to manage not time but energy: "We're facing an energy crisis, and this one's personal." Schwartz's op-ed serves not only to call attention to this growing "personal energy crisis" but also to advertise the services of the Energy Project, a consultancy and training firm that works "to create workplaces that are healthier, happier, and higher performing." Among the key ideas behind kinder, gentler capitalism would be energy management: "We're not meant to run at high speeds, continuously, for long periods of time. Science tells us we're at our best when we move rhythmically between spending and renewing energy—a reality that companies must embrace to fuel sustainable engagement and high performance."[27] Thus a workplace that satisfies the physical, emotional, mental, and spiritual needs of its employees "increases engagement, increases positive energy at work, builds employee loyalty, and increases life satisfaction."[28] Here then emerges one of the more vicious of the paradoxes of late capitalism. Capitalism exhausts and yet its deleterious tendency to exhaust becomes a spur to ever-greater productivity, not to mention product creation. We've witnessed the rise of personal or workplace energy consultation alongside ever-burgeoning energy bars, energy drinks, 5-hour energy, energy without jitters, energy without caffeine, quick bursts of energy, long-lasting energy.

There are more and more ways to power up, which seems only one more poisonous aspect of the capture of human flesh by capitalist urgencies.[29] "Sleep," Crary imagines, "poses the idea of a human need and interval of time that cannot be colonized and harnessed to a massive engine of profitability" and thus remains "an incongruous anomaly and site of crisis in the global present. In spite of all the scientific research in this area, it frustrates and confounds any strategies to exploit or reshape it. The stunning, inconceivable reality is that nothing of value can be extracted from it."[30] Yet the booming sleep industry suggests otherwise. In

2012 alone, products and services related to the cultivation of better sleep constituted a \$32.4 billion market.[31] While sleep may not be the utopia Crary imagines, the dilemmas posed by endless work and interrupted sleep ramify broadly. The very idea of "24/7," he argues, "is inseparable from environmental catastrophe in its declaration of permanent expenditure, of endless wastefulness for its sustenance, in its terminal disruption of the cycles and seasons on which ecological integrity depends."[32] At the heart of Crary's chilling analysis is the hope of a return to the natural—ecological integrity and sounder sleep. But how can we parse exhaustion in an age characterized by manic oscillation, like a child violently flicking a switch on and off. Power up or power down?

~

If the lights go down, how will you feel?

"They say it only took eight minutes for the world to go dark," announces a voice in darkness. But don't adjust your television set: it's only a show. On September 17, 2012, the television network NBC premiered *Revolution,* its named spelled with the familiar power icon (an o with a line through it) from electronics devices replacing the second *o.* It may come as little surprise, then, to hear that this postapocalyptic sci-fi–fantasy hybrid treats an energy apocalypse. What happens when a mysterious and seemingly permanent blackout seizes the earth, when "it turns off and it never turns back on?" The trailers for the series featured shots of darkness sweeping across highways, across cities, and across continents. Over a newly darkened city, an airplane spins out of control, crashes down, and explodes, providing the brightest of lights in the newly dimmed city. After the initial panic and chaos, what emerges are war and a regression to some fantasy of medieval life, complete with feudal power brokering, sword fights, threats of heads on pikes, and surprisingly immaculate costuming and styling. Apparently not all conveniences were lost. Far be it for any of us to expect realism from dramatic major network television. But that's not the point. The interesting question is not, if "it turns off and never turns back on" what would you do? After all, we've faced a decades-long glut of postapocalyptic science fiction, high and low, treating the practical, survivalist aspects of the regression and even disappearance of human civilization. The emergence of "cli-fi," or climate-change science fiction indicates a persistent inclination to reimagine the contours of life through the lens of disaster and decline.

If the lights go out, what will you feel? Many obvious answers emerge: terror, shock, and uncertainty represent merely a few. But for *Revolution*, a particular affective palate emerges as panic and mourning give way to anger and even a longing for revenge. In an early promo for the show, titled on YouTube "Our Hope," a central character, Charlotte "Charlie" Matheson admits, "Most people wish they could go back to before the blackout. Not me. When the world lost power, I found mine." What does enervation produce if not an alternative form of energy in the form of affect here captured or, more properly, hijacked by righteousness? A later promo showcases the program's freedom fighters, "They'll take back the power to take back the world." And here we have two of the more predictable fantasies articulated in this show—that in spite of the collapse of civilization and the emergence of powerful feudal lords (i.e., bad guys with power), families will be reunited and the people will rise up to conquer. Thus energy and ecology are secondary to human drama. Second, the blackout results not from overconsumption or careless stewardship but energy-zapping military nanotechnology gone wrong.

Righteous anger, community spirit, zeal for liberation: these are the affective tones that color a certain fantasy of the aftermath of energy collapse. In *Revolution* only one central character (who then dies or appears to die in the first episode) knows of the approaching darkness; he is, understandably, anxious, whereas the world blithely spins on, impending crisis unbeknown to the masses. And so it continues outside the frame of fictionalized crisis: some blithely spin on while others anticipate doom. What affective tonalities, somewhere between denial and apocalypse, might we associate with the anticipation of enervation? Can we veer from disaster without veering into false triumphalism?

Political change in the world of *Revolution* depends entirely on the violently oscillating affective patterns of the age of fossil fuels. One minute the world thrums with energic and economic activity and the next it collapses, complete with the terrifying images of airplanes falling from the sky. But energy that disappears from the grids and devices of the world cycles back in this fantastic, quasi-feudal world as people power. Such a fantasy distracts from the political paralysis, indecision, and denial that still dominate conversations about climate and energy. In the first days of 2016, the U.S. Senate, in its infinite wisdom, "voted virtually unanimously that climate change is occurring and not, as some Republicans have said, a hoax—but it defeated two measures attributing its causes to human activity."[33] This too is what progress looks like in an era in which the quasi-feudal longings of *Revolution* evade human responsibility for climate dilemmas by seeking an era apparently free of the enclosing apparatuses of late capitalism. The people will rise up, the lights will come back on, and the land will be cleansed. Powering down is really a form of powering up.

~

It's hard to turn away from rallying cries and the faith in human activism and community in an era such as ours, one characterized by extreme confusion, extraordinary paralysis, and an unwillingness to think and feel deeply about already-arriving futures. It is then ever-more important to ask what it might mean to power down. Perhaps not as a revolutionary or iconoclastic movement but as a critical pause, a ceasing of the engines of progress, a flattening out of the violent vicissitudes of petrocapitalism, a way of seeing and thinking and feeling differently. We exhaust ourselves in a late capitalist rat race of accumulation and inattention. Might we want to consider exhausting ourselves, being exhausted, as a way of making contact with different affective and corporeal realities?

To close, I want to turn to *Life* itself, that iconic instrument of American culture, and its coverage of the great 1965 blackout that struck the East Coast—and New York City particularly.[34] In a view from Times Square, looking east, a lightless Manhattan appears to sleep under a brilliant moon on the cover that also announces: "5:28 p.m., Nov. 9th: The Lights Went Out."[35] In "A Dark Night to Remember," Loudon Wainwright celebrates this suspension:

It shouldn't happen every evening, but a crisis like the lights going out has its good points. In the first place it deflates human smugness about our miraculous technology, which, at least in the area of power distribution and control, now stands revealed as utterly flawed. I have never regarded as trustworthy many of the appurtenances of modern America—automatic elevators, for example, have always chilled me with their smooth mindlessness—and it is somehow delicious to contemplate the fact that all our beautiful brains and all those wonderful plans and all that marvelous equipment have combined to produce a system that is unreliable.[36]

From the vantage point of 1965, Wainwright could not perhaps have imagined the regimes of "miraculous technology" to come. And yet he could already diagnose the paralysis that arises from being embedded in and exhausted by the tangled webs of technology and capital:

> Even better is the fact that *something happened*. With its virtually instantaneous transmission of news from all over, modern communication leaves people with the feeling that they are spectators in a world of action. The truth is that for most people, except for whatever quality of anguish or action exists in their minds, nothing much is going on, and it is exhilarating suddenly to become a performer in a drama, even if the cast has millions.[37]

To be sure, Wainwright's prose and *Life*'s photojournalism cite a series of heroic acts. In the faintest of illumination, a nurse cradles a newborn child, her face eerily defined by emergency lighting. Tales of other rescue abound. But what is perhaps most intriguing are the portraits of inactivity: the portraits of sleepers. Men sleep in a hotel lobby stretched across chairs. A crowd of men and women sleep in a rough circle in another hotel lobby, their bodies aligned with a round decorative medallion. Men recline in barbershop chairs, aprons drawn up as blankets. Men and women sleep on stairs clinging to railings, and each other, in the relentless dark. Even more fascinating than these snapshots of the powered down is the resulting awareness of the intimate entanglement of power systems and energy systems, of flesh and a world sustained by fuel. In explaining the disaster, Theodore H. White writes, "Whether a switching device had failed to operate, whether somewhere in the vast Northeast power grid a giant generator had suddenly gone out of phase, whether a computer experienced a nervous breakdown—no one would be able to determine exactly for some time afterward. But at that moment the electric pulse all up and down the great 345-kilovolt line surged, wobbled, and flickered; and, like a pinched aorta, it caused an entire civilization to flicker with it."[38] Still, in 1965, once could imagine the power grid as a body with "a pinched aorta." What difference might it make if we could still imagine the human body not merely captured by energy and power systems but entangled with them, even if we imagine power systems entangled with and not merely the ruination of larger ecologies? Would that render palpable relationships of mutual care and disaster?[39] Instead, energy systems seem scaled so far beyond the contours of the fragile human body that perhaps it's no surprise the fragile power grid doesn't feel like anything to us.

It sounds too good to be true. *Life*'s often-lyrical celebration of a world without energy belies those forms of terror and disaster that strike at the most vulnerable. This Wainwright admits, "After a few moments of darkness the other night, holding a lighted match near the face of my watch, I forgot about those people who might be endangered by the emergency and hoped the current wouldn't come back on too soon."[40] The consequence of such forgetting may be especially apparent to those, like us, who live in an era of exponentially strengthening superstorms, one in which the impacts of climate instability disproportionately impact the economically disadvantaged. Indeed it was not hope, solidarity, and community but fear, terror, and violence that would characterize the 1977 New York City blackout.[41] But there may be something more thoughtful and salubrious to retain from Wainwright's interest, which is not in some form of revolutionary people power that fantasizes its way out of energy and climate nightmares but in the forms of community and mutual care that arise when energy is in abeyance. "In the darkness," he notes, people "emerged, not as shadows, but far warmer and more substantial than usual. Stripped of the anonymity that goes with full illumination, they became humans conscious of and concerned about the other humans about them."[42]

It might seem that the appropriate response to being exhausted is empowerment. Yet if we assume that energy, everything from the galvanic pulses of bodies to the fantasy of cold fusion, can be sustained with respect to an ever-growing population, we thereby retain the endless-growth model of capitalism that conditions interlocking systems of energy, affect, and social power and renders sustainability impossible. What if we instead exhaust ourselves in a different manner, powering down not just our electronic devices but our affective states, responding to social and ecological crises (and apparent public indifference) not with fear, panic, anger, or false empowerment but with an aesthetics of exhaustion? For some proponents of sustainability, beauty is a lure that may still save us all. Robert E. King joins many other environmental activists in the hope of an economy of limitless abundance. "Nature," he argues, "produces the most profound, magnificent, and nurturing examples of beauty in endless abundance, and for free."[43] More importantly, he suggests, "returning to a sustainable way of life need not be thought of as sacrifice; instead it can be seen as an opportunity to increase aesthetic pleasure and the spiritual nourishment that comes from living in the midst of incalculable beauty."[44] In this vision, limitless natural abundance chases away the painful scarcities of economy. But is this another boom in search of its inevitable and subsequent bust? In an era of instability and indifference and in the anticipation of decline, a political aesthetics that poses the beauty of nature against the grotesqueries of petrocapitalism might not serve as well as an aesthetics of exhaustion. Imagine those bodies—asleep in the lobby, asleep in chairs, asleep on the floor, asleep on a train—powering down to dream.

Notes

Many thanks to Dominic Boyer, Jeffrey Jerome Cohen, Lowell Duckert, Tim Morton, Derek Woods, and Marina Zurkow for their thoughtful comments on a draft of this essay. Thanks also to the attentive and responsive audience to whom I first presented this work at the 2013 Culture of Energy Symposium, sponsored by Rice University's Center for the Study of Energy and Environment in the Human Sciences (culturesofenergy.com).

1. Richard Heinberg, introduction to *The Energy Reader: Overdevelopment and the Delusion of Endless Growth*, ed. Tom Butler, Daniel Lerch, and George Wuerthner (Sausalito, Calif.: Watershed Media, 2012), xvi.

2. Dominic Boyer, "Energopolitics and the Anthropology of Energy." *Anthropology News* (May 2011): 5.

3. That fossil fuels derive from lifeforms dead for millions of years is particularly the subject of new media artist Marina Zurkow. See particularly "Necrocracy" and "Outside the Work: A Tasting of Geological Time," http://www.o-matic.com/.

4. Of the many critiques of sustainability, see particularly the game-changing Alan Stoekel, *Bataille's Peak: Energy, Religion, and Postsustainability* (Minneapolis: University of Minnesota Press, 2007).

5. See particularly Matthew Schneider-Mayerson, *Peak Oil: Apocalyptic Environmentalism and Libertarian Political Culture* (Chicago: University of Chicago Press, 2015).

6. For a related approach to energy and affect, see Stephanie LeMenager's notion of "petromelancholia" in *Living Oil: Petroleum Culture in the American Century* (Oxford: Oxford University Press, 2014).

7. Regardless of how long the current oil glut lasts, we exist in what Michael T. Klare calls an era of "tough oil": "The world still harbors large reserves of petroleum, but these are of the hard-to-reach, hard-to-refine, 'tough oil' variety. From now on, every barrel we consume will be more costly to extract, more costly to refine—and so more expensive at the gas pump." "A Tough-Oil World," *Guernica,* March 13, 2012, https://www.guernicamag.com/daily/michael_klare_a_tough-oil_world/.

8. M. King Hubbert, "Nuclear Energy and the Fossil Fuels." Exploration and Production Research Division, Shell Development Company, publication no. 95 (June 1956), 1.

9. Ibid., 3.

10. Ibid., 9.

11. Ibid., 11.

12. Ibid., 27.

13. M. King Hubbert, "Energy from Fossil Fuels," *Science,* February 4, 1949, 108.

14. Robert Wooster and Christine Moor Sanders, "Spindletop Oilfield," *Handbook of Texas Online,* June 15, 2010, http://www.tshaonline.org/handbook/online/articles/dos03.

15. Ross Barrett and Daniel Worden, introduction to *Oil Culture,* ed. Ross Barrett and Daniel Worden (Minneapolis: University of Minnesota Press, 2014), xxvi.

16. Heidi Scott, "Whale Oil Culture," in Barrett and Worden, *Oil Culture,* 5.

17. Frederick Buell, "A Short History of Oil Cultures," in Barrett and Worden, *Oil Culture,* 71. For a much earlier account of the interpenetration of energy cycles and human cultural formations, see Leslie A. White, "Energy and the Evolution of Culture," *American Anthropologist* 45, no. 3 (1943): 335–56.

18. Clifford Krauss and Diane Cardwell, "Climate Pact Put to Test by Drop in Price of Oil," *New York Times,* January 25, 2016.

19. Karl Marx, *Capital,* vol. 1 (New York: International, 1967), 224.

20. Ibid., 245.

21. Francesco "Bifo" Berardi, *The Soul at Work: From Alienation to Autonomy* (Los Angeles: Semiotext(e), 2009), 61.

22. David McNally, *Monsters of the Market: Zombie, Vampires, and Global Capitalism* (Leiden: Brill, 2011), 3.

23. Berardi, *Soul at Work,* 24.

24. Francesco "Bifo" Berardi, *Heroes: Mass Murder and Suicide* (London, Verso: 2015), 55.

25. Jonathan Crary, *24/7: Late Capitalism and the Ends of Sleep* (London: Verso, 2013), 3–4.

26. Tony Schwartz, "The Personal Energy Crisis," *New York Times,* July 23, 2011.

27. http://theenergyproject.com/.

28. http://theenergyproject.com/key-ideas.

29. See also the rapidly expanding body of literature on speed and accelerationism, especially Robin Mackay and Armen Avanessian, *#ACCELERATE: The Accelerationist Reader* (Falmouth: Urbanomic, 2013); Benjamin Noys, *Malign Velocities: Accelerationism and Capitalism* (New York: Zero Books, 2013).

30. Crary, *24/7,* 10–11.

31. Maureen Mackey, "Sleepless in America," *Fiscal Times,* July 23, 2012, http://www.thefiscaltimes.com/Articles/2012/07/23/Sleepless-in-America-A-32 -4-Billion-Business.

32. Crary, *24/7,* 10.

33. John Lundin, "As 2015 Crushes Heat Record, Republicans Vote to Deny Climate Change," *Politicususa,* January 28, 2016, http://www.politicususa.com/ 2016/01/28/2015-crushes-heat-record-republicans-vote-deny-climate-change .html.

34. See David Nye's *When the Lights Went Out: A History of Blackouts in America* (Cambridge, Mass.: MIT Press, 2010).

35. See *Life,* November 19, 1965.

36. Loudon Wainwright, "A Dark Night to Remember," in *Life,* 35.

37. Ibid., emphasis in original.

38. Ibid., 46B.

39. For Jane Bennett, a North American blackout in August 2003 encourages a recognition of powers and agencies beyond the human. "How," she asks, "does recognition of the nonhuman and nonindividuated dimensions of agency alter established notions of moral responsibility and political accountability?" in "The Agency of Assemblages and the North American Blackout," *Public Culture* 17, no. 3 (2005): 446.

40. Wainwright, "Dark Night," 35.

41. See James Goodman, *Blackout* (New York: FSG, 2003).

42. Ibid., 35.

43. Robert E. King, "Cap the Grid," in Butler, Lerch, and Wuerthner, *The Energy Reader,* 327.

44. Ibid., 328.

Obsolesce

MARGARET RONDA

"There should be more nouns / For objects put to sleep / Against their will," begins "Phone Booth," a recent poem by Brenda Hillman that meditates on the lost pleasures of the phone booth.[1] These opening lines rue the impoverishment of our vocabulary for describing what happens to objects and materials that have been "put to sleep" before they are worn out and what remains of their obdurate, even "will[ful]" presence as they are rendered inactive by market forces or consumer whim. This not-yet-created language must account, Hillman suggests, not only for the complex material changes these objects undergo as they are placed into disuse but for the concomitant shifts in everyday phenomenological experience as new commodities and technologies replace older forms. It is this latter aspect that "Phone Booth" takes up, considering various dimensions of the phone booth as it once functioned in everyday life and its supersession by the cell phone. Exploring yet another turning point in what Walter Benjamin calls the "increasing atrophy of experience" that accompanies technological innovation under capitalism, Hillman's poem undertakes a retrospective, symbolic revaluation of a now-superseded technology and the slower, more ruminative encounters it afforded.[2] "We twisted the rigid cord / As we spoke / It made a kind of whorl," the poem ends, wistfully.

"Phone Booth" also points, in intriguing if undeveloped fashion, toward a continuing trajectory of the commodity as it is made to disappear from everyday use. Hillman's phrase "put to sleep" evokes a kind of spell cast on the object to render it outmoded. The mystical, fairy-tale-like language of this phrase echoes, in reverse, Karl Marx's description of the dynamic transfiguration that occurs to an object as it assumes the form of the commodity. As Marx writes in *Capital*:

> But as soon as it emerges as a commodity, it changes into a thing that transcends sensuousness. It not only stands with its feet on the ground, but, in relation to all other commodities, it stands on its head, and evolves out of its wooden brain grotesque ideas, far more wonderful than if it were to begin dancing of its own free will.[3]

The objects Hillman describes are not awakened into a new, topsy-turvy form but instead cast into a premature slumber. Yet in both cases, the object is transfigured into something strange—what Marx calls "suprasensible."[4] We might see Hillman's poem as pointing to the *end* of the process of commodification that Marx describes in *Capital*: the no-longer-circulating commodity can finally rest, "put to sleep" after its uncanny awakening. Its *social* life seems to end, as it begins, with a determination that inheres not in its use-value but its position within larger relations of production.[5] Hillman intimates, however, that this is not the end of the tale but another chapter, one that requires a new vocabulary. We need a richer language, she claims, for charting these commodity forms as they slowly alter into something unrecognizable—as they *obsolesce*.

The abstract injunction placed on the artifact so that it is framed as no longer significant, able to be forgotten, is of course a key productive logic of capitalism. The innovative advances that drive growth and the pursuit of profit in capitalist economies ineluctably create obsolescence in various forms, from "built-in" obsolescence in technological and other consumer products to outdated forms of industrial infrastructure. Tied to the dynamics of commodity production and circulation, the logic of obsolescence highlights the essential quality of "creative destruction" in capitalist economies. As Joseph Schumpeter writes in his 1942 work *Capitalism, Socialism, and Democracy*, "The fundamental impulse that sets and keeps the capitalist engine in motion comes from the new consumers' goods, the new methods of production and transportation, the new markets, the new forms of industrial organization that capitalist enterprise creates."[6] This ceaseless innovation, Schumpeter argues, "incessantly revolutionizes the economic structure *from within*, incessantly destroying the old one, incessantly creating a new one." As we will see in the following section, which highlights the rise of obsolescence as a method of staving off overproduction and generating new growth after World War II, obsolescence necessitates not only the foreshortened duration of a product—often cheaply made, quickly superseded—but a consumer sensibility oriented toward the new and disdainful of the outmoded. Describing this new "rhythm of consumer life," Arjun Appadurai writes that it requires new affective and imaginative practices that undertake "the social discipline of the imagination, the discipline of learning to link fantasy and nostalgia to the desire for new bundles of commodities."[7] Obsolescence as a productive logic thus involves both a supply-side and a demand-side dimension, profit strategy and consumer ideology working in sync.

In all these senses, obsolescence entails a strategic evacuation of value, social circulation, and formal recognizability. Hillman's insight that we lack words for what happens to outmoded goods reveals the extent to which this process is characterized, for the consumer, by negation and erasure. With the introduction of the new commodity, the consumer is "disciplined," in Appadurai's phrase, to regard the older version as worthless and disposable. Hillman's poem evokes this shift as sudden and culturally pervasive:

One day we started to race past
& others started racing
Holding phones to their ears
Holding a personal string
To their lips[8]

As these lines describe, the newer product is often associated not only with greater ease, efficiency, or beauty, but with increased speed. Older means slower, clunkier; newness promises mobility, lightness, seamless interface. The heavy "whorl" of the cord attaching the phone to its rooted circuitry is replaced by the "personal string" invisibly connecting the active speaker with a communications network. Left behind, the cast-off good falls prey to what Michael Thompson, in his groundbreaking study of the anthropology of waste, *Rubbish Theory* (1979), calls "a conspiracy of blindness."[9] This "blindness" is not only a consumer orientation, generated and maintained by market strategems, but also a wider set of relations with regard to the storage, transport, management, and disposal of these goods. Consigned to the landfill, moldering in attics, or sent overseas to be "recycled," these materials lose salience in their prior configuration as first-order commodity but importantly *remain*. Some retain a semblance of their former shape and function; others are reworked into new forms in order to wrest more value. Reminiscent of the spiraling motions of the "whorl" that Hillman evokes in her poem, these various materials enter into newly proliferating cycles beyond their appearance and use.

To attend to the socioecological activity of *obsolescing* thus necessitates thinking against the grain of this economic logic that would consign the obsolete material to invisibility and considering what happens next. As a commodity falls into disuse, how might we understand its modes of existence and its larger impacts as it undergoes transformation? From these remaindered, disordered forms, what new aggregates or extracts might be found? What do these materials reveal about the larger dynamics that foster them and the sites, spaces, and practices developed to manage them? In this chapter I consider how obsolescing indexes, in distinctive ways, the complex relations between economy and ecology organized by the concept of *second nature*. I argue that obsolescing does not always exemplify the endgame of the capital accumulation process but can also embody another phase of its productive relations, one that reflects many of its most exploitative features. Differentiating obsolescing from the larger category of waste, I claim that obsolescing remains tied to capitalism's ongoing pursuit of value. Yet obsolescing also illuminates the uneven temporalities and unintended ecological consequences that accompany this pursuit. Following the trajectory of obsolescing, I examine the ways material cast-offs of a given present convey essential insights into the historicity of this present beyond or below its progressive motions. As Theodor Adorno writes in *Minima Moralia*, "In an order which liquidates the modern as backward, this backwardness, once condemned, can be invested with the truth over which the historical process obliviously rolls."[10]

"Just Throw It Out the Window / Onto Somebody Else"

In the opening of his best-selling 1960 book, *The Waste Makers*, Vance Packard offers a satirical vision of Cornucopia City, a future paradise of consumerism. Cornucopia City produces myriad consumer goods meant to be quickly replaced, such as cars "made of a lightweight plastic that develops fatigue and begins to melt if driven more than four thousand miles."[11] Unveiled alongside these marvels of design is an entire infrastructure for quick disposal: a mart full of "receptacles where the people can dispose of the old-fashioned products they bought on a previous shopping trip," factories on a cliffside that toss unneeded products "directly to their graveyard," and weekly "Navy Days" where a warship is stocked with a variety of surplus goods—"playsuits, cake mix, vacuum cleaners, and trampolines"—to dump in the sea.[12] Organized entirely around harmonious cycles of production, circulation, consumption, and disposability, Cornucopia City is revealed to be the hopeful dream of "marketing experts" who fear the "haunting problem of saturation" in the booming postwar American economy. This is, of course, a problem endemic to capitalist productive relations, as Marx and Engels point out in their 1848 *Communist Manifesto*: "In these crises, there breaks out an epidemic that, in all earlier epochs, would have seemed an absurdity— the epidemic of over-production."[13] These crises are temporarily overcome, according to Marx and Engels, by "enforced destruction of a mass of productive forces" and "the conquest of new markets." For his part, Packard identifies the postwar turn toward planned obsolescence and an intensifying ethos of disposability as following these strategies in order to forestall such an epidemic.[14] The developments of Cornucopia City were already well underway in 1950s American consumer capitalism, Packard argues, creating a culture of profligate and seemingly isolated "waste-makers." Mid-century American poet Lorine Niedecker describes this sensibility as the "Time's buying sickness," accompanied by heedless waste: "needn't clean anything," she notes sardonically, "just throw it out the window / onto somebody else."[15]

As an economic strategy, obsolescence comes into its own amid the postwar economic growth that Niedecker and Packard chronicle and becomes an increasingly central organizing principle of commodity production in the global economy by the late twentieth century. The idea is often traced to economist Bernard London's 1932 idea for ending the Great Depression. London argues for an expiration date to be added to products and a system of taxes created to spur consumers not to reuse goods that are "legally dead."[16] As Packard's book makes vividly clear,

London's idea, while never adopted in law or tax code, becomes general practice in postwar American product development and marketing, employed with relation to goods ranging from cars to household appliances to fashion items. Disposability becomes increasingly built into consumer goods to encourage repeat buying, and many products are made not for durability but for limited use, with features designed to break down or lose functionality. Industrial designer Brooks Stevens popularized the concept in the early 1950s, arguing that "our whole economy is built on planned obsolescence.... We make good products, we induce people to buy them, and then next year we deliberately introduce something that will make those products old fashioned, out of date, obsolete."[17] A potential feature of virtually any commodity (excepting those of wholly organic composition), obsolescence becomes most centrally associated with electronics and technological products in the shift, by the mid-1960s, toward an "information society."[18] In *Made to Break: Technology and Obsolescence in America,* Giles Slade details the accelerating technological innovations in hardware and circuits that produced new products, from digital watches to computers, leading to unprecedented levels of product turnover. "By 1965," he writes, "the ground was prepared for America's e-waste crisis."[19] Today, computers and cell phones might be placed at the head of a long list of products oriented by the command to obsolesce.

Part of the disconcerting quality of these materials, as Packard's startling descriptions of Cornucopia City's untouched or barely used goods dumped on land and sea suggest, is that they appear to *resist* the decompositional qualities of waste. They often maintain their shape, texture, and utility after they have been discarded, largely because of their industrially produced, synthetic composition—made to break, but not to *break down.*[20] As Michael Thompson writes, "One of the most striking features of rubbish is that we all instantly recognize it when we see it, hear it, read it, smell it, or, horror of horrors, touch it.... Just as dogs are undoubtedly dogs, so rubbish is undoubtedly rubbish. There are no questions of degree."[21] By contrast, obsolescing materials seem to occupy a strange middle ground. This median category is exemplified in portrayals of obsolescing materials that highlight their identifiable shape, from Packard's description of "millions of tons of motorcars, refrigerators, alarm clocks, and metal play wagons cluttering attics and landscape" and "in town dumps or wayside gullies," to the old videocassettes and Zippo lighters in the 2008 Pixar film *WALL-E*.[22] These portrayals reveal a *different* sense of "matter out of place" than Mary Douglas's key anthropological definition of dirt, drawn on by various scholars of waste studies—matter still imbued with the uncanny, lively qualities of the commodity that Marx describes.[23] To observe these materials in their obsolescing condition is thus to confront the limit of an idea of virtuous return to the soil; rather than disappearing, they linger in *undead* form. In their lingering, they embody what Adorno calls *ciphers,* where nature and history dialectically interfuse as second nature.[24]

Second Nature

Second nature as a historical–materialist concept first emerges in Georg Lukács' 1914 *The Theory of the Novel* and is taken up not only in his subsequent work on reification but in the negative dialectics of Adorno and Benjamin, who each turn toward "ruins and fragments" to discern the hidden history of the present. For Lukács, second nature emerges from the distortions endemic to capitalist society that obscure the objective nature of social relations. If *reification* is the process by which "even the individual object which man confronts directly, either as producer or consumer" becomes "distorted in its objectivity by its commodity character," *second nature* is the greater world this process produces.[25] Lukács writes: "This second nature is not dumb, sensuous and yet senseless like the first: it is a complex of senses—meanings—which has become rigid and strange, and which no longer awakens interiority; it is a charnel-house of long-dead interiorities."[26] At once real and estranged, second nature is what surrounds us as artifice that has paradoxically become naturalized. In an early reading of Lukács, Adorno describes second nature as a "world of things created by man, yet lost to him," a world that "encounters us as *ciphers.*"[27] Adorno's reading introduces what he calls a "natural-historical" dimension into the concept of second nature. Adorno argues that the ciphers of second nature present the "petrifying" of history into nature as well as the "petrified life of nature as a mere product of historical development."[28] The problem that Lukács unfolds, Adorno writes, is "how it is possible to know and to interpret this alienated, reified, dead world"—how, that is, to confront these "ciphers" in order to discern their *natural–historical* content rather than their estranged unknowability.

In his *Arcades Project,* Benjamin provides one answer, turning to the relics of nineteenth-century consumer capitalism as "fossils" preserved in the decaying arcades of modern Paris. An image of the uncanny times and materials of second nature is apprehensible in the arcades' petrifying commodities and their process of obsolescing, Benjamin claims. He writes:

> On the walls of these caverns, their immemorial flora, the commodity, luxuriates and enters, like cancerous tissue, into the most irregular combinations. A world of secret affinities: palm tree and feather duster, hair dryer and Venus de Milo, prosthesis and letter-writing manual come together here as after a long separation.[29]

Though their form bears witness to their prior life, these obsolescing materials have come unfastened from the productive circuits of capitalism, while remaining irreducible to the cyclical temporalities of biotic life. Instead, they take on strange new configurations, a wanton tangle of "organic world and inorganic world" as the cast-off "commodity intermingles and interbreeds."[30] Through these uncanny forms emerge a sense of second nature defined *not only* as the reified world of commodity production but as the material recombinations that emerge in its wake. Indeed, if their cast-off condition points to capitalism's ongoing forms of creative destruction—the "charnel-house" of Lukács' description— this obsolescing also reveals the *nondestruction,* the estranged persistence, that accompanies it. Carrying forward into an extended future time-frame in their slow decomposition, these materials remain figures of discontinuity rather than progressive succession. In so doing they illuminate the uneven and expanded temporalities of second nature as generated through—but not wholly *controlled by*—the transformative energies of capitalism.

For these Frankfurt School philosophers, second nature is a way of naming the ongoing and naturalized "catastrophe" of capitalist modernity.[31] In their natural–historical dialectic of progress and ruin made apprehensible in the ciphers of decaying commodities, these theorists offer a prefiguration of the ecological dimensions of second nature that becomes increasingly obvious by the later twentieth century. From tires, refrigerators, air conditioners, and synthetic fabrics like nylon, to PBDE-impregnated seat cushions, mattresses, and car seats, to aluminum cans and other metals, a host of commodified materials present complex environmental challenges in their toxic, non-biodegradable obsolescing.[32] Plastic is perhaps the best-known example of a commodity material whose postconsumer persistence creates a wide variety of disturbing ecosystemic effects. In *The World Without Us,* Alan Weisman quotes the marine scientist Tony Ardrady discussing the unknowable duration of plastic after it is disposed: "Except for a small amount that's been incinerated, . . . every bit of plastic manufactured in the world for the past 50 years or so remains. It's somewhere in the environment."[33] Cast-off plastic has a particularly outsized impact on the world's oceans, collecting in great garbage patches and being pulverized into microplastics, which are in turn ingested by various forms of marine life.[34] Plastic in landfills also creates an extended risk of chemical contamination of groundwater, soil, and larger ecosystems.[35] The material effects of obsolescing materials present one example of what geographer Jason Moore calls "negative-value"—the undesirable counterpart to the generation of surplus-value—proliferating in the contemporary "capitalist world-ecology."[36]

One of the key forms of obsolescing materials is e-waste, which is the fastest-growing global waste stream, according to the UN.[37] E-waste—computer monitors, laptops, cell phones, printers, scanners, tablets, TVs, electronic toys, and other gadgets—mostly ends up in landfills, as with other forms of postconsumer materials.[38] E-waste contains various hazardous materials, from heavy metals such as lead, mercury, and cadmium to brominated flame retardants, which can cause damage to waterways, soil, humans, and various other organisms. However, e-waste, more than any other form of non-biodegradable detritus, also points to another dimension associated with obsolescing: *recycling* and *salvage.* For many U.S. consumers, the concept of recycling is a powerfully positive one, even operating as what psychologists Jesse Catlin and Yitong Wang call a "'get out of jail free card,' which may . . . signal to consumers that it is acceptable to consume as long as they recycle the end product."[39] A significant portion of these materials is taken by consumers to recycling centers each year in the hopes that these discarded machines will be reused in environmentally sustainable ways.[40] Yet the actual processes of e-waste recycling bear little resemblance to the fantasy of a waste-free end result to consumption. Jennifer Gabrys details the extended afterlife of e-waste in her fascinating study, *Digital Rubbish: A Natural History of Electronics*:

Typically, electronics are first collected by recyclers in North America or Europe, who salvage high-grade machines for resale and extract valuable metal from devices for scrap or who alternately bundle defunct machines in shipping containers. In either case, at some stage down the line of processing, the electronics are usually sent to developing countries for scrap and salvaging of components.[41]

The salvage process extends the productive circuit of capital further, as value is appropriated from former commodities that now operate as raw material for extraction. A complex global infrastructure, including warehouses and other facilities, mechanical shredding devices and smelters, shipping containers and other transportation units, and myriad informal processing and trading operations, has been developed to facilitate this industry built on obsolescing goods.

This process retraces, in reverse, the global itinerary of commodity production and circulation, sending e-waste back to countries such as China, India, Nigeria, and Ghana (sometimes using the same shipping containers both ways).[42] Plastic that is made in Shenzhen for companies like Foxconn—producer of electronics products for companies like Apple and Hewlett-Packard—finds its way into mountains of e-waste in the informal workshops of Guiyu. Discarded tablets, laptops, and cell phones containing heavy metals mined in Africa end up in cities like Accra and Lagos, where people scavenge in dumps for material to sell to local scrap markets. In these and other cities, informal workers, including children, labor in toxic conditions to dismantle e-waste and extract metals. According to a recent Indian chamber of commerce report, Indian informal e-waste workers generally cannot work after the ages of thirty-five to forty because of their poor health.[43] In turn, the local waters, air, soil of these areas are severely contaminated, reflecting the conditions of what critical geographer Mike Davis calls "slum ecology."[44] The expanded temporality of obsolescing involves such key dimensions of global political ecology—the growing proportions of the global precariat and the harmful conditions in which they live and eke out a subsistence.[45] In many ways, to highlight obsolescing is thus to return to a familiar story—the production of surplus-value through the exploitation of labor-power—and to see its ramifications extend into new phases and intensifying, if uneven, material effects on humans and ecosystems.

Coda: Wish Images

Yet if obsolescing unveils the workings of second nature as the propulsive progress of capitalist production and the alienated and petrifying forms it leaves in its wake, obsolescing materials also operate, according to Benjamin's *Arcades Project,* as collective "wish images."[46] Benjamin argues that wish images are "images in the collective consciousness in which the new is permeated with the old."[47] Obsolescing objects, with their strange sense of nonsynchrony, come to embody utopian longings for another mode of communal existence, as in Hillman's description of the slow, intimate connectivity that the phone booth and the phone cord's tactile "whorl" evokes. If, as the French group the Invisible Committee claims, human and nonhuman life in the present is "*held captive* by the general organization of the commodity system," we might see the wish image of obsolescing as a figuration of collective *release* from this captivity.[48] Left alone, obsolescing materials evoke a sense of liberation from what Benjamin calls the "drudgery of being useful," but perhaps more significantly, from the tyranny of value creation for profit—a vision that extends to subjects and objects alike.[49] In Benjamin's description, these ciphers in their natural–historical entangling evoke "elements of primal history," including dimensions of "classless society"—a utopian glimmer drawn from decaying things.[50]

Such a wish image might bring to mind truly posthuman imaginings, such as Weisman's vision of the planet's life after humans or postapocalyptic narratives like *Earth Abides*—a world in which production has come to a cataclysmic standstill. Or it might evoke a return to peasant-based agriculture on a global scale, as ecofeminists such as Mariarosa Dalla Costa propose.[51] But obsolescing might also express desires that are more immediately attuned to alternative socioecological speeds and times. Above all, obsolescing evokes *senescence*: the processes of aging, slowing down, not keeping up. As a wish image, this offers a different orientation toward ecological being-in-relation, one that honors slowness and deceleration. To turn our attention to this wish image is perhaps to reorient our thinking toward finitude—ecological and economic—and to denaturalize our ethos of speed and novelty. As Hillman ruefully asks: "Why did we live so fast[?]"

Notes

1. Brenda Hillman, "Phone Booth," in *Practical Water* (Middletown, Conn.: Wesleyan University Press, 2009), 17.

2. Walter Benjamin, "On Some Motifs in Baudelaire," in *Walter Benjamin: Selected Writings, Volume 4: 1938–1940*, ed. Howard Eiland and Michael Jennings (Cambridge, Mass.: Harvard University Press, 2003), 316.

3. Karl Marx, *Capital*, vol. 1, trans Ben Fowkes (London: Penguin, 1990), 163–64.

4. Ibid., 165.

5. Marx writes, "The mystical character of the commodity does not therefore arise from its use-value." Ibid., 164.

6. Joseph Schumpeter, *Capitalism, Socialism, and Democracy* (London: Routledge, 1994), 82–83.

7. Arjun Appadurai, *Modernity at Large: Cultural Dimensions of Globalization* (Minneapolis: University of Minnesota Press, 1996), 82.

8. Hillman, "Phone Booth," 17.

9. Michael Thompson, *Rubbish Theory: The Creation and Destruction of Value* (Oxford: Oxford University Press, 1979), 2.

10. Theodor Adorno, *Minima Moralia: Reflections on a Damaged Life* (London: Verso, 2005), 221.

11. Vance Packard, *The Waste Makers* (New York: David McKay, 1960), 4–5.

12. Ibid.

13. Karl Marx and Friedrich Engels, *Manifesto of the Communist Party, Marxists.org*, https://www.marxists.org/archive/marx/works/1848/communist-manifesto/.

14. For a detailed discussion of the dynamics of overcapacity and overproduction in postwar economies, see Robert Brenner, *The Dynamics of Global Turbulence* (New York: Verso, 2006), 34–40.

15. Lorine Niedecker, *Collected Works* (Berkeley: University of California Press, 2006), 157, 119.

16. Cited in Bill Brown, "The Obsolescence of the Human," in *Cultures of Obsolescence: History, Materiality, and the Digital Age*, ed. Babette Tischleder and Sarah Wasserman (New York: Palgrave Macmillan, 2015), 22.

17. Giles Slade, *Made to Break: Technology and Obsolescence in America* (Cambridge, Mass.: Harvard University Press, 2006), 153.

18. Ibid., 187.

19. Ibid., 198.

20. As Daniel Abramson writes: "Obsolescence connotes a terminal process, an emptying of usefulness and value in competition with something new and better. But the suddenness and externality of obsolescence leaves the obsolete object intact, as opposed to slow, intrinsic, physical decay. The problem then

becomes what to do with the superceded yet more or less whole artifact." "Boston's West End: Urban Obsolescence in Mid-Twentieth-Century America," in *Governing by Design: Architecture, Economy, and Politics in the Twentieth Century,* ed. the Aggregate Collective (Pittsburgh: University of Pittsburgh Press, 2012), 56.

21. Thompson, *Rubbish Theory,* 94.

22. Packard, *Waste Makers,* 209.

23. Mary Douglas, *Purity and Danger: An Analysis of the Concepts of Pollution and Taboo* (London: Routledge, 1984), 35.

24. Theodor Adorno, "The Idea of Natural-History," in Robert Hullot-Kentor, *Things beyond Resemblance: Collected Essays on Theodor Adorno,* trans. Robert Hullot-Kentor (New York: Columbia Press, 2006), 261.

25. Georg Lukács, "Reification and the Consciousness of the Proletariat," in *History and Class Consciousness: Studies in Marxist Dialectics,* trans. Rodney Livingstone (Cambridge, Mass.: MIT Press, 1972), 93.

26. Georg Lukács, *The Theory of the Novel,* trans. Anna Bostock (Cambridge, Mass.: MIT Press, 1971), 64.

27. Adorno, "Idea," 260–61.

28. Ibid., 261.

29. Walter Benjamin, *The Arcades Project,* trans. Howard Eiland and Kevin McLaughlin (Cambridge, Mass.: Belknap Press of Harvard University Press, 1999), 827, 874.

30. Ibid., 827.

31. Ibid., 473.

32. See Alan Weisman, *The World Without Us* (New York: Thomas Dunn Books, 2007) 131, 205.

33. Ibid., 126.

34. Ibid., 123.

35. Jessica Knoblauch, "Plastic Not-So-Fantastic: How the Versatile Material Harms the Environment and Human Health," *Scientific American,* July 2, 2009, http://www.scientificamerican.com/article/plastic-not-so-fantastic/.

36. Jason Moore, *Capitalism in the Web of Life* (London: Verso, 2015), 274.

37. John Vidal, "Toxic 'E-waste' Dumped in Poor Nations, says United Nations," *Guardian,* December 14, 2013, http://www.theguardian.com/global-development/2013/dec/14/toxic-ewaste-illegal-dumping-developing-countries.

38. An unknown amount of e-waste is also found in stockpiles. See Ian Urbina, "Unwanted Electronic Gear Rising in Toxic Piles," *New York Times,* March 18, 2013, http://www.nytimes.com/2013/03/19/us/disposal-of-older-monitors-leaves-a-hazardous-trail.html?pagewanted=all&_r=2&.

39. Jesse Catlin and Yitong Wang, "Recycling Gone Bad: When the Option to Recycle Increases Resource Consumption," *Journal of Consumer Psychology* 23, no. 1 (January 2013): 122–27.

40. According to the nonprofit group Electronics Take-Back Initiative, citing figures from the EPA, the United States had a 27 percent recycling rate of overall e-waste in 2010. See Electronics Take-Back Initiative, "Facts and Figures on E-Waste and Recycling," June 25, 2014, http://www.electronicstakeback.com/wp-content/uploads/Facts_and_Figures_on_EWaste_and_Recycling.pdf.

41. Jennifer Gabrys, *Digital Rubbish: A Natural History of Electronics* (Ann Arbor: University of Michigan Press, 2011), 91.

42. Heather Rogers, *Gone Tomorrow: The Hidden Life of Garbage* (New York: New Press, 2005), 201.

43. Matt Wade, "Inside Delhi's Gadget Graveyard Where the West's E-waste Ends Up," *Sydney Morning Herald,* January 9, 2016, http://www.smh.com.au/world/inside-delhis-gadget-graveyard-where-the-wests-ewaste-ends-up-20160107-gm1h7z.html.

44. Mike Davis, *Planet of Slums* (London: Verso, 2007), 121.

45. A key source of e-waste labor in the United States is prisoners in federal penitentiaries. See Leslie Kaufman, "Toxic Metals Tied to Work in Prisons," *New York Times,* October 26, 2010, http://www.nytimes.com/2010/10/27/science/earth/27waste.html.

46. Benjamin, *Arcades Project,* 4.

47. Ibid.

48. The Invisible Committee, *To Our Friends,* trans. Robert Hurley (Los Angeles: Semiotext(e), 2015), 208.

49. Benjamin, *Arcades Project,* 19.

50. Ibid., 4.

51. Mariarosa Dalla Costa, "Food Sovereignty, Peasants, and Women," *Commoner,* June 21, 2008, http://www.commoner.org.uk/?p=42.

Decorate

DANIEL C. REMEIN

What decorates? And how? Why should the ecologically minded practitioner of the humanities concern herself with decorating—an activity of expenditure, of waste, an activity that resonates more with the theoretical invocations of *oikos* that mark an exclusively human household economy and the excesses of the *domus* (the household, yes, but also the unsustainable extravagance of the *dominus,* the lord, whose decorating displays his sovereignty and ownership) than the invocations of *oikos* in an ecological thinking that would mark the etymology of *eco-* in order to better think the earth as a much larger and complex household?[1]

I come to these questions under the sign of *ecopoetics.* "Any writer," writes ecopoetics practitioner Jonathan Skinner (in an effort, like this one, to inject more of the *-poetics* into the thought of *eco-*), "who wants to engage poetry with more-than-human life, has no choice but to resist simply, and instrumentally, stepping over language."[2] To decorate, as an act of ecological theory or criticism, will intensify attention to the rhetoric of ecocriticism and the *poetics* of ecocritical argument.

This essay is old fashioned that way: it privileges literary poetics as a mode for thinking a generalizable techne (a transliterated Greek term for *art* or *skill* as, for example, in rhetoric or architecture in their contrast to classically defined "unworked" nature—deployed widely as a term for a mode of practice, working, or making, in a variety of twentieth-century "theory," especially in association with phenomenology and deconstruction), in this case, as an ecology that is only thinkable first as a cosmology.

In this I follow mid-twentieth-century poet Robert Duncan, who conditions the ecopoetical idea of the poem as a living system determined as "a form that maintains a disequilibrium," on the notion that the poem is structured by/as an apprehension of cosmology: "The most real, the truth, the beauty of the poem is a configuration, but also a happening in language, that leads back into or on towards the beauty of the universe itself."[3] If, as Duncan writes elsewhere, "from whatever poem, choreographies extend into actual space,"[4] this is possible because of a receptivity to different orders of *oikos*: "It is not that poetry imitates but that poetry enacts in its order the order of first things."[5]

¶

Joining the above-cited invocations of the term *oikos*, Skinner formulates his ecopoetics as "a house making."[6] In its recurrent investment in the concept of *oikos*, ecopoetics would need to be able to think how to decorate its house. But an ecopoetics that decorates will mean that the nonstructural activities of *making-house* will supplant the activity of *making a house: without taking the initial emergence of any given individual architectonic structure for granted, its supposedly secondary decoration will still appear as prior.* Via an ecopoetics that decorates, building will be supplanted by decorating as the primary paradigm of making; and, insofar as it decorates, an ecology will become a cosmological *ecotecture*.

I want to suggest this term because *to decorate* can mean to elaborate surface (or nonspatiality) into the textures that catch, articulate, hook, entangle, and mix as a spatiality where different perceptual systems overlap: gems that blind, details that distract, tendrils that tangle. To decorate is to build, but not in such a way as to distinguish between the built and the wild.

But who decorates anymore? We expect poets and architects will avoid it—especially the avant-garde, the minimalist, the conceptualist— at least since Modernism. By 1964 Marxist critic Ernst Bloch was able to articulate a reserved approval that modernist architecture had dedecorated certain nineteenth-century and early twentieth-century practices, and yet he could still lament "an iconoclasm against decoration unheard of thus far."[7] "It is certain," Bloch observes, "we suffer from a little too much of glass and steel and a little too little of that which decorates."[8]

And anyone who has seen the high-end glass condos soar over the skies of Brooklyn over the last decade amid a capital-driven nostalgia for a certain minimalism could say something quite similar to today.

Yet the command not to decorate by both architectural and poetic Modernisms encodes an untenable blend of motivating racism, misogyny, and homophobia.[9] Ezra Pound's exhortations to poets to "use absolutely no word that does not contribute to the presentation" suggest that his recommendation to "use either no ornament or good ornament" be read as a dogmatic prohibition of ornament altogether.[10] Despite Pound's early experiments with pre-Raphaelite decorative medievalism, Pound's later turn to the medieval in search of the poetics of a vital outside world (a sort of ecomedievalist Pound) champions what he calls the *Tuscan aesthetic*: "the medieval clean line, as distinct from medieval niggle."[11] Pound explains that the "clean line" means an avoidance of pejoratively characterized anal, fecal, or hemorrhoidal acts: "ornament flat on the walls, and not bulging and bumping and *indulging in bulbous excrescence*."[12] The "clean line" is a program of racialized misogynist hygiene that elides backwardness, orientalism, femininity, passivity, and weakness as exemplified in the Northern and Gothic architecture of "these European Hindoos [sic]"; for "mess, confusion in sculpture, is always symptomatic of supineness, bad hygiene, bad physique (possibly envy)."[13]

Touting a similar program in his landmark, and transparently titled, essay, "Ornament and Crime," twentieth-century critic and designer Adolf Loos explains the urge to decorate as a threat to health, a symptom of degeneracy, and an undesirable fecopoetics:[14]

> The man of our day who, in response to an inner urge, smears the walls with erotic symbols is a criminal or a degenerate. . . . A country's culture can be assessed by the extent to which its lavatory walls are smeared.[15]

For Loos, an urge to decorate takes its place as a symptom within a malignant racial and cultural development narrative: "What is natural to the Papuan and the child is a symptom of degeneracy in the modern adult"; for "*the evolution of culture is synonymous with the removal of ornament from utilitarian objects*."[16]

But the call to decorate bulges in on us, excretes all around us, blinds us with shine—breaking into the city, creeping into the hearth, encircling thought and writing. Even Bloch, partly satisfied with Modernism's expurgations, still yearns for us to begin to decorate again:

> The scabby and ulcer-like ornaments, as Adolf Loos called them, fell off. They could not be pasted back on again, and the plain transportation box or egg box as the building, which had "Hurrah, nothing comes to our minds anymore!" as its only head decoration, does not have to be the final word against the former scab.[17]

As queer theorist José Esteban Muñoz argues, "[Utopian] potentiality for Bloch is often located in the ornamental."[18] For Bloch, to decorate can mean to register new forms of collectivity: especially what he theorizes as the "Gothic style" heralds "a storm brewing on the horizon of collectivity."[19] Even as Bloch's twentieth-century Marxism circumscribes his sense of this "collectivity" firmly within the human,[20] Bloch's account of what decorates anticipates and potentializes larger collectivities in the activities of physical phenomena and their processes:

> There already where nobody is yet, this vital clue smolders upward toward us. It is the same force that takes effect in lava, when molten lead is poured into cold water, in wood veining, and first and foremost in the quivering, bleeding, tattered, or curiously conglobate forms of the internal organs.[21]

And Bloch's account of how the "Gothic line" decorates explicitly celebrates a nonhuman bulging, perhaps perversely associating an escape from the strictly human household with its traditionally most central structure:

> The *Gothic* line, on the other hand, retains the hearth. This line is restless and uncanny in its figures: the bulges, the serpents, the animal heads, the watercourses, a tangled criss-cross and twitching where the amniotic fluid and the incubation heat sit, and the womb of all pains, all lusts, all births, and of all organic images begin to speak.[22]

There is no "interior" decorating. For feminist theorist, Elizabeth Grosz, too, when we decorate with furniture, we follow a pattern generalizable to all biological life—furnishing entrances and exits between the *oikos* of bourgeois domestic life and the *oikos* of a larger sphere.[23] The line that decorates is a line of flight from the human house in complicity with the morphologies of an already decorating nonhuman *oikos*: an interface not renouncing the human home but entangling it with snakes and waterfalls, bulges organic and mineral, living and nonliving.

Hooked on the lure of that utopian line, contemporary Canadian poet Lisa Robertson rehabilitates the pro-decorating insight of nineteenth-century architect and art historian Gottfried Semper—as elaborated in his magisterial historical study, *Style in the Technical and Tectonic Arts; or, Practical Aesthetics*—that what decorates buildings is not ancillary to the history of tectonic structure.[24] In theorizing her interest in Semper's project, Robertson points to the importance accorded to *surface* in Semper's approach:

> Semper . . . proposed a four-part unsubordinated architectural topology, where surface was in non-hierarchical dynamic relationship with molded plasticity, a framework of resistance, and foundational qualities. The transience and non-essential quality of the surface did not lessen its topological value.[25]

To decorate is to build, as in the ordinary functions of fiberwork: "1. to string and to bind; 2. to cover, to protect, and to enclose."[26] For Semper, it is a domestic elaboration on, or folding of, *surface* or fiber and not a priestly manipulation of architectonic structure, by which humans make their first architectural gestures.[27] This is possible because humans can recognize how the nonhuman world already decorates itself,[28] folding surfaces into covers:

> Human beings first learned to recognize the essential purpose of natural covers (shaggy animal skins, protective tree bark) and began to use them for their own ends according to their correctly perceived natural use. Later, they imitated them with synthetic weaving.[29]

The weaver, the interior decorator, was the first wall fitter.[30] The structural masonry wall "was an intrusion into the domain of the wall fitter by the mason's art."[31] It is difficult to tell here if the earliest human architecture consists of huts (following Rousseau) or something more akin to decorating the body with clothes. To decorate is to build, but to build is different than we thought. So to *make house* is again to make a house/ *oikos*.

But this ecotectural spatiality does not emerge from decorating because of a secret depth or because of a simple opposition to "depth" or "symptom" such as some literary critics have worried about in recent years.[32] Robertson produces, as critic Lytle Shaw contends, "a carefully staged refusal of opposition between surface and depth as an immersive utopian experience."[33] Rather, it is *on account of* superficiality that surface yields to an ecotecture. This is perhaps why Robertson loves to decorate with the torsions of rhetorical surfaces, writing as the "Office for Soft Architecture," whose practices "will reverse the wrongheaded story of structural deepness," on the premise that "that institution is all doors but no entrances."[34]

On the other hand, writing about the capacity for color to decorate, Robertson considers that "the medium is also an economy"[35]—that is, rhetoric remains as a kind of turnstile even once we find an entrance or an exit. How then to entangle one *oikos* with another? In one instance Robertson decorates an interface between human perceptual economies ("the day is our house," she writes, of our temporal *oikos*) and that flutteringly evasive nonhuman economy, the weather: "Consider that we need to drink deeply from convention under faithfully lighthearted circumstances in order to integrate the weather, boredom utopic, with waking life."[36] To decorate the weather with rhetoric can yield a relay switch between different economic/ecological and perceptual orders because surface phenomena (like rhetorical convention) can work as a Mobius strip to disarticulate inside and Outside at specific curves of its slipstream.[37] Thus, Robertson contends:

> Artifice is the disrespect of the propriety of borders. Emotion results. The potent surface leans into dissolution and disrupts volition. . . . This is the *pharmakon*: an indiscrete threshold where our bodies exchange information with an environment.[38]

This is why the way that the invasive blackberry bush *(Rubus armeniacus)* decorates teaches us that "if architecture is entombed structure or *thanatos* [as a vulgar shorthand for the very personification of the death drive], ornament is the frontier of the surface . . . where lively variability takes place."[39]

But this variability is indeed a *pharmakon* (a Greek term for *drug,* as the deadly cure, the healing poison that Plato's Socrates uses to describe writing and that Jacques Derrida famously explores as the "movement, the locus, and the play" that philosophy is constantly trying to arrest),[40] and in this way to decorate as ecotecture will mean—with the "Swimmer Poetics" of Steve Mentz—to give up fantasies of the sustainability of any *oikos*.[41] Robertston asserts this ineluctable and rhetorical imbalance in adumbrating the often-menacing grammar of such ecotecture, as expressed by how the blackberry bush decorates:

> The limitless modification of the skin is different from modernization— surface morphologies, as *Rubus* shows, include decay, blanketing and smoothing, shedding, dissolution and penetration, pendulous swagging and draping, as well as proliferative growth, all in contexts of environmental disturbance and contingency rather than fantasized balance.[42]

And imbalance too requires work (i.e., the poem as a *disequilibrium*). To decorate is to elaborate—it is undeniably a form of labor and subject to all the same exploitations: from the Latin *elaboro,* a working out from, possibly with both care and exertion, sometimes to the point of "overdoing it."[43]

No two lines of poetry that I know of decorate such an elaborate ecotecture so much as two lines that constitute the preamble to two poems from the compilation conventionally known as the Old English Riddles, preserved in the late tenth-century Exeter Book manuscript:[44]

> Is þes middangeard   missenlicum
> wisum gewlitegad,   wrættum gefrætwad
> [this middle earth is in a variety
> of ways decorated/beautified, decorated with jewels/ornaments].[45]

Although these lines may present a formulaic way to open a riddle-poem,[46] they are interesting precisely in that they formulaically or conventionally *decorate* the poem. Clearly "overdoing it," they elaborate a rhetoric that decorates on the level of cosmos itself. Although critics have a long history of disagreeing on everything from how many separate riddles there are in this manuscript to how to solve the riddles, most can agree with the early editor of the collection, Frederick Tupper, whose 1910 edition explains that the Old English riddle "arises out of the desire to invest everyday things and thoughts with the garb of the unusual and the marvelous."[47]

This returns us to my opening questions: what decorates, and how? For Tupper, what the Old English Riddles *do* is *decorate*—as part of the economy of interacting with the everyday, the domestic—as a mode of entanglement with an *oikos*. This was, however, already our first answer: *poetry decorates*. But what has done the decorating?

The world is decorated with decorations: decorations decorate. Elaborating such a tautology is important because a historicizing reading of these lines might answer simply that *God decorates with ornaments.* After all, the trope echoes the Vulgate itself: "igitur perfecti sunt caeli et terra et omnis ornatus eorum" [then the heavens and the earth were completed in all of their ornamentation] (Gen. 2:1). And Old English vernacular poetry often explicitly identifies God as the interior decorator of the cosmos in a manner that recalls these verses from Genesis.[48] Far from a secular ecopoetics, this might herald a cosmology in which all decorating is subordinated to explicitly theological significance.

But here, these two lines disrupt significance, disrupt narrative, delay the riddles that follow. *To decorate is to disrupt narrative* and the anti-environmentalist economy that instrumentalizes and capitalizes on each syntactic unit to be reinvested as clauses of Value in a story.[49] To decorate is to disrupt signification and instead to set gems that will distract, to carve curlicues that will catch. In any case, with lines like these, one should perhaps decorate before signifying. Medievalist Mary Carruthers laments an "overtheologized" medieval studies in which "every colour and ornament is said to conceal a lesson for the improvement of the viewer or listener."[50] This is because the style, or aesthetics—a term that here indicates rhythms and procedures of physiological perception rather than abstract form—of a medieval artifact, as Carruthers argues, interacts with a medieval body thought of as "porous . . . not a waterproof sack, but permeable to influences from all sorts of external agents, benign

and harmful."⁵¹ The famous Anglo-Saxon "interlace" aesthetic, in literary form or specific styles of stonework, cloissoné, or open-work buckles
or brooches, is perhaps better thought as an "interface" aesthetic, embroi-
dering porous bodies into an active *oikos*.⁵²

Stylistically, the rhetorical surface of the lines from the Old English
Riddles constitutes an excellent specimen of the figure of *variation*: as
grammatical complements to *þes middangeard* (this middle-earth), the
phrases *missenlicum/ wisum gewlitegad* (in a variety of ways/decorated
or beautified) and *wrættum gefrætwad* (decorated or adorned with ornaments) are syntactically apposite variations of each other. The variation
that is such a hallmark of Old English verse here yields a style that we
might call *mixed*: mixing the *altus* (high) style of words like *wræt* (orna-
ment/jewel) with the prosaic adjective *missenlic* (various, diverse, dissim-
ilar). As Carruthers explains, while classical *varietas* was to result from
placing each thing in its proper place, determined as *dignitas* (fittingness),
over the course of the first millennium, medieval rhetoric developed a
different approach to variety as the result of both *diversitas* and *mixtura*
(diversity and mixture).⁵³ Medieval variation conflated the idea of the
variation of qualities in temporal succession and the idea of simultaneous variegation so that the style "was regularly linked causally to 'change'
in affect and feeling" rather than to a notion of appropriateness.⁵⁴

These aesthetics underscore an alternative lexicon of what decorates.
The Present-Day English verb, *decorate*, first enters the language in the
early modern period (as a transitive verb with more or less the meaning
it has now), the past participle of the Latin *decoro* having been borrowed
first as a participial adjective and subsequently taken as the stem of a
verb.⁵⁵ Classical Latin uses *decoro*, in remarkably the same way, to decorate, adorn, embellish, beautify (and tropologically, to distinguish or
honor, in the way that the army might now "decorate" a soldier).⁵⁶ The
term descends from Proto-Indo-European root *dek̑-*, which has something to do with the actions of taking or receiving, leading to Greek and
Latin reflexes that pertain to a capacity to be taken or received (i.e., fittingness, dignity, elegance, beauty, and, hence, decorating and ornamenting).⁵⁷

The Old English verb *gewlitigian* (to decorate, to beautify), from the
riddle above, can function as a gloss on the Latin *decoro*,⁵⁸ yet means
more specifically to make beautiful or radiant with reference to a particular *wlite,* a radiance or brightness belonging to a face or a countenance:

an "aspect."[59] Most prosaically, to *decorate* in this sense might involve the use of jewelry or makeup—to accentuate the already decorative aspect of a countenance. The verb phrase that is apposite to *gewlitigad*, that is, *wrǣttum gefrǣtwed* (decorated with ornaments), elaborates this sense. *(Ge)-frǣtwian* means to adorn, ornament, or add luster to something, usually by means of the addition of *frǣtwe* (the verb's root noun). Without veering too far from the verbal focus of this volume, allow me to decorate this essay with attention to just two nouns—*frǣtwe* and *wrǣt*— with the defense that these nouns both ineluctably veer into action. A *frǣtwe* may be an adornment or ornament of any kind, but the term seems to suggest accoutrements that come in the plural, that multiply: treasure, coins, jewels, ornaments, and so on.[60] A *wrǣt*, on the other hand, is one possible outcome of someone *doing* decorating: an ornament that is either a work of art (i.e., human made) or in fact a precious stone or a "jewel"[61]—the common element here a degree of variation and complexity recalling the plastic arts of early medieval Britain.

Yet lexicographers of Old English also traditionally associate the related adjective *wrǣtlic* and adverb *wrǣtlice* with the *wonderous* as well— eliding surface and more phenomenologically robust phenomena.[62] I do not mean here to leap to meaning before thinking about style. But the complicity of the mixed and varied style with this vocabulary of decorating inherently harbors the idea that appearance itself, in a phenomenological sense, somehow already decorates. According to these lines, to decorate is an affirmation of an aesthetic hooked into the structure of the cosmos itself. They recall us to the capacity for the Greek term κόσμος (cosmos), stemming from its primary sense as related to the beautiful order or form of a thing, to refer to ornament *before* it refers to the universe.[63]

✍

To ask about what decorates is to ask a question about ecotectural style; this is an important ecopolitical question. These lines of Old English poetry concern the structure of what the cosmos feels like. In a cosmos that feels varied, in which things appear to each other inasmuch as they decorate each other, the surfaces of different entities are always co-elaborating each other, compromising volitions, and so on—but I think, crucially, without necessarily foreclosing the possibility of meaningful and radical differences between them, without closing that opening between *bio* (life) and *gea* (earth) that Michel Serres considers at length.[64] To decorate is not necessarily to relate every thing to every other thing, and indeed presupposes its impossibility. Even in Anglo-Saxon interlace patterns, some tendrils lead to a dead end. To decorate can mean to resist saturated, fascist, totalities—including that of global capital and its ecological disaster. To decorate in this way is to affirm an ecological thinking that can conceptualize what amplifies or dilates (inescapably rhetorical terms) the interrelations of multiple beings, species, systems, structures, topographies, cosmologies, and so on, without veering from a secular ecology toward a theology amnesiac of difference. As Shevek, the utopian physicist hero of Ursula Le Guin's novel *The Dispossessed,* argues, "We don't want purity but complexity . . . our model of the cosmos must be as inexhaustible as the cosmos."[65]

To decorate is not the most obvious of ecopoetics, because, like phenomenology, it radicalizes itself as ecotecture "not by taking its place in absolute consciousness without mentioning the ways by which this is reached, but by considering itself as a problem"—not by "leaping" outside of the problems of a species-specific perceptual *oikos,* but by convoluting inside and Outside into mutual decorating.[66] To decorate is to maintain the human as a decoratable question, to question how the activities of the humanities can veer toward those of the nonhuman *oikos.* And in the same way that Robertson, as Shaw argues, accords a literal quality to discursive sites, which have their own "sedimentary history,"[67] to decorate a site that is "merely" discursive is no more or less ecological than nature writing.

But as rhetoric—that is, a *techne*—governed by a generalized logic of writing that menaces the very possibility of contemporaneous relation,[68] to decorate may unnerve environmentalist culture: it will not—poof!—render up ready-to-hand vibrant materials or agential objects as instrumentalizable phenomena for our writing. As Serres reminds us, "swirlings" can be both *against* and *with.*[69] Yet to decorate is still always to lavish and entangle, to spend without reserve on the inessential. This is neither a gnomic paradox nor a performance of hipster irony. Utopian plenitude decorates even with patterns from echoes of the echoes of absent cosmic forces and, like the weather, needs no agent to be present in the present. For:

Is þes middangeard   missenlicum
wisum gewlitegad,   wrættum gefrætwad
[this middle earth is in a variety
of ways decorated/beautified, decorated with jewels/ornaments].

In a cosmos in which to decorate is the basic mode of relation—the question of *what* decorates is superseded by the question of *how*—as if to suggest an impersonal usage for the verb *to decorate*. As with the weather that Robertson decorates, of which it can be said, *it rains, it snows,* and so on, one can ask about how it's going between this or that order of *oikos,* hoping for the answer, *it decorates.*

Notes

1. On the former usage of *oikos,* see Michel Foucault, *The History of Sexuality, Volume 2: The Use of Pleasure* (New York: Vintage, 1990), 153–84. For representative citations of the latter use, see Gary Snyder, *Earth House Hold: Technical Notes and Queries to Fellow Dharma Revolutionaries* (New York: New Directions, 1969); Jeffrey Jerome Cohen and Lowell Duckert, "Howl," *postmedieval: a journal of medieval cultural studies* 4, no. 1 (Spring 2013): 5: "The *oikos* is our home, but our home is not ours alone." For an example that renders the two usages less distinguishable, see Jennifer Scappettone, "Garbage Arcadia: Digging for Choruses in Fresh Kills," in *Terrain Vague: Interstices at the Edge of the Pale,* ed. Manuela Mariani and Patrick Barron (New York: Routledge, 2014), 149: "Trash serves as an acrid reminder of the elements of the *oikos* cast off in the interests of economy."

2. Jonathan Skinner, "Why Ecopoetics?" *ecopoetics* 1 (Winter 2001): 105. Skinner argues that "environmentalist" culture and criticism "may be eco . . . but it certainly comes up short in 'poetics.'" (7). "Editor's Statement," *ecopoetics* 1 (Winter 2001): 7. This has been changing radically in the years since, e.g., Lowell Duckert, "Glacier," *postmedieval: a journal of medieval cultural studies* 4, no. 1 (2013): 68–79.

3. Robert Duncan, "Towards an Open Universe," in *Robert Duncan: Collected Essays and Other Prose,* ed. James Maynard (Berkeley: University of California Press, 2014), 128–30.

4. Robert Duncan, introduction to *Bending the Bow,* qtd. from *Robert Duncan: The Collected Later Poems and Plays,* ed. Peter Quartermain (Berkeley: University of California Press, 2014), 297.

5. Duncan, "Open Universe," 131.

6. Skinner, "Editor's Statement," 7.

7. Ernst Bloch, "On Fine Arts in the Machine Age," in *The Utopian Function of Art and Literature: Selected Essays,* trans. Jack Zipes and Frank Mecklenburg

(Cambridge, Mass.: MIT Press, 1988), 200–201. In contrast, Walter Gropius and his associates did not initially spurn medieval decoration per se. See, for example, Gropius et al., "New Ideas on Architecture," in *Programs and Manifestos on 20th-Century Architecture,* ed. Ulrich Conrads, trans. Michael Bullock (Cambridge, Mass.: MIT Press, 1971), 46–48.

8. Bloch, "On Fine Arts," 200.

9. For a foundational ecofeminist discussion of the gendering of the household and the process of domestication, see Stacey Alaimo, *Undomesticated Ground: Recasting Nature as Feminist Space* (Ithaca: Cornell University Press, 2000).

10. Ezra Pound, "A Retrospect," in *Literary Essays of Ezra Pound,* ed. T. S. Eliot (New York: New Directions, 1935), 4–5.

11. Ibid., 151–52.

12. Ibid., 151, emphasis added.

13. Ibid., 150.

14. One of the best theorizations of fecopoetics has come, not insignificantly, out of medieval studies. See Susan Signe Morrison's *Excrement in the Late Middle Ages: Sacred Filth and Chaucer's Fecopoetics* (New York: Palgrave, 2008) and *The Literature of Waste: Material Ecopoetics and Ethical Matter* (New York: Palgrave, 2015).

15. Adolf Loos, "Ornament and Crime," in Conrads, *Programs and Manifestos,* 19–24.

16. Ibid., 20, emphasis in original.

17. Bloch, "On Fine Arts," 201.

18. José Esteban Muñoz, *Cruising Utopia: The There and Then of Queer Futurity* (New York: New York University Press, 2009), 7.

19. Ernst Bloch, "Art and Utopia," in *The Utopian Function of Art,* trans. Jack Zipes and Frank Mecklenberg (Cambridge, Mass.: MIT Press, 1989), 94.

20. For example, see Bloch's "Art and Utopia," 96, 101–102, and "Building in Empty Space," 198, in *The Utopian Function of Art.* Bloch locates architectural utopianism in "the anticipation of a space adequate for human beings."

21. Bloch, "Art and Utopia," 78.

22. Ibid., 96.

23. Elizabeth Grosz, *Chaos, Territory, Art: Deleuze and the Framing of the Earth* (New York: Columbia University Press, 2008), 15.

24. Gottfried Semper, *Style in the Technical and Tectonic Arts; or, Practical Aesthetics,* trans. Harry Francis Mallgrave and Michael Robinson (Los Angeles: Getty Research Institute, 2004), 106.

25. Lisa Robertson, "*Rubus armeniacus*: A Common Architectural Motif in the Temperate Mesophytic Region," in *Occasional Work and Seven Walks from the Office for Soft Architecture,* 3rd ed. (Toronto: Coach House, 2011), 111.

26. Semper, *Style,* 113.

27. Cf. Martin Heidegger on the temple as the paradigm of architecture, in *Poetry, Language, Thought,* trans. Albert Hofstadter (New York: Harper Perennial, 1971), 38–56.

28. For a sympathetic argument, see Roger Callois, *The Writing of Stones,* trans. Barbara Bray (Charlottesville, University of Virginia Press, 1985).

29. Semper, *Style,* 123.

30. See Gottfried Semper, *The Four Elements of Architecture,* trans. Harry Francis Mallgrave and Wolfgang Herrmann (Cambridge: Cambridge University Press, 1989), 103–4.

31. Ibid.

32. Cf. essays collected in a special issue of *Exemplaria* on "Surface, Symptom, and the Future of Critique," *Exemplaria* 26, nos. 2–3 (2014): 127–298.

33. Lytle Shaw, *Fieldworks: From Place to Site in Postwar Poetics* (Tuscaloosa: University of Alabama Press, 2013), 254.

34. Lisa Robertson, "Soft Architecture: A Manifesto," in *Occasional Work,* 21.

35. Lisa Robertson, "How to Colour," in *Occasional Work,* 125.

36. Lisa Robertson, "Introduction to *The Weather,*" in *Occasional Work,* 60.

37. I borrow here from Miriam Nichols's language in discussing the poetry of Robin Blaser. See her *Radical Affections: Essays on the Poetics of Outside* (Tuscaloosa: University of Alabama Press, 2010), 193.

38. Robertson, "How to Colour," 123.

39. Robertson, "*Rubus armeniacus,*" 110.

40. Jacques Derrida, *Dissemination,* trans. Barbara Johnson (Chicago: University of Chicago Press, 1987), 127.

41. Steve Mentz, "After Sustainability," *PMLA* 127, no. 3 (2012): 586–92.

42. Robertson, "*Rubus armeniacus,*" 122.

43. Charleton T. Lewis and Charles Short, *A Latin Dictionary* (Oxford: Clarendon, 1879), s.v. *elaboro* 1, and see especially 2b.

44. For a general introduction to the riddles for nonspecialists, see Craig Williamson, ed. and trans., *A Feast of Creatures: Anglo-Saxon Riddle-Songs* (Philadelphia: University of Pennsylvania Press, 2011). These riddles are increasingly understood as part of a larger Anglo-Latin Riddling tradition. See Mercedes Salvador-Bello, *Isidorean Perceptions of Order: The Exeter Book Riddles and Medieval Latin Enigmata* (Morgantown: West Virginia University Press, 2015). Unlike many Latin manuscript counterparts, the Old English Riddles do not include mention of their solutions within the extant manuscript.

45. Riddles 31 and 32, in George Philip Krapp and Elliott Van Kirk Dobbie, eds., *The Exeter Book,* ASPR 3 (New York: Columbia University Press, 1936), lines 1–2. All translations mine.

46. For example, see Marie Nelson, "The Rhetoric of the Exeter Book Riddles," *Speculum* 49, no. 3 (1974): 421–40.

47. Frederick Tupper, ed., *The Riddles of the Exeter Book* (Boston: Ginn, 1968), xiv.

48. See, for example, "The Panther," in *The Exeter Book,* lines 1–3; and the Old English Psalm 101:22, in *The Paris Psalter and the Meters of Boethius,* ed. George Philip Krapp, ASPR 5 (New York: Columbia University Press, 1932).

49. See Steven McCaffrey, "From the Notebooks," *L=A=N=G=U=A=G=E* 9, no. 10 (October 1979): n.p.: "Words . . . are invested into the sentence, which in turn is invested in further sentences. Hence, the paragraph emerges as a stage in capital accumulation. . . . [Thus,] meaning finds its place in bourgeois epistemological economy as a consumed surplus value . . . found wholly as surplus value at the end of a reading."

50. Mary Carruthers, *The Experience of Beauty in the Middle Ages* (Oxford: Oxford-Warburg, 2013), 8.

51. Ibid., 33.

52. John Leyerle, "The Interlace Structure of *Beowulf,*" in *Interpretations of Beowulf: A Critical Anthology,* ed. R. D. Fulk (Bloomington: Indiana University Press, 1991), 146–67; Rosemary Cramp, *Grammar of Anglo-Saxon Ornament: A General Introduction to the Corpus of Anglo-Saxon Stone Sculpture* (Oxford: Oxford University Press, 1991).

53. Carruthers, *Experience of Beauty,* 152–54.

54. Ibid., 136.

55. *Oxford English Dictionary,* s.v. *decorate* 1.

56. Lewis and Short, *A Latin Dictionary,* s.v. *decoro.*

57. Jonathan Slocum, et al., eds., *Indo-European Lexicon,* Linguistics Research Center at the University of Texas, Austin, 2009–11, http://www.utexas.edu/cola/centers/lrc/ielex/, sv. deḱ-.

58. J. R. Clark Hall, *A Concise Anglo-Saxon Dictionary,* 4th ed. (Toronto: University of Toronto Press, 1960), s.v. *wlitigian;* Joseph Bosworth and T. Northcote Toller, *An Anglo-Saxon Dictionary* (Oxford: Oxford University Press, 1972), s.v. *wlitigian* 1–2. Bosworth and Toller do not invoke terms of adornment or decoration explicitly in its definition but do indicate that the present participle *wlitigende* occurs as a gloss of the Latin participle form *decorans.*

59. See Bosworth and Toller, *Anglo-Saxon Dictionary,* s.v. *wlite* 1, 2, and *wlitig;* Clark Hall, *A Concise Anglo-Saxon Dictionary,* s.v. *wlitig.*

60. See *Dictionary of Old English,* ed. Cameron et al. (Toronto: Toronto University Press, 1986–), s.v. *frætwe.*

61. Bosworth and Toller, *Anglo-Saxon Dictionary,* s.v. *wræt.*

62. Ibid., s.v. *wrætlic, wrætlice;* Clark Hall, *Concise Anglo-Saxon Dictionary,* s.v. *wrætlic, wrætlice.*

63. See H. G. Liddell and Robert Scott, *An Intermediate Greek-English Lexicon* (Oxford: Oxford University Press, 1889), s.v. κόσμος, A, A.203, II, IV.

64. Michel Serres, *Biogea,* trans. Randolph Burkes (Minneapolis: Univocal, 2012). See esp. 45–51.

65. Ursula Le Guin, *The Dispossessed* (New York: Harper Voyager, 2011), 266.

66. Maurice Merleau-Ponty, *Phenomenology of Perception,* trans. Colin Smith (London: Routledge, 2002), 73.

67. Shaw, *Fieldworks,* 255.

68. I do refer to Jacques Derrida's *Of Grammatology,* corrected, ed., trans. Spivak (Baltimore: Johns Hopkins University Press, 1997) and its reading of "the trace," that movement anterior to the distinction of presence/non-presence, living/nonliving—"which opens appearance and *[l'apparaître]* signification" (65). To decorate will not make easy joinings in the present. I take my cues for ecotheory with a Derridean backdrop from Julian Yates, especially "Sheep Tracks—A Multi-Species Impression," in *Animal, Vegetable, Mineral: Ethics and Objects,* ed. Jeffrey Jerome Cohen (Washington, D.C.: Oliphaunt, 2013), 173–209.

69. Serres, *Biogea,* 188.

Remember

CORD J. WHITAKER

Love is like a cancer, three lyrics drawn from a representative collection of popular musical genres suggest. For hip-hop's part, Lloyd Banks offers, "Love is like a cancer when you don't let go."[1] Post-hardcore punk band letlive offers, "And love is like a cancer."[2] For the part of the blues, from which hip-hop and punk ultimately derive, Son Seals growls, "Your love is just like a cancer."[3] Cancerous love consumes its subject until nothing remains. But love also gives life: in the Middle Ages to the lusty squire of Chaucer's *Franklin's Tale* "yong, strong, right vertuous, and riche, and wys . . . oon of the beste faryng man on lyve" (933, 932) and to any modern reader who has known the joys of new love, eschewing sleep in order to converse with her beloved through the night yet rising in the morning feeling "fressher . . . and jolyer of array . . . than is the month of May" (928).[4] Cancer also gives life, though at least since the 1971 advent of the U.S. "war on cancer" it has not been thought of this way.[5] Though modern Americans are inundated with the language of battle and haunted by images of emaciated victims, though cancer can result in losing a loved one, cancer does not begin in death.[6] Tumors *grow*; they do not, of their own accord, shrink. Living, cancerous cells multiply in number as a tumor grows, and cancer can be construed as, like love, vivifying. Of course, in cancer, and sometimes in love, things go terribly wrong. The dual edge of vivification that paradoxically leads to reduction—whether in grief, loss, or death—might seem to be the (not insignificant) extent of the similarity between cancer and love. It is not.

This essay takes the role of memory as a point of commonality, as a crossroads where fourteenth-century elegiac love lyric productively informs twenty-first-century cancer theory: love leads to grief and mourning only when it has been lost and is fondly remembered, while new cancer research suggests that malignant cells multiply relentlessly only because they remember, as a function of their genetic codes, how they were designed to reproduce a billion years ago. This essay argues that Chaucer's *Anelida and Arcite,* an elegy in which a jilted lover mourns love lost, proffers lessons about memory and progress as well as loss, repetition, and reconstruction that can help compel the study of cancer beyond the yes-and-no, right-and-wrong arguments that often hinder scientific research. At the same time, *Anelida* offers lessons in the cultural and personal roles of memory, even for those who are not falling in love—or dealing with cancer.

The dynamics by which memory spans generations are an oft-treated theme in literature and they drive the theoretical physicist Paul Davies's groundbreaking and controversial claim that cancer is the result of a "genetic subroutine programmed into [cells'] ancestors" around a billion years ago.[7] The novel about colonialism's traumatic legacy *Le Livre d'Emma* by the Haitian-Quebecoise writer Marie-Célie Agnant offers a literary example: Lesley S. Curtis writes of "l'aspect intergénérationnel du trauma colonial" (the intergenerational aspect of colonial trauma) that informs the coalescence of the main character, Emma, and Flora, her translator. In Curtis's analysis of characters who are thirled by history's horrors, whether they remember them through personal or inherited experience, she concludes that "le passé va toujours refuser d'enfanter l'avenir si l'on n'accepte pas son existence" (the past will always refuse to give birth to the future if we do not accept its existence).[8] Curtis's analysis asserts the multiple roles of intergenerational memory—as a fastener to the past and a driver of the future. *Anelida and Arcite* offers a medieval analogue in its recognition of memory's ability to transcend an individual's life and death. The text is perhaps most famous for the opening and closing lines of the main character's complaint:

So thirleth with the poynt of remembraunce
The swerd of sorowe . . .
. . . .

How that Arcite Anelida so sore
Hath thirled with the poynt of remembraunce (211–350)

Thirl, from Old English *þyrlian,* is to pierce or run through as a sharp sword does to a body or to bore a hole as a drill does.[9] Lee Patterson points out that "the poynt of remembraunce" derives from Dante's *Purgatorio* canto 12 in which Dante and Virgil walk on sculptured paving stones that depict the prideful. They are compared to the memorial stones that cover the tombs of those buried in a church nave. The stones "spur" mourners to weep for the dead "at the prick of memory" *(per la punctura de la rimembranza).*[10] Anelida weeps for the death of her love relationship with Arcite. In modern literature, in *Anelida,* in Dante, and in Davies's theory, memory traverses the border between life and death. Memory's intergenerational nature means that it gives new life to the dead—whether love, an individual, or bygone modes of cellular reproduction—even as it ushers those who live and mourn toward emotional or physical death.

In an attempt to stabilize the alternately vivifying and deadly roles of memory, its powers to animate love and cancer and to turn them equally to creative and destructive ends, I strive to arrest the concept long enough to discuss it by enfolding it in the term *remember.* This enfolding is a fantasy of stasis (of memory's submission to critical inquiry), and using what is properly a verb as a noun draws attention to the effortfulness of its reification and the falseness of its stasis. Memory resists standing still.

In addition to mobility among past, present, and future, and vacillation between creation and destruction, *remember* encompasses the unity of the literary experience of intergenerational memory with memory's real-world effects. The modern story of writer and teacher Kristine Morrow, her father, and the search for his lost childhood home offers a real-world case study in the role of memory and will serve as a touchstone in this essay.[11] Morrow was a teenager when she first drove west with her father from their apartment in a busy Philadelphia suburb into the Appalachian Mountains. He was on a quest to recover his past. She was learning what it meant when he would finish gardening on the little plot of land he had outside their building. He would stand, his knees darkened with soil, and slap his hands against his thighs. It was a farmer's gesture: all done, it proclaimed, as the earth fell from his fingers and palms. To Kristine, it seemed out of step with the multistory buildings crammed together on the gravelly, overused thoroughfare. The sun barely reached the ground for the shadows cast by the gray, institutional apartment houses, the stores, the tall strip-mall shopping plaza signs. But this man dusted himself off as if he were in the midst of wide open fields, as if the only things making noise nearby were cattle or sheep. She would learn on that drive that they were not wide open fields for which her father longed. Instead, he yearned for the dense tree cover of the Appalachians, for the rush of Jacob's Creek as it descends from the Laurel Highlands until it reaches the Youghiogheny River, for the babbling of some 177 miles of tributary streams.[12] Comprising one of these streams and the valley it supports is the place he longed for: the unique ecosystem in southwestern Pennsylvania known as Morrow Hollow. Morrow's yearning, exemplified by his farmer's gesture in an urban/suburban garden, always had most of the hallmarks of *remember*—grief even unto despair brought on by looking to the past and the spurring onward that comes of necessity yet is shaped by desire for that which is unrecoverable—and its intergenerational nature became apparent on the drive west.

The elder Morrow was "thirled with the poynt of remembraunce." Years later, Kristine would come to understand that her father was not only pierced by his memories but his condition corresponded with the other elements of thirling, too. He was held in place, fixed to a spot, to Morrow Hollow, as if it were a spindle. His adult experiences spun out from and circulated around his memories of Morrow Hollow. He could spin around it but not away. Whenever he heard of environmental degradation, local or global—deforestation in Brazil, for instance—he would grieve deeply. The Morrow Hollow stream and its larger recipient Jacob's Creek carved—one might say *thirled*—their way into the mountain around them over eons. The Hollow had thirled Morrow by the time he was nine when he was wrenched away because his family could no longer make a living off the land.

It is no coincidence that Morrow was thirled by the memory of an unrecoverable home, and his troubles elucidate Anelida's. Nothing may animate the notion of *remember* as effectively as the topos of home, a place of origin, comfort, and belonging. *Remember,* as it functions in the Morrows' story, illustrates in modernity the affecting dynamic of exodus, memory, and desire for return that often drives medieval romances, especially "family romances," "where rediscovery and reconciliation of parents and children constitute the recovered equilibrium of the happy ending."[13] In comparison, the function of *remember* in *Anelida and Arcite* significantly complicates the notion of home: Thebes is a city that is equally thirled and thirling, run through and running through, and Chaucer's use of the city illuminates the process by which destructive and creative potentials inhere and continually supplant one another in *remember*. Citing the renowned story of Oedipus, the Theban who kills his father and unwittingly commits incest with his mother, Lee Patterson writes that "the Theban story is itself about disordered memory and fatal repetition, about the tyranny of a past that is both forgotten and obsessively remembered."[14] Patterson goes on to describe "the profound circularity of Thebanness, its inability ever to diverge from the reversionary shape ordained in and by its beginning." According to Ovid's *Metamorphoses,* Thebes is founded when Cadmus slays a serpent sacred to Mars and is told that he will be made into a serpent as punishment. He plants the slain serpent's teeth and dragon warriors rise up who then kill one another. These "first Thebans" instantiate a recursive "economy" that haunts the Theban line all the way down to Oedipus.[15] It is a city thirled by exile, murder, and cursedness, and it is always poised to thirl again: in one medieval recounting, the short Latin work *Planctus Oedipi,* Oedipus says that he "struck iron through [his] father's loins."[16] Literally, he thirls his father. Metaphorically, by boring a hole in that which formed him he thirls together his father and himself in a cycle of destruction that portends toward recursive repetition but instead veers. Each destruction, rather than reducing Oedipus's narrative and history down to nothing, produces something new and unexpected, another narrative: an Oedipus who strikes his father in a new and unexpected way, an Anelida whose mourning constitutes a new narrative.

Anelida and Arcite is particularly elucidating of *remember's* creative and destructive capacities because of the moment in Thebes's history at which it occurs. As Dominique Battles points out, Chaucer's predecessors in the *Roman de Thèbes* and Boccaccio's *Teseida* represent the city as a war-ruined shell of its former self, while Chaucer sets *Anelida* in a functioning Thebes that nevertheless naively teeters on the edge of destruction.[17] While Thebes is alive in Chaucer's text, it is also already dead, or at least dying: Chaucer positions Anelida's story after the city's two ruling, and warring, brothers, Etiocles and Polynices, have been killed. Creon steps in to rule. Though tyrannical, he reinvigorates the city, drawing the friendship of regional nobles with entertainment and the like. Anelida, queen of Armenia, is one of Creon's guests. When she falls in love with and is jilted by the Theban knight Arcite, she describes her grief as "dedly adversyte" (258). She exclaims, "I am so mased that I deye" (322). Like Thebes, whose destruction was by Chaucer's time a literary commonplace, Anelida is "always dying but never dead."[18] Though Thebes and Anelida seem to march toward death, the veering cycle on which Thebes is founded and that informs its continual literary redeployment means that they are continually given new life.

Remember's life-giving properties are manifest in the earliest experimental studies of personal, or autobiographical, memory. Sir Frances Galton's late nineteenth-century experiments tested for memory associations. During a stroll along Pall Mall in central London, he noted how the objects he saw drew to mind a great variety of memories. He decided to test how quickly his memory reacted to a given stimulus and what various stimuli drew to mind by observing a word, allowing his mind to recall several memories in response, and recording what he remembered. He found that no fewer than 39 percent of his memory associations went back to his youth. Galton was the first to demonstrate the "reminiscence" effect—that as we approach the age of sixty, our associations tend to turn back to our youth.[19] Far from respecting linear progression, which would dictate that a person remember most recent events first, the reminiscence effect evokes something similar to Theban circularity: the closer one is to death, the closer she is to the beginning of life.

As he felt himself drawing nearer to the beginning of his life, the elder Morrow became more adept behind the wheel. These roads challenged other drivers, drivers *from away,* with hairpin turns and blind bends hugged by steep drop-offs to the right and hulking boulders teetering precariously to the left. To Morrow, these roads felt like home. Kristine had no idea where they were going. She felt lost. The boulders, drop-offs, and blind turns threatened her. She was *from away.*

Morrow had been away too long. The roads felt like home, but they did not lead him to it. For hours, he could swear he was close. Wasn't that the kind of tree that grew in the Hollow? Wasn't that a Hollow flower? They did not find Morrow Hollow. As they returned to the home Kristine knew, she saw Morrow differently. He bore an air that was somehow like the sense of determination and accomplishment he displayed in the garden. But this was sadder. In his eyes, she saw that determination had turned to grief, yet it was a grief that she knew would turn to determination once again. Studying her father's profile, with the deep green Appalachian pines whizzing by for a backdrop: this was how Kristine came to remember Morrow Hollow, a place she had never been.

Kristine's new memory of her father's past demonstrates the extent to which intergenerational memory has real implications for the present. Medievals thought of memory similarly. Considering mnemonic devices, thirteenth-century Italian rhetorician Boncompagno da Signa defines memory as that faculty by which humans "assiduously remember the invisible joys of paradise and the eternal torments of hell."[20] Mnemonic devices are also discussed in the *Rhetorica ad Herennium*, the standard text for rhetorical theory in the late Middle Ages:

> We ought, then, to set up images of a kind that can adhere longest in the memory. And we shall do so if we establish likenesses as striking as possible; if we set up images that are not many or vague, but doing something *[agentes imagines]*; if we assign them exceptional beauty or singular ugliness; if we dress some of them with crowns or purple cloaks, for example, so that the likeness may be more distinct to us; or if we somehow disfigure them, as by introducing one stained with blood or soiled with mud or smeared with red paint, so that its form is more striking.[21]

The *agentes imagines*, "beautiful or hideous human figures," are what Frances Yates has called "'corporeal similitudes' of 'subtle and spiritual intentions."[22] Though obtaining heaven or avoiding hell at first appear to be future-oriented goals, the *imagines agentes* have a profound effect on lived experience: they remind him who remembers them how to act in the present, how to treat other people ethically, and how to behave prudentially in the present and in the future.[23] Modern experimentation also demonstrates memory's profound effect on the present. Citing Willem Albert Wagenaar's interpretation that "our memory is particularly good at storing those events that are hardest to reconcile with our self-image," Douwe Draaisma asserts that the memory of humiliation and other affronts "gives your body every chance to re-experience the physical reactions you had at the time."[24] Hurtful events are etched into our memories and brought into the experience of the present in order to direct us toward prudential and ethical behavior in the future. The memory of Morrow Hollow thirled Kristine and would become an integral part of her present.

Medieval and modern theory affirm Paul Davies's deployment of memory in the etiology of runaway cellular reproduction that characterizes cancer. Davies, a theoretical physicist, began working on cancer when Anna Barker, deputy director of the National Cancer Institute, called him for help. She cited his work for asking, "What are the hidden assumptions we are making, how can we reconceptualize this problem, how can we make progress?"[25] The result is twelve physical sciences-oncology centers around the United States, in which cancer biologists and oncologists work closely with physicists, physical chemists, astrobiologists, engineers, and nanotechnologists. Davies has sought to develop a theory of cancer by situating it within the evolutionary story—an approach germane to his career in theoretical physics and astrobiology. The evolutionary framework asks, how did life get here? How has life developed? Davies's work asks, as a subset of those questions, how did cancer get here? How did cancer develop? The results of his questions are a testament to the creative and destructive power of *remember* as well as its wide reach.

The veering circularity that defines *remember* is borne out in Davies's theory. He and his colleagues posit that cancer's "genetic subroutine" is "really just a re-run of an embryonic developmental program." Embryonic cells, according to Davies and Charles Lineweaver, represent the most ancient and highly conserved cells.[26] They "envisage a collection of ancient conserved genes driving the cancer phenotype, in which the metastatic mobility of cancer cells and the invasion and colonization of other organs merely reflects the dynamically changing nature of embryonic cells and their ability to transform into different types of tissues."[27] In other words, cancer arises when cells revert to the expression of their most ancient genes, which were developed for the reproduction of single-celled organisms about a billion years ago. This form of expression continues to have a role in modernity as embryonic reproduction, but in multicelled organisms cellular differentiation occurs in order to form the variety of necessary biological structures. Differentiation did not occur in the single-celled organisms from which humans evolved. In the evolutionary approach to cancer, *remember's* destructive power is revealed: what once led to immortality—in the form of the unchecked and exact replication of a cell—now leads instead to death of the organism *and* the otherwise immortal replicating cell. As with Anelida's mourning, the vibrance of life (the love relationship in Anelida's case) gives way to death.

Remember's status in the etiology of cancer is a contested issue. Davies's theory has received a lot of press: in addition to numerous interviews with such global outlets as the *Guardian* and more local publications such as the Australian Broadcasting Corporation's *The Science Show,* in July 2013 *Physics World* magazine published a special issue titled "The Physics of Cancer." It celebrated physicists' interventions in oncology and featured Davies's work.[28] Not everyone, however, is on board. Davies's critics include the outspoken biologist Paul Myers, whose blog *Pharyngula* has been ranked by *Nature* as the most popular blog written by a scientist.[29] Myers charges that Davies and Lineweaver misunderstand gene suppression:

> Genes that are suppressed decay and are lost, not lurking. Genes that remain and are tinkered with by mutations acquire new functions, lose old ones, and are assembled into new, coevolving networks. . . . They're not in a separate "layer," they're not waiting, unmodified, to blithely re-enact ancestral states.[30]

In other words, according to Myers, genes are not subject to *remember's* veering, thirling circularity; they simply progress unidirectionally in a straight line. Myers's blog is known for sarcasm and harsh criticism and is true to form when he judges it *"embarrassing* to see two smart guys with a measure of legitimate prestige in their own specialties charging off into another discipline with such crackpot notions."[31] For Myers, it is ludicrous to imagine cells reverting to historical behavior based on memory functions embedded within their genetic codes.

The debate about *remember's* status in cancer's etiology is far from over. In a 2015 paper, a team led by Amy Wu and including Davies showed that certain genes mutate more and others less in the development of chemotherapy drug resistance. They chose multiple myeloma, a bone-marrow cancer that is usually incurable, because its cells develop resistance quickly. The researchers designed and implemented a "micro-environment, with drug gradients and connected microhabitats" in order to significantly speed up resistance's development. Once they had drug-resistant cells, they were able to identity "hot" or "hypermutated" genes that had at least one single-nucleotide variant and "cold" genes that underwent no mutation. Of the 785 genes that were fully sequenced, 163 of them, or 21 percent, were cold. They found that the mutating hot genes were involved in various biological functions, including cell division, metabolism, and apoptosis, or "programmed cell death."[32] Irregularities in apoptosis are an important driver of malignancy's progression: cancer cells proliferate but do not die in order to make way for new cells. What's more, at an average age of 1.7 ± 1.0 billion years "the cold, zero-mutation genes . . . are older on average than all human genes." Cold genes represent "ancient genes associated with spermatogenesis, oxidation–reduction, and glycolysis," fundamental processes for reproduction and energy production. Wu et al. postulate that these zero-mutation genes are "possibly protected" in the development of drug resistance. These genes also seem to be "abnormally upregulated"; they are expressed at far greater levels in drug-resistant cells, possibly to compensate for the mutation of hot genes. Wu et al. answer Myers's criticism by positing that there are "two broad components of information dynamics in cancer evolution. One involves permanent changes in which genes are subject to gain or loss-of-function substitutions" and the other "is the information in the human genome that is not mutated and in fact is protected from mutations." Wu et al. conclude that "the cancer cell . . . can upregulate or downregulate any number of strategies used for survival and proliferation during embryogenesis, development, and normal adaptation to environmental stresses."[33] In other words, Wu et al.'s analysis demonstrates that Myers is right some of the time: some gene functions are substituted and lost. But a portion of the human genome is "highly conserved," and *remember*—the veering and thirling circularity of creation, destruction, and re-creation, of experience, desire, and reexperience—is characteristic of, and perhaps a function of, humans' genetic code.

The thirl of circularity—the propensity to nearly but not quite repeat, to veer—is intrinsic to *remember*. In Davies and Lineweaver's theories, cosmology reveals the multidirectionality that characterizes evolution when they "envisage cancer progression within a host organism as like running the arrow of biological evolution backward in time at high speed."[34] The "arrow of time," or the "one-way slide of the universe towards total disorder" leading to inevitable death, has also been described as the assumed continual rise in biological complexity.[35] The arrow, however, does not point one way. After all, who is to say that increasing biological complexity is quite the same thing as an increase in disorder, or entropy, on the cosmological level? Indeed, Davies's work in cosmology, his specialty, shows that "arrows" of development cut across one another in ways that produce summary effects different from that of a single arrow. While the universe's matter and radiation started out in thermodynamic equilibrium, a maximum-entropy state, the universe's gravitational field started in a low-entropy state of near uniformity. Together, their entropy has always been rising as the gravitational field becomes increasingly distorted. Constituent of this rising, however, is the *decrease* in matter and radiation's entropy as gravity causes them to form increasingly ordered systems.[36] Increase and decrease inhere within one another, and cosmology demonstrates that *remember* is a multidirectional system characterized by back and forth, pushing and pulling motions that create the whirling vortex that thirls.

Anelida might seem stuck, but she in fact demonstrates *remember's* thirling multidirectionality. Toward the end of her complaint, Anelida moves from sorrow to anger:

> Wher is the trouthe of man? Who hath hit slayn?
> Who that hem loveth, she shal hem fynde as fast
> As in a tempest is a roten mast.
> Is that a tame best that is ay feyn
> To fleen away when he is lest agast? (311–16)

Anelida compares Arcite to a rotten mast and a wild beast. This a far cry from her earlier claim that she will always love him and that she will forgive him (274–80). What's more, in Chaucer's "apparently unfinished" ending, Anelida heads to the temple of Mars, god of war. Aranye Fradenburg argues that the text registers "anxiety about, even hostility towards . . . *endings*" and notes that a Derridean approach deems erroneous the notion that "some texts *are* finished, that there are ever absolute boundaries between texts and other texts."[37] Indeed, if a reader knows the *Knight's Tale* or the *Teseida,* the text spills over into its analogues. But *remember* exceeds intertextuality. Anelida's turn to anger suggests that she moves forward; perhaps the healing process is underway. She abruptly retreats from anger with apology: "Now merci, swete, yf I mysseye? . . . my wit is al aweye" (317–19). The reader remains mired in *remember's* circularity, in its thirling veer. At the same time, she is shown glimmers of the way forward: a change in emotion, healing and empowerment, an ending that opens onto other texts and a variety of possibilities.

The dynamics of cancer mimic Anelida's lament, and the text offers a lesson that will benefit further cancer research. Anelida's lament demonstrates stasis even as the poem progresses; it moves forward by being stuck in the mode of lament. Cancer, similarly, progresses when cells reproduce themselves exactly and in an unregulated fashion; it progresses by being stuck. Wu's study, in showing that cancer depends on mutating and static genes in order to drive cellular reproduction, suppress apoptosis, and increase cellular perseverance under stresses such as metastasis, bears out the coexistence and interdependence of stasis and change. So does Kristine's reception of Morrow's memory. Decades later, still thirled by Morrow Hollow, Kristine ran across it on a United States Geological Survey map. She drove west once again. She replayed in her memory Morrow's farmer's gesture, their drive, and her disorientation. His grief. His determination. This time she found it: the rich green tree cover and the dappled mountain light filtering through it, the tributary's water rushing down the Hollow to meet Jacob's Creek, varieties of mountain flowers she did not know. Birdsong. His determination had become hers. His grief her joy. The static memory of nine-year-old Morrow's home persisted intergenerationally, and *remember* served its recuperating purpose.

The place had changed, though in a way that permits stasis its role. Kristine found the wells: 11 hydraulic fracturing, or *fracking,* wells now mar Morrow Hollow's landscape. It should have come as no surprise. Natural gas extraction from the Marcellus Shale, the 380 to 400 million-year-old rock formation on which much of Pennsylvania is situated, has made the state one of the fastest-growing U.S. natural gas producers.[38] In Westmoreland County, where Morrow Hollow is located, there are 251 active wells out of 7,788 statewide.[39] The wells represent change, but their express purpose is to tap into history. Change in this case is the means by which stasis, in the form of preserved ancient natural gas and seawater, is revealed. Fracking is thirling of the earth.

And it is a modern manifestation of *remember*. Water with any number of chemical additives is injected down into shale at extremely high pressures in order to open fissures through which natural gas is drawn to the surface. The water that comes back up is called *flowback,* but it brings with it "produced water," the remnants of ancient seawater. Produced water's salts have reached extreme concentrations and contain radionuclides such as radium, which can cause lymphoma, bone cancer, and leukemias. It also contains bromide, which can cause liver, kidney, and nervous-system problems.[40] Though measures are taken to contain and treat produced water, mechanisms have been known to fail and contaminate drinking water.[41] Morrow Hollow is thirled by the "poynt of remembraunce" that is the injection well.

Remember is systemic, and when humans thirl the earth, the earth can respond in kind by thirling the human genome. The best cancer research practices are those recognizing the interdependence of stasis and change. They recognize that development, genomic and otherwise, is a multidirectional process in which cycles suggest circular repetition while their vectors veer away from completion. Thirling, they drill down as each point moves toward its former position in one plane and beyond it in another. Seeming stuck, they create that which is new. Sometimes the new, like a fracking well to ancient seawater, recovers and reinvigorates the old. Sometimes new life, even unto cellular immortality, is thirled together with death.

If only for a moment, Anelida gets angry. Perhaps this portends a post-elegiac future for her. Regardless, her story in a Thebes thirled and thirling is already new. I was thirled by the Morrows' story. Hurtling along mountain roads and descending down into the Hollow again and again in my mind long after Kristine had finished, I could not shake Morrow's face set with sad determination. I cannot shake his farmer's gesture in a small suburban garden. I do not want to leave the deep green tree cover of Appalachian pines and the dappled mountain light filtering through it. This is how I have come to remember Morrow Hollow, a place I have never been. *Anelida* and the Morrows' story demonstrate that, while the songs say that love is like cancer, it is more proper to say that, in the context of *remember,* cancer is like love.

Notes

This essay has been made possible by a Wellesley College Faculty Award and the efforts of my indefatigable research assistant, Sarah Michelson.

1. Lloyd Banks, "Where I'm At," Christopher Lloyd, Matthew Samuels, Matthew Burnette, Marshall Mathers III, *H.F.M. 2,* Eight Mile Style Music, Sony/ATV Tunes LLC, 2010, CD.

2. letlive, "Virgin Dirt," Jason Aalon Butler, Jean Nascimiento, Jeff Sahyoun, Ryan Jay Johnson, *The Blackest Beautiful,* Peermusic, 2013, CD.

3. Frank "Son" Seals, "Your Love Is Like a Cancer," *The Son Seals Blues Band,* Alligator Records, 1973, CD.

4. Geoffrey Chaucer, *Riverside Chaucer,* ed. Larry D. Benson (Boston: Houghton Mifflin, 1987). All references to Chaucer are taken from this text.

5. The "war on cancer," or the National Cancer Act of 1971, was designed to fund research that would identify and eradicate the retroviruses that were believed to cause cancer. Robert A. Weinberg, "Coming Full Circle—From Endless Complexity to Simplicity and Back Again," *Cell* 157, no. 1 (March 2014): 267–71 (267–68).

6. For a discussion of bellicose language in cancer discourse, see David J. Hauser and Norbert Schwarz, "The War on Prevention: Bellicose Cancer Metaphors Hurt (Some) Prevention Intentions," *Personality and Social Psychology Bulletin* 41, no. 1 (2015): 66–77 (74).

7. Paul Davies and Charles Lineweaver, "Cancer Tumors as Metazoa 1.0: Tapping Genes of Ancient Ancestors," *Physical Biology* 8, no. 1 (2011): 15001–8.

8. Lesley S. Curtis, "'Vite elle se referme': L'opacité dans *Le livre d'Emma* de Marie-Célie Agnant," *Women in French Studies* 21 (2013): 68–78 (73, 77). Intergenerational trauma in literature often responds to histories of racism. See also Derek Walcott's *Omeros* and Toni Morrison's *Beloved* and *Home.*

9. *OED,* s.v. *thirl*; *MED,* s.v. *thirl*: Though the *OED* asserts that modern English *drill* comes into English independently from the Dutch *drillen,* an old strong verb *þrell-/þrall-/þrull-* may have influenced both the Dutch and Old English, informing the modern words' identical meanings.

10. Lee Patterson, *Chaucer and the Subject of History* (Madison: University of Wisconsin Press, 1991), 78–79. Dante's poetry is taken from *Purgatorio* canto 12, lines 16–21.

11. Kristine Morrow, Untitled sermon, Calvary United Methodist Church, Philadelphia, Pa., November 8, 2015.

12. Jacob's Creek Watershed Association, http://www.jacobscreekwatershed .org.

13. Helen Cooper, *The English Romance in Time: Transforming Motifs from Geoffrey of Monmouth to the Death of Shakespeare* (Oxford: Oxford University Press, 2004), 333.

14. Patterson, *Chaucer,* 75.

15. Ovid, *Metamorphoses,* trans. and ed. Charles Martin (New York: W. W. Norton, 2004), book 3, lines 13–162; Patterson, *Chaucer,* 75–76.

16. Patterson, *Chaucer,* 76, translates the Latin from the poem's witness in Berlin MS lat. 34, fol. 113, printed as *Edélestand de Méril, Poésies inédites du moyen âge* (Paris: Franck, 1854), 310–13.

17. Dominique Battles, *The Medieval Tradition of Thebes: History and Narrative in the OF Roman de Thèbes, Boccaccio, Chaucer, and Lydgate* (New York: Routledge, 2004), 87.

18. "Always dying but never dead" is Patterson's pithy formulation for the representation of despair in medieval writing and the *Pardoner's Tale* in particular. Lee Patterson, "Chaucer's Pardoner on the Couch: Psyche and Clio in Medieval Literary Studies," *Speculum* 76, no. 3 (2001): 638–80 (657).

19. Francis Galton, "Psychometric Experiments," *Brain* 2 (1879): 149–62; Douwe Draaisma, *Why Life Speeds Up as You Get Older: How Memory Shapes Our Past,* trans. Arnold and Erica Pomerans (Cambridge: Cambridge University Press, 2010), 3–7.

20. Frances Yates, *The Art of Memory* (Chicago: University of Chicago Press, 1966), 59–60.

21. *Rhetorica ad Herennium,* trans. Harry Caplan, Loeb Classical Library 403 (Cambridge, Mass.: Harvard University Press, 1954), book 3, chapter 22.

22. Yates, *Art of Memory,* 76.

23. Ibid., 53–61; and Mary Carruthers, *The Book of Memory: A Study of Memory in Medieval Culture,* 2nd ed. (Cambridge: Cambridge University Press, 2008), 81–89.

24. Draaisma, *Why Life Speeds Up,* 46–47.

25. Paul Davies, "Rethinking Our Approach to Cancer," interview by Robyn Williams, *The Science Show,* Australian Broadcasting Corporation, February 15, 2014, http://www.abc.net.au/radionational/programs/scienceshow/rethinking -our-approach-to—cancer/5246414#transcript.

26. Davies and Lineweaver, "Cancer Tumors," 15001–8.

27. Paul Davies, "Exposing Cancer's Deep Evolutionary Roots," *Physics World,* July 2013, 39–40.

28. Paul Davies, "Cancer Can Teach Us about Our Own Evolution," *Guardian,* November 28, 2012, http://www.theguardian.com/commentisfree/2012/nov/18/ cancer-evolution-bygone-biological-age; Zeeya Merali, "Did Cancer Evolve to Protect Us?" *Scientific American,* October 2, 2014, http://www.scientificamerican .com/article/did-cancer-evolve-to-protect-us/; Marcus Strom, "Cancer Theorist Paul Davies to Speak on the Disease's Evolutionary History," *Sydney Morning Herald,* December 2, 2015, http://www.smh.com.au/national/health/controver sial-cancer-theorist-paul-davies-to-speak-on-the-diseases-evolutionary-his tory-20151124-gl6lod.html; "Physics of Cancer," special issue, *Physics World,* July 2013, http://physicsworld.com/cws/download/jul2013.

29. Declan Butler, "Top Five Science Blogs," *Nature*, July 6, 2006, 9, http://www.nature.com/nature/journal/v442/n7098/full/442009a.html.

30. Paul Myers, "Aaargh! Physicists! Again!," *Pharyngula* (blog), November 20, 2012, http://scienceblogs.com/pharyngula/2012/11/20/aaargh-physicists-again/.

31. Ibid., emphasis in original.

32. Davies, "Exposing," 39.

33. Amy Wu et al., "Ancient Hot and Cold Genes and Chemotherapy Resistance Emergence," *Proceedings of the National Academy of Sciences of the United States of America* 112, no. 33 (August 18, 2015): 10467–72 (10467, 10470–10471).

34. Davies, "Exposing," 40.

35. Paul Davies, "Directionality Principles from Cancer to Cosmology," and Charles Lineweaver, "What Is Complexity? Is It Increasing?" in *Complexity and the Arrow of Time,* ed. Charles Lineweaver et al. (Cambridge: Cambridge University Press, 2013), 1–3.

36. Davies, "Directionality Principles."

37. L. O. Aranye Fradenburg, "Voice Memorial: Loss and Reparation in Chaucer's Poetry," *Exemplaria* 2, no. 1 (March 1990): 169–202 (178).

38. Valerie Brown, "Radionuclides in Fracking Wastewater: Managing a Toxic Blend," *Environmental Health Perspectives* 122, no. 2 (February 2014): A51. The Jacob's Creek Watershed in particular is situated on bedrock estimated to be 290 to 330 million years old. Jacob's Creek Watershed Association, *Jacobs Creek Watershed Implementation and Restoration Plan* (Pittsburgh: A. D. Marble, 2009), 2, http://files.dep.state.pa.us/Water/Watershed%20Management/WatershedPortalFiles/NonpointSourceManagement/ProgramInitiatives/ImplementationPlans/JCWA%20319%20report%206-17-09.pdf.

39. Chris Amico et al., "Shale Play: Natural Gas Drilling in Pennsylvania," StateImpact Pennsylvania, National Public Radio, 2011, http://stateimpact.npr.org/pennsylvania/drilling/.

40. Brown, "Radionuclides," A51–A52, A54. Environmental Protection Agency, "Radionuclides: Radium," March 6, 2012, http://www.epa.gov/radiation/radionuclides/radium.html#affecthealth.

41. Amico et al. report 4,006 known violations in Pennsylvania wells in 2011. See also Laura Legere, "Hazards Posed by Natural Gas Drilling not Always Underground," *Scranton Times-Tribune*, June 21, 2010, http://thetimes-tribune.com/news/hazards-posed-by-natural-gas-drilling-not-always-underground-1.857452#axzz1l9sBu6j4; Environmental Protection Agency, *Assessment of the Potential Impacts of Hydraulic Fracturing for Oil and Gas on Drinking Water Resources* (Washington, D.C.: Office of Research and Development, U.S. Environmental Protection Agency, 2015), ES-11.

Represent

JULIAN YATES

Veering is what living creatures do, human and otherwise.

—NICHOLAS ROYLE, *Veering: A Theory of Literature*

Anthropology is . . . faced with a daunting challenge: either to disappear as an exhausted form of humanism or else to transform itself by rethinking its domain and its tools in such a way as to include in its object more than the *anthropos*: that is to say, the entire collective of beings that is linked to him but is at present relegated to the position of a merely peripheral role.

—PHILIPPE DESCOLA, *Beyond Nature and Culture*

On the face of it, it seems hard to imagine a less likely candidate for inclusion in a lexicon of verbs vital to ecological thinking than the word *represent*, rubbished as it comes by a history of bad mediations, infidelities, ideological freighting, reduction, and redaction. The word sets in motion a string of approximating substitutions almost as if it concedes, from the beginning, that what matters, what it hopes to convey, shall simply slip through its fingers.

To represent means to remediate with words, images, bodies, with whatever technical resources you care to name, one encoding of information in a manner that renders that information present elsewhere, hosted by this or that differently configured arrangement of matter. An elected official represents some portion of an electorate. I represent my family at a wedding or a funeral. A picture of a rose represents the flower. The word carries with it the possibility of a faithful demotion of he or she or it that represents to that which is represented. Yet by advertising the fact of mediation (of its secondary, after the fact, distance from what it re-presents), the word raises the possibility of a catastrophic or compromised reduction. Does the representative who speaks for make the absent thing speak to ends other than its own? Does the verb itself embed and compound this confusion?[1]

In fact, the more I write, the more I am tempted, like the proverbial rat on a sinking ship, to jump, before things get worse. Perhaps I might find shelter in a word that slides off the tongue with less plosive unpleasantness, with less of an aura of botched catch up. To translate might provide a nice alternative. The task of the translator, he or she or it that helpfully carries something across a distance (whether or not it wants to go), may be impossible, but translators have learned to mind (or curate) the gaps and to acknowledge the static interference of the stepping-stones that permit their transmissions. "No translation without transformation" might then become my happily optative slogan as I make my nest within translation's endless passages.[2] Choose to translate and the world might happily become my rat's oyster. But no, even as translation might prove an effective because seemingly more neutral alternative, I shall stick with *represent*, even if (and perhaps because) the word remains hostage always to its evil twin, *re-present*, to present again, belatedly, probably badly, something that this verb had a hand in making appear absent in the first place. Better a piece of dry land, so many territorializing representations, so many emphatic here's and now's, however Balkanized, than the endless passages and transfer tickets offered by an optimized theory of translation.[3]

Yes, I know. My verb has already lost its vigor and become a noun (something our editors warned us against). But that makes me happy also. *To represent* means to embark on a process of making images and copies, stand-ins that necessarily sport good and bad representations (products, mediations). Almost immediately, from before the verb is written or said aloud, *represent* designates the action by which the verb transforms into nouns that then serve as inputs to successive acts of making or which provoke objections and accusations, interrupting the process or opening it to what appears to have been lost by and in and through the representations that have been made. Poor old *represent* seems only incidentally a verb. It refers distantly to activities long passed, overwhelmed, as it is, by a mob scene of nouns that it casts off only then to attract.

Ultimately, it is this after-the-fact relation to the process of making, to the erring, wild-card, phenomenalizations of *poiesis* that I like about the word. Rooted as it is in a sense of territory, a zone of appearance, here and now, for me, for you, for this "us" that takes a share in these representations of absent things—or who disputes them, *to represent* always also means *to apologize*. The representations the verb sets in motion must somehow manage the fact of their belatedness—well, badly, hospitably, or with violence. The verb retains a technical relation to ethics and politics, even as we may want to sort out or strategically separate the senses of the verb—to speak for and to re-present. In rhetorical terms, then, *to represent* also means to justify and defend (well or poorly, explicitly or implicitly) the ethos (animal, vegetable, mineral) and so ecology that its poetic acts posit or construct. Never sufficient, endlessly embattled, representations remain sites of contact and negotiation, of rhetorical gamesmanship, of danger, risk, but also, therefore, possibility.

Much like the word *environment,* which, as Nicholas Royle observes, seems primed always by a turning or circular movement that both "inscribes" and "effaces" its veering middle syllable, *represent* designates an activity whose orientation to process seems immediately to default to its products.[4] "There is no environment," offers Royle, "without veering," without "the haphazard or unpredictable non-teleological de-centering," or veer. But the word *environment* manages this errancy, marshals it to a set of anthropic ends. "Wherever there is 'environment,'" he counsels, "certain kinds of centrism—logocentrism, anthropocentrism, egocentrism—are at work." For good reasons, then, Royle proposes that we stick with the verb. His concern "is to try to think about veering in ways that question—and veer away from—" the shelter from an aleatory exposure to the inhuman that the word *environment* seeks to broker. The word *human,* likewise, always more an "infrahuman" or folding of the outside in, as Jacques Derrida offers, than a thing unto itself, "represents" an equivalent shelter, an ongoing prescription and drawing down of limits.[5] *To veer,* then, means not having to decide. To veer refuses to curtail or to draw down boundaries. Sticking with the verb enables you to prorogue the madness or violence of decision (in Latin, *to decide* means *to cut*), of demarcating boundaries. To veer undoes decisions (cuts) that have been made. Veering restores the movement that an environment stabilizes into a routine orbit. It renders an environment into an environing movement—refuses to forget that movement or to allow the routines to settle.

The verb *to represent*, in contrast, offers a differing set of possibilities that is harder to inhabit by way of a present participle or a gerund. The word offers no verbal vector that will sustain an orientation to process over and against its products. The word signals instead a chain of making predicated on the decision, "cut," or termination that produces a representation.[6] Tripped up by nouns, bumping into familiar faces before it even begins, stay-at-home *represent* does not enable an equivalent sense of mobility or movement as does *veer*. Instead, the verb places us in an archival relation to the representations we have made, the environments or shelters that have been built. This archival relation proves key. For, as anthropologist Philippe Descola puts it, "representations [remain] a privileged means of accessing the intelligibility of the various structures which organize the relations between humans and non-humans."[7] Though, as he would be the first to point out, the subject of the verb *to represent* has now to include a range of beings otherwise than human. It's not just you and I any longer (if ever) who represent, who produce representations.

My aim in this essay is not then to rehabilitate the verb *to represent* exactly, or, for that matter, the noun *representation*. I should make for a poor advocate. Instead, ratlike, in the manner of a parasite, I want to explore what might happen—how we might read and write, and so construct shelters now—if we were to pluralize access to the verb such that not only other animals, plants, and other living creatures but also the inorganic world of minerals are understood also to represent, to produce representations.[8] Descola retains the word *representation* in order to designate an archival contact zone between and among "humans and non-humans." This move orients one approach that would posit human writing systems as merely a reduced form of a general relation to coding or inscription. This orientation has generated an allied series of moves in multispecies anthropology, moves that attend to the media ecologies, translational relays, and modes of inscription (or representation) deployed in order to render impressions of deep sea worlds and to ask in what way we might understand how forests think and how vying or complementary multispecies alliances or associations might make possible different worlds.[9]

With this renewed attention to questions of mediation comes a more fundamental parsing of the verb *to represent* that would go further than either the liquidation of the nature/culture binary or its hyphenation, as in natureculture. What if we were to extend something like "culture" to otherwise than human beings, a modeling of writing that freed the "pluri-dimensionality" of the term, in Derrida's sense, from the regimen of linearization not merely for us but for other entities also?[10] Key contributions to this endeavor come from a disparate set of scholars distributed among various disciplines: animal behavior studies, ethology, anthropology, and philosophy, all of whom might be said to ponder or address themselves to the metaphysics of other beings—not necessarily with the idea of their retrieval and ready processing but as a tariff on our own ways of being. How do we proceed if cetaceans are understood to have a sense of group belonging or group identity, if sheep have opinions and even rhetoric, or if a troupe of baboons has more in common with children on a preschool field trip than with red-tailed monkeys?[11]

The aim here is not to enfranchise all beings within some renewed sense of an-at-last-achieved universal category or to assume that the writing or representations of others means that we can know the difference between what we name a (quasi-automatic) reaction or welcome as a response.[12] On the contrary, the assumption that beings otherwise than human represent, or write, has to recognize crucial differences in scale, modes of embodiment, and cognition that impede such an explosion of categories, even as it might muddy them further. Differences in scale canalize the modes of relation between proximal forms of being and decide the extent to which we may be said to share (or not) a common world, as Jakob von Uexküll's famous modeling of what seems to us the highly delimited relation of a tick to its environment *(Umwelt)* makes plain.[13] Differences among beings traditionally understood to derive from ontology (animal, vegetable, mineral, and so on) cease to have any necessary explanatory or classificatory function once all beings are understood to be differently scaled encodings of information (technical, bio-informatic, social). Singular difference plays out instead as plural differences—differences in scale, form, connectivity, and mobility—that derive from how these beings are distributed and differentiated, in relation to each other, within themselves, and within their own groups. The morphologies of different entities and the extent to which they may be said to share a world make possible but also limit the transversal relations that develop between beings even as they offer resources for possible identifications.

The assumption that many more entities than us represent their worlds tends also to have the salutary effect of provincializing the representations we set in motion as merely one small set of a much larger repertoire of images. It invites us to consider the ways in which (following Descola), for certain writers and artists, the act of representation exists not as a moment of capture so much as a ceding of our writing systems to others, a hosting not of the other but of something even stranger, the representations of the human that we receive from beings otherwise than human. Depending on the scale of the encounter, this orientation to the activity of representation might induce you to traumatic counter-identifications, such as when the so-called chickenhead, J. R. Isidore, of Philip K. Dick's *Do Androids Dream of Electric Sheep?* summons up what bounty hunter Rick Deckard must be like from the perspective of the andys (androids) he helps to find shelter. J. R. shudders as he receives "an indistinct, glimpsed darkly impression: of something merciless that carried a printed list and a gun, that moved machine-like through the flat, bureaucratic job of killing."[14] The issue here is not one of simple identification but of a time-bound, sympathetic encounter. J. R. parts company from these same andys as he watches them dismember a spider. But his nightmare vision of Deckard as a robotic killing machine promotes his understanding of them even if he concludes that they are not like him.

At other times, the effect might remain even more curious or remain hidden in plain sight, routed through more familiar, even mundane, representations, such as what it means to eat something—say, to eat an orange. As a first step, then, this multispecies orientation to the word *represent* would have to pause so that it might estrange things that seem so familiar to us that we reach for them, peel, and segment them into edible parts. For what exactly is an orange? What, for that matter, does it mean to eat?

Prior to its assimilation to the human (though this qualification shall prove erroneous), an orange is a type of berry, part of the reproductive technology of a species of plant. However fond some of us may be of them, however much you or I may enjoy their scent or find ourselves distracted by their color, the orange itself remains indifferent to us. It unfolds according to a program that, technically, does not discriminate on the subject of species. For an orange addresses itself not to us exactly but to the great variety of differently animated entities (human and otherwise) that might render it mobile, whom it rents or recruits as it makes use of their bodily capacity for movement, their physiology, offering itself to be eaten in and as the process by which it moves. It is this moment of consumption as animal recruitment, the moment at which we peel and eat an orange that, in *Le parti pris des choses (Siding with Things)*, prose-poet Francis Ponge chooses to dramatize in his poem "*L'orange.*" In what remains of this essay, my rat shall seek to inhabit this poem, or perhaps, this orange—to trace or travel with the veering whirl of orange that Ponge sets in motion.

The poem begins with what remains after the orange has been eaten, with the now archival peel, which retains an "elasticity" that seems to wish to return to its previous shape, to mimic the *thing* it was before:

Like the sponge, the orange, after undergoing the ordeal of expression, longs to recover its composure. The sponge always succeeds, though, the orange never: for its cells have burst, its tissues are torn apart. Only the peel, thanks to its elasticity, to some extent retains its shape. Meanwhile an amber liquid has been spilled, which, refreshing and fragrant as it may be, often bears the bitter consciousness of a premature expulsion of pits.[15]

The orange never manages to reverse its ordeal, to recover itself to itself. But still, something of its former self remains in the responsiveness of its skin, of this remainder that somehow remembers. This remainder shall prove instructive.

But first Ponge asks us to wrestle with the fact that in the poem the orange cohabits with Ponge's namesake and emblem, the sponge *(l'éponge)*, the zoophyte or plant-animal that defies categories, that loops the inside and outside of being, so as to appear self-grounding. In contrast to the shape-changing reversibility of the sponge, the orange figures an expenditure that is all "too passive." It lets us "off," as Ponge elaborates, "much too easily," rewarding us with its "perfume" as the skin is broken and the "glorious color of the resulting liquid" that, "as is not the case with lemon juice," requires "no apprehensive puckering of the taste buds." The orange makes so few demands. It requires only that "the larynx . . . open wide to pronounce the name as well as to ingest it." Both the word and the fruit go down with an easy equivalence. The opening of the mouth necessary to eat the orange corresponds to the movement of the mouth necessary to saying its name. The orange and its "torturer" become mutually animating presences in the poem. Both are subjects. Both are objects. The orange gives itself to be eaten—and by that giving comes to be named, and so recruits or convokes us as its ally. We eat and name the orange. Or, the orange, in leading us to do so or in not resisting us, names itself, in our mouths.

The poem ends by approaching what it calls the innermost core of the orange, its indigestible pip, or seed, that will pass through us. At bottom, with the pip, the "wood," it seems as if we hit something like a referent or an outside to Ponge's poem, the limit to what he can do as a prose-poet of substance or what Jacques Derrida punningly names "signsponge."[16] But the poem permits no such exit. Instead, it proposes something stranger still. "In concluding," he winks, "this all too summary study, carried out as roundly as possible," Ponge finds that he has still to

> cope with the pip. This seed, shaped like a miniature lemon, is the color of the white wood of the lemon tree, and inside a pea or tender sprout. After the sensational explosion of that Chinese lantern of tastes, colors, and scents—the fruity balloon itself—we here recognize the relative hardness and acidity (not entirely insipid) of the wood, the branch, the leaf: a very small sum but unquestionably the *raison d'être* of the fruit.

As Ponge "copes with" or, more literally, "has to come to" *(il faut en venir),* the pip, which appears indigestible, unusable, and so resists the orange's otherwise seemingly gratuitous assimilation to the human, we encounter not the limit to what he can say but, instead, an unfolding of the orange as it takes hold of his language and this poem. The pip sprouts,

turns tree, grows and sheds its leaves, buds, flowers, fruits, and so exfoliates the multiplicity that is orange within *"L'orange,"* within *this* orange
that we have tortured, eaten, and read.

At the moment of conclusion at which Ponge might be said to exhaust
the orange, he encounters neither a referent nor an outside to language,
so much as the point at which the expressivity of the orange comes to
take hold of the poem, of him, of his language, and so of us. The innermost pip of the orange reveals the orange to exist as what we might call,
already, the expression or writing of the plant, an expression housed in
pith and peel, a dormant *kairos,* that waits, cut and tortured already, a
pre-cut and segmented entity, the reproductive technology of a particular genus of plant that goes mobile in and by its recruitment or rental of
the differently animated entities we name *animals* (human and otherwise). So it is that Ponge, the sponge, the sovereign zoophyte, takes in
and expels the orange. As he does so, he passes from active subject to
passive, sensing orangey substrate or screen, from animal to vegetable
and back, as the looping motion of his muscle for writing passes from
animal enjoyment, his heady, violent, expression or explosion of the
orange, to registering the process by which, to "conclude," to reach an
end, or make an "end" of *this* orange and of *orange,* requires that we
encounter the telos, "end," or vectoring of the orange itself, of the way in
which all that we take from it, all that he registers under the term *expression,* all that we enjoy, is made possible by the orange as itself a form of
writing that impresses itself on us.

Yes, the poem "can be read," as Sianne Ngai offers, "as a figure for a
number of personification strategies."[17] Yes, the poem serves as an exem-
plum on the reversibility of *prosopopoeia,* the trope that means to give
face and so voice to things and which, as Paul de Man writes, "implies
that the original face can be missing or non-existent," that it exists only
because of its being figured or by the program of figuration itself."[18] Yes,
the orange may be "cute," an object that both attracts and repels, whose
passivity engages the reader, viewer, eater, in a reciprocal passivity that
dominates both—it's certainly handy, lends itself to the hand, befriends
handedness, naturalizes (itself to) the fact of hands. And yes, this cuteness functions a little like that of the commodity form in that when
and as we consume it, we are, in fact, put to use, made use of, even as, in

Ponge's poem, we appear to be in control, our agency maxing out as we eviscerate and torture the orange, using it all up—except for the pip. "*L'orange*" may serve to set in motion a *mise en abîme* by which *to eat* becomes *to be made to swallow* and *to be made to speak,* but beyond or beside offering a template for either the glamour of the commodity that we passively consume or the status of a modernist work of art, the poem registers the existence of the orange as itself a form of writing in which the orange tree exfoliates itself.

What is an orange? The answer might be that it is itself the result of an act of representation—a representation of what the plant assumes that we sighted, smelling, tasting animals might desire. To posit the orange as already a form of writing or, better still, a representation, from the perspective of the plant, of what it is that we want, will be pleasing to our senses, enables us to return to where Ponge begins his poem and to understand the longing or "aspiration" of the orange's skin to "recover its composure," or "*contenance,*" bearing, attitude, or face. The automatic motility or automimesis of the peel functions as an archival remnant of the orange as already an expression of orange writing—of, what Michael Marder might name the plant's "proto-writing." In *Plant Thinking,* Marder offers the iterability of the leaf or fruit as a gloss on Ponge's own frequently stated admiration for the ability of plants to "'repeat the same expression, the same leaf, a million times' and, 'bursting out of themselves,' to produce 'thousands of copies of the same . . . leaf.'"[19] Like the leaf, each and every orange "will vary with the material expression of the being of plants, that 'is' the being of plants," a fact that, as Marder points out, is as true of the leaf or orange as it is of every "word or concept that will carry slightly different semantic overtones depending on the singular event of its enunciation."[20] If both the leaf and the orange serve as an "ephemeral register for the inscription of vegetal time as the time of repetition, a register not archived [exactly] but periodically lost and renewed," then with "*L'orange*" we encounter a record or inscription of that ephemeral archive on Ponge's language and our senses, the two coming to serve as a substrate for the expression that is the orange.[21]

Orange is merely the name we give to the plant's dispersal strategy that produces (for us) a series of orangey flowerings or irruptions. But, however fragrant, however colorful, however sweet, however joyful the

putting to use, the orange remains entirely indifferent. It does not discriminate. Except that that's not quite right either. The orange remains
indifferent, but that is not to say that it is uninterested. It functions as
something more than a purely grammatical or material–semiotic program. Because it is a representation, because it offers itself to us in a
manner that we might find attractive, the orange reveals itself to be a
thoroughly rhetorical entity. It inclines toward the trope, offering itself
as a mode of hyperoccasional multispecies address that seeks, posy-like,
to interpellate you and me as the plant plays its own form of bio- or
zoopolitics—the orange, as it were, always in the vocative case, the orange
itself an act of nomination that designates the multispecies polity of animate beings that shall eat and so name and so disseminate it.

To whom does Ponge address his poem? Who is it that joins him to
become the "we" that expresses the orange? This collective subject need
not, from the perspective of the orange, be an exclusively human entity.
It might refer just as well to any passing animal or animating vector as
to you or me. The "we" that the poem proposes continues just as well
by way of tooth and claw or beak, by way of this or that prosthesis, as by
the way of hands and fingers—however handy the orange. And this "we,"
accordingly, does not refer to a stable category so much as it designates
merely the range of beings convoked by the orange, the group of animals,
plants, fungi, and bacteria that all "eat" and so, in different ways, "name"
and express the orange.

In their profound neutrality, Ponge's prose-poems stand as points of
contact with what he calls the "side of things," and what, in this instance,
we might name "orange-being," the serialized multiplicity that accounts
for or which expresses orange and oranges, punctuating the object world
with particles of sweetness and color. After this encounter neither we
nor the orange shall recover our former bearing or faces, even as the
shared elasticity of our skins mimic the *things* that we once were. Ponge's
poem suggests the way in which what we take to be our archives, archives
backed with this or that animal, vegetable, or mineral matter, rely on and
are no different from the differently situated ephemeral registers of the
leaves and fruits of plants that remember by forgetting. With Ponge, we
come to recognize the way in which the "proto-writing" (or simply "writing") of plants exfoliate within our own poetic acts, marking them and
inscribing on us and through us the movements of vegetal time.

So it is I choose to side with the verb *to represent*—awash as the world is with representations to us from beings otherwise than human, representations that code our desires, that please or traumatize an "us" that exceeds the limits of the human. Who knows how such a renewed sense of the verb might cause us to veer? Who knows also what might happen if we were to accept that the verb *to represent* and the activity it designates no longer (and never exactly did) refer exclusively to the preserve of a human subject?

Notes

1. For a now-classic exploration of the ideological stakes of this confusion and the desire to do without representation as a conceptual difficulty, see Gayatri Spivak, "Can the Subaltern Speak?" in *Marxism and the Interpretation of Culture,* ed. Cary Nelson and Lawrence Grossberg (Urbana: University of Illinois Press, 1988), 271–313. Famously, Spivak tracks a strategic confusion of the terms *vertreiten* (speaking for) and *darstellen* from Karl Marx's *The Eighteenth Brumaire of Louis Napoleon* as they migrate through a conversation between Michel Foucault and Gilles Deleuze (274–77). The essay ends with a rallying cry of sorts that sides with Derridean reading practices against busting through the problem of representation/s: "The subaltern cannot speak. There is no virtue in global laundry lists with 'woman' as a pious item. Representation has not withered away. The feminist intellectual has a circumscribed task, which she must not disown with a flourish" (308).

For a differently staged conservation of representation as a constitutive problem, see Bruno Latour's ongoing exploration of what he names a "parliament of things," initially in *We Have Never Been Modern,* trans. Catherine Porter (Cambridge, Mass.: Harvard University Press, 1993), and then crucially in *Pandora's Hope: Essays on the Reality of Science Studies* (Cambridge, Mass.: Harvard University Press, 1999) and *An Inquiry into Modes of Existence: An Anthropology of the Moderns,* trans. Catherine Porter (Cambridge, Mass.: Harvard University Press, 2013).

2. Bruno Latour, *Aramis; or, The Love of Technology,* trans. Catherine Porter (Cambridge, Mass.: Harvard University Press, 1996) 48.

3. On the not entirely pessimistic impossibility of translation, see Walter Benjamin, "The Task of the Translator," *The Selected Writings of Walter Benjamin: Volume 1, 1913-1926,* ed. Michael W. Jennings (Cambridge, Mass.: The Belknap Press of Harvard University Press, 1996), 253–63; and, among others, commentaries by Paul de Man, "'Conclusions': Walter Benjamin's 'The Task of the Translator,'" in *The Resistance to Theory* (Minneapolis: University of Minnesota Press, 1986), 73–105; Jacques Derrida, "Des Tours de Babel," in *Psyche,* ed. Peggy Kamuf and Elizabeth Rottenberg (Stanford: Stanford University Press, 2007), 195–225.

For an alternate, optative, if not always optimistic path, see the philosopher Michel Serres, whose life's work might be named a "general theory of relations," or translation and the translational networks adopted and advocated for by the sociologist Bruno Latour. See, as a start, Michel Serres and Bruno Latour, *Conversations on Science, Culture, and Time,* trans. Roxanne Lapidus (Ann Arbor: University of Michigan Press, 1995), and for this quotation, 66.

For a consideration of Serres's model of translation alongside a deconstructive problematic, see Julian Yates, "Wet?" in *Elemental Ecocriticism: Thinking with Earth, Air, Water, and Fire,* ed. Jeffrey Jerome Cohen and Lowell Duckert (Minneapolis: University of Minnesota Press, 2015), 183–208.

4. Nicholas Royle, *Veering: A Theory of Literature* (Edinburgh: Edinburgh University Press, 2011), 69.

5. "'Eating Well'; or, The Calculation of the Subject: An Interview with Jacques Derrida," in *Who Comes after the Subject?* ed. Eduardo Cadava, Peter Connor, and Jean-Luc Nancy (New York: Routledge, 1991), 116.

6. On the violence of decision as cutting or the creation of an edge, see, in different registers, Michel Serres, *The Natural Contract,* trans. Elizabeth MacArthur and William Paulson (Ann Arbor: University of Michigan Press, 1995), 55; Jacques Derrida, *The Gift of Death,* trans. David Wills (Chicago: University of Chicago Press, 1992), 53–82.

7. Philippe Descola, *The Ecology of Others,* trans. Geneviéve Godbout and Benjamin P. Luley (Chicago: Prickly Paradigm Press, 2013), 85. The quotation from Descola that appears as an epigraph to my essay comes from his earlier *Beyond Nature and Culture,* trans. Janet Lloyd (Chicago: University of Chicago Press, 2013), xx.

8. On this relation to parasitism, see Michel Serres, *The Parasite,* trans. Lawrence R. Schehr (Minneapolis: University of Minnesota Press, 2007).

9. Donna Haraway's *When Species Meet* (Minneapolis: University of Minnesota Press, 2008) serves as a crucial point of entry here, but see also Eben S. Kirksey and Stefan Helmreich, "The Emergence of Multispecies Ethnography," *Cultural Anthropology* 25, no. 4 (2010): 545–76; and, less programmatically, Stefan Helmreich, *Alien Ocean: Anthropological Voyages in Microbrial Seas* (Berkeley: University of California Press, 2009); Eduardo Kohn, *How Forests Think: Toward an Anthropology beyond the Human* (Berkeley: University of California Press, 2013); John Hartington Jr., *Aesop's Anthropology: A Multispecies Approach* (Minneapolis: University of Minnesota Press, 2014); the collection of essays in cultural anthropology, *The Multispecies Salon,* ed. Eben Kirksey (Durham: Duke University Press, 2014); Anna Lowenhaupt Tsing, *The Mushroom at the End of the World* (Princeton: Princeton University Press, 2015).

10. Jacques Derrida, *Of Grammatology,* trans. Gayatri Chakravorty Spivak (Baltimore: Johns Hopkins University Press, 1974), 84.

11. It is not possible to provide an exhaustive list of contributions here, but some highlights include Vinciane Despret's *Quand le loup habitera avec l'agneau* (Paris: Editions du Seuil, 2002) and *Que diraient les animaux, si . . . on leur posait les bonnes questions?* (Paris: La Découverte, 2012), and, in English, "Sheep Do Have Opinions," in *Making Things Public: Atmospheres of Democracy*, ed. Bruno Latour and Peter Weibel (Cambridge, Mass.: MIT Press, 2005), 360–68; Haraway, *When Species Meet*; and primatologist turned sheep observer Thelma Rowell's essays, such as "A Few Peculiar Primates," in *Primate Encounters: Models of Science, Gender, and Society*, ed. Shirley C. Strum and Linda Fedigan (Chicago: University of Chicago Press, 2000), 57–70.

12. On the difficulty of parsing the difference (or producing a knowable demarcation) between a so-called organic "response" and a machinic "reaction," see Jacques Derrida, *The Animal That Therefore I Am*, trans. David Wills (New York: Fordham University Press, 2008), 29.

13. Jakob von Uexküll, *A Foray into the Worlds of Animals and Humans*, with *A Theory of Meaning*, trans. Joseph D. O'Neil (Minneapolis: University of Minnesota Press, 2010), 44–52 and 219–21.

14. Philip K. Dick, *Do Androids Dream of Electric Sheep?* (New York: Del Rey Books, 1968), 158.

15. Francis Ponge, *Selected Poems*, trans. C. K. Williams and John Montague (Winston-Salem: Wake Forest University Press, 1994), 23.

16. Jacques Derrida, *Signéponge/Signsponge*, trans. Richard Rand (New York: Columbia University Press, 1984).

17. Sianne Ngai, "The Cuteness of the Avant-Garde," *Critical Inquiry* 31, no. 4 (Summer 2005): 832. See also her work *Our Aesthetic Categories: Cute, Zany, Interesting* (Cambridge, Mass.: Harvard University Press, 2012), 91. In addition to Ngai, my reading of the end of Ponge's poem owes a debt to Helen Deutsch's reading of the poem as one in a succession of anecdotal fragments produced by the orange as itself an anecdotal entity. See Helen Deutsch, "Oranges, Anecdotes, and the Nature of Things," *SubStance* 38, no. 1 (2009): 31–54.

18. Paul de Man, *The Resistance to Theory* (Minneapolis: University of Minnesota Press, 1986), 44. Ngai draws on de Man's "Autobiography and Defacement" in *The Rhetoric of Romanticism* (New York: Columbia University Press, 1984), 67–81.

19. Michael Marder, *Plant-Thinking: A Philosophy of Vegetable Life* (New York: Columbia University Press, 2013), 115 (quoting Ponge, *Selected Poems*, 71).

20. Ibid., 116.

21. Ibid., 114.

Compost

SERPIL OPPERMANN

An essential process of biodegradation, composting occurs through the decomposition of nutrient-rich organic materials to enhance the soil properties. Since time immemorial nature's invisible organic engineers have been composting to keep the soil alive, making it porous and aerated. If we zoom into this *living canvas* (to use Dorian Sagan's definition of soil),[1] we encounter a throng of organisms composting in an eccentric landscape of diffractive relations.[2] An epic of life is being played out in this extraordinary terrain of microentities internalizing one another from their own residues without completely eliminating the traces of their origins. Fraught with its own remnants, nature recycles itself this way within that vortex of ecological spin-off that bears material repercussions for every living thing. *To compost* means being part of a fertility cycle, whether we acknowledge it or not, as our bodies compost with agents we always live with but don't always see. It means being part of things that are dead but are in a way undead, which makes Schrödinger's ambivalently limbo cat seem quite ordinary. Different from this unnamable cat who is bluntly poised at the strange domain of non/existence where she can neither decompose nor recompose, composting entities couple into an undivided creation, a continuous system interlinking all bodily natures with soil, air, and water. One can even call the system *compost poiesis*. Although *poiesis* originally signifies *making* (as in making material objects like tables, shoes, and artifacts, as well as story-making, or art-making), *compost poiesis* amplifies the word's originary sense by reconciling *making* with *unmaking* to give a sense of the twofold

136

condition of composting entities who undergo continuous decomposition and recomposition. Through metamorphosing with one another, they mesh with the earth's evolutionary systems to become part of a dervish whirl of environmentality.

From the scientific vantage point, compost is a source of nitrogen and phosphorous, and to a lesser extent of potassium for enriching organic matter; it is a material resource of nature's biocycles. From a poetic viewpoint, however, it is a metaphor of change in which a continual storying of the world is unalterably scripted. There are "new kinds of possibilities for relating and understanding"[3] in this burrowing story that writes itself out in the composting medium. Heeding the stories buried in compost piles as an invitation to think the world anew, I reflect on these possibilities that constellate around *veer ecology*—a more expansive understanding of the "non/human and in/organic things that suddenly, and unpredictably, go off course" and transform other beings "in ways catastrophic, pleasurable, life-changing," as the editors of this volume have specifically underlined in their call for papers. With these veer deliberations in mind, I ask: How does *compost* (as verb) capture the alternative meanings of human–nonhuman enmeshments? Does it help reinterpret life? How do visions of the anthropos transform in its light? I approach these questions through composting stories that entice us to think about the strangeness of our ecological destiny tied to many discomforting nonhumans. As such, composting can be said to fertilize veer ecology in quite a realistic way of theorizing the enmeshment of the human within the natural world. As a detrital ecoactivity, then, composting is also a theoretically fertile ground to endorse the oftentimes bewildering "relationalities of becoming of which we are parts."[4]

Although it is a vexing site punctured with ferocious entanglements, composting landscapes are rich with alternative ecologies, images, and stories that are not yet encumbered by anthropocentric potencies and can propel change by deepening our understanding of environmentalism. To compost is to veer off course in ex-centric swerves toward new ecological trajectories; or put differently, to compost—to make and compose and even dine with the nonhuman—veers us away from anthropocentricity by transforming sites of decay into vibrant sites of fecund imagination.

To understand this process, let us picture an ecology of human and nonhuman detritus swirling together: animal manure, twigs, woodchips, biological litter like plant debris, rotting leaves, vegetables, fruits, egg-shells, and other food scraps as organic waste compost with microorgan-isms (beneficial bacteria), which feed on and digest decaying matter to restore the soil's vitality with rich nutrients, such as nitrogen, phospho-rous, and potassium, which are essential for the cycle of life to continue. By producing carbon dioxide, water, heat, and humus, microorganisms play a crucial role in the life-sustaining cycles of biodegradation. The vanguards of biodegradation are the mesophilic bacteria, which produce heat up to 40°C to break down the soluble compounds. When the tem-peratures rise above 50°C, thermophiles take over to decompose proteins, fats, and complex carbohydrates, as well as to destroy human or plant pathogens. In the final phase of composting, the temperature decreases and mesophilic microorganisms reappear to cure the remaining organic matter.[5] This is an irreversible labyrinth of endings and beginnings, a cauldron of biotransformations for the earth's miniscule beings. As expressed by Sagan's witty lines, if we come closer to this living canvas, we see "organisms blend[ing] into a pointillist landscape in which each dot of paint is also alive."[6] Without such a re-creational capacity, life would cease to exist. Composting is, therefore, a rescripting of life in the "sedimented historialities"[7] of decay and fecundity. It is as if every life-form is harvested in one another while composting and bringing some-thing essential into existence with each decaying molecule in the soil's manifold mouths.

With this in mind, let us now imagine a savoring of compost as food in the nonhuman mouth as if it were *komposto*, a sweet delicacy for the all-too-human palate. It is because *komposto*, as a pun from Turkish for compost, provides a palpable example, which is, *in fact*, compote made of pieces of fruit, such as sour cherries, quinces, grapes, dried apricots, and apples cooked in water with sugar, cinnamon, and cloves. *Komposto* is, therefore, a juicy tempter to ponder about compost in the mouth—but not our all-too-human mouths. Like the effect of *komposto* in our mouth, the soil is refreshed when it composts in the other-than-human mouths filled with the decomposed food scraps that become soil nutri-ents. While the scraps are broken into their metamorphosing compo-nents, a relishing of compost is crystallized and liquefied in the soil's gut. Maggots, worms, microbes, bacteria, and other tiny organisms feast on this plate of organic remnants consigned to the compost pile, making it a palate-veering reflection. If the soil's microcosmic gut (always too many) is full of petrified remains of lifeforms digesting their own dissolving DNA, transcending their own life, and rising from their own funeral with renewed life, then this gut peremptorily anatomizes rejuvenation from annihilation. Like the phoenix, the soil's multigut goes through yet another cycle of life in innate modes of self-transformation. There is no getting off this merry-go-round once decomposed lifeforms begin composting in a sustainable biomass[8] of becoming-together in the grid of becoming-other. Here, decaying things that are bacterial, fungal, mineral, vegetal, and animal compost to rejuvenate the soil. No messy saturation in this telluric vitality, however, promises an easy rebirth; so lively matter has to go through the circuits of an acrimonious dissolution to ensure a new fusion. Its greatest curse is also its greatest miracle; because, mixed-up, broken down, and reassembled through various stages of metamorphosis, biodegradable matter wrests death free from itself and connects loss to renewal. In sum, composting is nature's method of conveying how life-composing decomposition emerges in convoluted ways to guarantee its generativity. As this tangly process affords both liveliness and demise, rotting, decaying, putrefying, and disintegrating can be simultaneously terrifying and magical.

Composting, therefore, is a matter not only of realistic ecology but of the poetics of transformative matter in a self-sustaining structure in which it draws its existential meanings from its relation to other matter. This is what Susan S. Morrison calls "metaphoric composting,"[9] conductive to material imagination. In his innovative book *This Compost: Ecological Imperatives in American Poetry,* Jed Rasula similarly offers an inventive ground to think this way. Quoting Eugene Jolas, the editor of the avant-garde journal *transition,* Rasula contends that "compost is not conceptually restricted to the decay of organic matter; it affords a commanding prospect of correspondences, resonant parallelisms, glimpses of independent figures participating in a fortuitous isomorphism."[10] Conceived this way, to compost metaphorically is much like "spinning off dark incalculable rhythms"[11] that revitalizes the past in the present whereby, as Rasula imagines it, ideas collapse into each other and constantly recycle their energy. To spin actually means to whirl around quickly, but it also means to make threads by drawing out and twisting fibers, indicating revitalization. By drawing out and twisting the fibers of new life from the mortifying effects of decomposition, the decaying and fermenting things crowd in a space fundamental to the making and the unmaking of their stories and entail a radical rethinking of human/nonhuman/inhuman coalitions and mutual interactions or, in Karen Barad's terms, *intra-actions*. Sagged to the dirt, the earth's crumbling denizens compost to transcend their local decomposition and recapitulate stories of resilience and resurgence. Although they invoke "something offensive, muddled, or dirty,"[12] these stories demand to be read as instances of provocative environmentalism, encouraging us to "enter the perspective of every other creature—human and nonhuman."[13] A maggot, for example, would tell the story of savoring the primordial compost soup while being consumed in it, whereas a bacterium's story would recount the pleasure of disbanding the maggot. As if on a surreal artist's canvas, this image may dredge up a vision of a disjunctive creativity compressed in death and dissolution; but the same image is also somewhat recuperative, as it is suggestive of "food production and consumption and, coextensively, the infrastructure of human habitats."[14]

Compost piles are all-too-inhuman, yet they emerge from and deeply permeate the all-too-human lives through many bodies "collapsing into each other . . . by constant recycling,"[15] the purpose of which is of course sustenance through nourishing substances. From this perspective, whether or not composting is understood as returning in generative nurture or putrefying abandonment, it embodies environmentalism as a visceral venture, and however repellent its materialization and deviant its meanings, edible (or not) vegetation depends on it. In effect, these recalcitrant narratives potentially fuel *compost poiesis* when the disseminated beings, returning through the ventilated backdoor for renewed bodily clusters, thread their stories in dirt, which is the very "weather of existence" and the "climate of our being," as succinctly described by Charles Olson, the major architect of postmodern American poetry, in "The Chiasma."[16] But even when registered figuratively, to compost is to disclose "the modes of being together with the dirty side of the green"[17] that compel us to think about an alternative vision of worldedness, one "that *grounds* humans within the continuum of life."[18]

Obviously, composting is "imaginatively fertile;"[19] for the "couplings and decouplings"[20] of composting organisms sharing a "predatory heritage" and a "biocentric lineage"[21] not only ensure constitutive relations between parting elements for the stabilization of organic residues but also kindle our imagination to drift toward their disanthropocentric meanings in a shared world of recurrent interactions and transformations. Humberto Maturana and Francisco Varela, for example, would explain these relations in terms of (a) "co-ontogenic coupling(s)"[22] to cultivate the soil and our imagination. Hence, compost is "adventurous veering" toward "a shared worldedness,"[23] as Jeffrey Cohen would say, to see matter revitalizing as if out of its own memory, from what is backdropped by naturecultures. In this sense, to compost is to witness the past impressed on the future through the vanishing traces of the present. This is an ecological karma where every thing and being leaves a trail even if *biodegradable* means leaving no trace behind. In fact, there is no absolute biodegradability. Matter that composts is not really completely destroyed when it biodegrades. It is metamorphosed. One might argue along with Jacques Derrida here that just like the recycling stories in cultures, composting matter does not decay into oblivion; it simply "submits to composition, decomposition, recomposition."[24]

When matter composts, it is transformed into a rich humus (amorphous earth), contributing to the soil's nutrient retention and fueling plant growth. We can think of humus as the signature of metamorphosed matter in its recomposed phase. Moving in and out of poignant stages of mutation, the food we eat is earthing itself this way, as playfully recounted by postmodern author Raymond Federman in *Return to Manure*:

> You see, first you cultivate the earth to make things grow. Hay, wheat, vegetables, corn, beets, grain, all the stuff you feed animals to fatten them up so that afterwards you can slaughter them and stuff yourself with their meat. But before those animals die, wow can they shit. Of course, we humans also shit after we devour the animals, and eventually all that shit is returned to the soil to enrich it, to make the things that grow in the ground bigger, more nourishing, more fattening because of the manure. And so it goes until the end of time. What a beautiful system nature is. It keeps jump-starting itself with shit.[25]

The inbuilt story of this drama is not only distressing but also edifying for the human, whose name comes from humus produced by composting matter. Bruno Latour also notes that the word *human* "derives from the same humus that constitutes the earth itself,"[26] which is to say that if the anthropos is humus, this arrogant species's identity as a geological force propagating deadly narcissism dissolves along with other decaying matter. Caught in the geo-corporeal intimacies as humus-body, the human must compost back into the earth, fold into the soil, which is the only cure for the Anthropocene's "mass hallucinatory fantasy."[27] The reason is simple. "What is to be composed may, at any point, be *de*composed,"[28] says Latour; hence the human must compost to re-compose as an earthbound being like all others. Latour is right in insisting that the word *composition* "carries with it the pungent but ecologically correct smell of 'compost,' itself due to the active 'de-composition' of many invisible agents."[29] The body is thus commensal and amorphous, symbiotic and pathogenic; it is storied humus as a knot of biological and imaginative journeys. Its compost stories are inscribed in dirt, in the food we eat, and in poetry. How can it not *de*-compose and compost both ecologically and conceptually in order to re-compose in different versions? Walt Whitman lyrically expresses it in *Song of Myself*:

> I bequeath myself to the dirt to grow
>   from the grass I love,
> If you want me again look for me under
>   your bootsoles.
> You will hardly know who I am or what I
>   mean,
> But I shall be good health to you
>   nevertheless,
> And filter and fibre your blood.[30]

As suitably discerned by Whitman, neither ignominious nor glorious, when humans compost with germs and bacteria *in* the earth (or "dirt," to be precise), they are dynamically enfolded in each part of the grass from which, as Whitman says, they grow. To enfold means to fold inward, and in the ecosystems everything is folded into everything else, and the planet is in principle enfolded into all of its components. Composting is a dynamic activity that reveals this fundamental life-principle and is inherently creative, however peculiar it may be. Poets can compost of such creative disclosures and show that, to quote John Fowles, "all the special privileges we claim for our species, all the feathers in our cap, might seem as absurd as the exotic ceremonial finery of some primitive chieftain."[31] If we are all enmeshed, as Cohen puts it, "in a vortex of shared precariousness and unchosen proximities,"[32] the feathers in human caps representing anthropocentric arrogance seem ridiculous indeed.

As Whitman's poem also illustrates, to compost ecologically is to be contained and unbounded at once, because all organisms are composed of each other's boundless atoms while composting *through one another* (diffractively) in a fertility cycle of anonymity. To compost, in this sense, is to be mutually environed in nature's cyclical rhythms, which awakens a "composting sensibility"[33] for poets like Whitman whose literary imagination conjectured a poetic corollary of composting. The compost medium is a "strangely familiar realm of estrangements"[34] both for the poets who fertilize the literary canon and for the microorganisms that literally split and splinter to fertilize the soil. This innate biotic reflex is an enactment of the nexus between death and life, a generative concern for life sustenance, and a ritual for guaranteeing resurrection. Expressed differently, this is *creative veering*.

There is also *destructive veering* in composting activity that needs to be addressed as well if we want to schematize a fair disanthropocentric vision of veer ecology. When heavy metals like lead, chromium, copper, nickel, mercury, and zinc compost with microbial communities, this destructive facet surfaces to confront us with more than concerted metaphors of toxicity. Composting heavy metals are abiotic factors that cause environmental and epidemiological risks; but biotic factors, like infectious pathogens (virus, microbe, fungus, parasite, viroid) in organic waste, are also major actors of environmental and health-risk scenarios.[35] According to environmental hygiene professor Reinhard Böhme, contaminated soil infects the microflora and fauna and introduces infectious pathogens and zoonotic agents into the food chain, because pathogens, he explains, "may survive for a remarkable period of time in manure, wastes, waste related materials, sludges and the environment. This is the base of the resulting epidemiological risks."[36] If the outcomes of composting can be so risky, how do we deliberate on veer ecology, of which compost is a part, and unearth its disanthropocentric meanings? The answer lies in acknowledging the word *veer* as a deviation from human exceptionalism and thus from the entrenched notion of humanity's privileged status as if it exists outside of the earth systems. Veering is turning away from the standard formulation of the anthropos imprinting a lasting signature on the planet's geomorphological processes. Despite the risk scenarios it may involve, composting offers an alternative narrative to veer us off this deified anthropogenic course and opens both material and poetic potentials for building a more ecofriendly life by resuscitating the diffractive relations we seem to have replaced with hierarchical ontologies.

Composting not only signifies these relations as a metaphor for disrupting anthropocentrism as a dysfunctional discourse, but by materially enmeshing things together it also underscores the complex dynamics of our internal entanglements with the nonhuman. This understanding heals our shortsighted vision, awakens us to our responsibility to the nonhuman and, as Barad emphasizes, results in embracing "an ethics of worlding."[37] It is important to understand, as Levi Bryant also argues, how "beings are knotted to one another," and if "they internalize one another, being affected and affecting each other," then we, along with the inhuman particles, compost of this world, and "our flesh is intertwined with its flesh."[38] When the human and the nonhuman compost *through one another* in fecund cycles of such interbeing, environmentalism manifests in its defamiliarized codes and navigates us toward more ecologically feasible standpoints, vocabulary, discursive strategies, and textual forms, which are, to some extent, incipient determinants of life. From this vantage point, composting is a negation not of the category of nature in its *double-coding* that implies both continuity and temporality but of nature consigned to the anthropocentric mindset. Its "patterns of diplopia"[39] (double vision) provide a true veering off course to explore the extent to which veer ecology can help us configure what Stanley Cavell calls "companionable thinking."[40]

A way of envisioning "companionable thinking" is to mentally experience composting in order to reach *ecological anagnorisis,* a realization of our earthbound identity. We then become aware of the meshwork of connections woven into the fabric of this earth, which is a recognition that melts anthropocentric illusions like the mesophilic bacteria breaking down the soluble compounds in compost piles. Understood this way, composting the world anew in the orbit of veer ecology consequently leads to making bold changes in our ontological, epistemological, and ethical frames to overcome the myth of dualism. It is not a plea but a necessity in crisis-laden ecologies of which we are the complicit party. Each of us will eventually compost to be liberated from the "permanency of boundaries"[41] and become a mélange *of* the world. We will veer off to a place where the barriers between the interior and the exterior will cease to matter. Veer, therefore, also implies doing away with the boundaries between the inside and the outside. To veer in this sense is environing inside out, being "rolled round in earth's diurnal course, / With rocks, and stones, and trees," as in William Wordsworth's poem "A Slumber Did My Spirit Feel."[42] Like the poem's persona, veering off course is actually being part of the earthly spin, encircling, turning, swerving in reciprocal being with other Earth agencies. But unlike the girl who has no motion, no force, and no feelings, the spinning agencies, including humans, even in stages of composting, swarm with life, which is ever unfolding in many currents, colors, and shapes. Life always undergoes continuous birth "in earth's diurnal course" and exceeds death. Even though "the touch of earthly years" is not to be experienced by the composting agencies, spinning with the earth *is*.

In this sort of encircling without boundaries, what prevails is the memory of ancient kinship that makes us, as David Abram puts it, "interdependent constituents of a common biosphere, each of us experiencing it from our own angle, and with our own specific capabilities, yet nonetheless all participant in the round life of the earth."[43] And here the compost vortex spins while we, along with all matter, transmute into "petrified swimmers"[44] covered with signs and slime, like the composting eukaryotic organisms, such as protozoans, algae, oomycetes, and slime molds creating life in their various states of change. As expressed in Olson's famous poem "The Kingfishers": "To be in different states without a change / is not a possibility."[45] If we reformulate this line as "to compost without a change is not a possibility," we realize that life itself composts as an iterative journey in a sporadic frequency, often ushered in cries, whimpers, and gasps, but also wonders, miracles, and mysteries. That is why composting of traumas, catastrophes, and fears materializes a compost pile of more anthropocentric practices, whereas composting creatively veers us off course in ex-centric swerves toward new ecological trajectories that allow us to be more responsible for our relations with other beings. Life, for example, can be imagined to compost in a storied creation always blossoming with revised stories and unfolding histories of so many human/nonhuman/inhuman mouths savoring their own *komposto*. This *komposto*, however, is not so acerbic; it moves us toward mental transmutation. Spinning with the flows of composting relations, whether dissonant or concordant, is swimming in the tides of matter's dissolving and assembling memories and forming existential cohesions and partitions with each collapsing bodily boundary in the flux of the earth.

If we take *veering off course* through compost as a means of such reflections, we should not be caught in a conceptual and temperamental melancholia but rather engage in a solution-oriented sense of trans-formation that allows composting together. Like the thermophiles that destroy the pathogens, different solutions also compost in the unpredict-able maelstrom we call *environing*. Therefore, we should "Behold this compost! Behold it well,"[46] as Whitman exclaims in his poem "This Compost," both in its beneficial and detrimental effects. Like Whitman, seeing "endless successions of diseas'd corpses,"[47] we can be terrified at the earth, but we can also see how the same earth "distills such exquisite winds out of such infused fetor," and "grows such sweet things out of such corruptions."[48] Reminiscently, Félix Guattari wrote that "we cannot conceive of solutions . . . without a mutation of mentality."[49] The ecology of compost offers a platform for such mental mutations to reflect on how the material imagination confronts an existential crisis that composts ecological vicissitudes. "The resurrection of the wheat" that "appears with pale visage out of its graves"[50] in Whitman's poem is a discernible example for understanding how this imagination works in following the stories of composting elements working to create new tracks that may not always echo hallelujahs, but they do not necessarily carry infectious meanings either.

Composting, then, echoes a sense of ecological liberation from the Anthropocene narratives that enmesh the human in destructive fanta-sies of overpowering. Even imagining ourselves as humus is a humbling experience. Such an inner ecology teaches the human to be unpreten-tious as the human also composts in the earth, and with the earth. What we learn through composting is the vanity of praising ourselves as the self-appointed governors of the other-than-human environments. Fur-thermore, to speak of the human as composting *in* the earth is to be reminded that being biodegradable is a better imprint in the environment than the one forced by the anthropos. It is to be reminded that "the rotted man inside, who used to seem archetypal, is biological, and his 'language environment' is amniotic and porous."[51] It is not easy to let go of the anthropos identity, but once we do, it is intellectually and ecologically liberating. If veering off course means to imagine the human as *humus*, it affords a moral and mental swerve toward a counternarrative embedded in *compost poiesis*.

In consequence, to story the world in terms of an undivided existence is important because, as the compost activity illustrates, there are no rigid boundaries even in life and death. If compost sustains a better worldview than the one constructed by the glorified anthropos, we can ultimately cease to give human-centered reality the significance of utilitarian ways of thinking, which have evidently proved to be disastrous, and instead begin to see and experience ourselves as parts of a fundamentally interdependent planetary existence. Veering off course in this sense will be conductive to the overall health of the biomass as it enters into our discourses, intentions, actions, and practices. Hence, a major change of direction is imperative in material–discursive practices, social and political choices, and economic tactics that are born of the anthropocentric mindset. Let us then compost into emancipatory processes and stories that need to be emplotted, now more than ever.

Notes

1. Dorian Sagan, *Cosmic Apprentice: Dispatches from the Edges of Science* (Minneapolis: University of Minnesota Press, 2013), 167.

2. By "diffractive relations," I refer to the relational nature of different organisms that compost *through one another* to make a beneficial difference in the soil's texture. Like the entangled subatomic particles, composting organisms can be envisioned as agents of interference to sustain vegetation.

3. Eduardo Kohn, *How Forests Think: Toward an Anthropology beyond the Human* (Berkeley: University of California Press, 2013), 7.

4. Karen Barad, *Meeting the Universe Halfway: Quantum Physics and the Entanglement of Matter and Meaning* (Durham: Duke University Press, 2007), 393.

5. For more information about the phases of composting, see Nancy Trautmann and Elaina Olynciw, "Compost Microorganisms," at http://compost.css.cornell.edu/microorg.html.

6. Sagan, *Cosmic Apprentice,* 167.

7. Barad, *Meeting the Universe Halfway*, 180.

8. The total mass of living matter.

9. Susan Singe Morrison, *The Literature of Waste* (New York: Palgrave Macmillan, 2015), 10.

10. Jed Rasula, *This Compost: Ecological Imperatives in American Poetry* (Athens: University of Georgia Press, 2002), 2.

11. Ibid., 14.

12. Janelle A. Schwartz, "Introduction: Vermiculture," in *Worm Work: Recasting Romanticism* (Minneapolis: University of Minnesota Press, 2012), xiii–xxv (xix).

13. Morrison, *Literature of Waste*, 174.

14. Amanda Boetzkes, "Techniques of Survival: The Harrisons and the Environmental Counterculture," in *West of Center: Art and the Counterculture Experiment in America, 1965-1977*, ed. Elissa Auther and Adam Learner (Minneapolis: University of Minnesota Press, 2012), 306–23 (309).

15. Rasula, *This Compost*, 17.

16. Charles Olson, "The Chiasma; or, Lectures in the New Sciences of Man," *OLSON* 10 (Fall 1978): 3–113 (36).

17. Sebastian Abrahamsson and Filippo Bertoni, "Compost Politics: Experimenting with Togetherness in Vermicomposting," *Environmental Humanities* 4 (2014):125–48 (125).

18. Tim Ingold, *Being Alive: Essays on Movement, Knowledge, and Description* (New York: Routledge, 2011), 49–50.

19. Matthew Griffiths, "'Rummaging behind the Compost Heap': Decaying Romanticism in Wallace Stevens and Basil Bunting," *Green Letters: Studies in Ecocriticism* 18, no. 1 (2014): 36–47 (41).

20. Ian Bogost, *Alien Phenomenology; or, What It's Like to Be a Thing* (Minneapolis: University of Minnesota Press, 2012), 7.

21. Rasula, *This Compost*, 77.

22. Humberto R. Maturana and Francisco J. Varela, *The Tree of Knowledge: The Biological Roots of Human Understanding* (Boston: Shambhala, 1987), 235. By "co-ontogenic coupling," the authors primarily refer to language, which they claim constitutes us "in a continuous becoming that we bring forth with others" (235). Although their focus is on the linguistic world built with other human beings, I have adapted their phrase to explain the links between biological relations and environmental imagination.

23. Jeffrey Jerome Cohen, *Stone: An Ecology of the Inhuman* (Minneapolis: University of Minnesota Press, 2015), 134, 50.

24. Jacques Derrida, "Biodegradables: Seven Fragments," trans. Peggy Kamuff, *Critical Inquiry* 15 (Summer 1989): 812–73 (815).

25. Raymond Federman, *Return to Manure* (Tuscaloosa, Ala.: FC2, 2006), 78. This is the story of a narrator named Federman who spent three years on a

French farm during World War II. Following his family's 1942 Nazi roundup in Paris and murder at Auschwitz, thirteen-year-old Federman shovels manure on this farm working for an abusive farmer. As in all of his novels, Federman plays with ambiguous relations between memory and imagination when he revisits the realm of the unspeakable in his distinctive surfictional mode.

26. Bruno Latour, "The Climate to Come Depends on the Present Time," trans. Tim Howles, *Le Monde*, 14 November 2014, http://www.bruno-latour.fr/sites/default/files/downloads/14-11-ANTHROPO-transl-GB.pdf.

27. Barbara Kingsolver, "A Good Farmer," *Nation*, November 2003, 13.

28. Bruno Latour, "An Attempt at a Compositionist Manifesto," *New Literary History* 41, no. 3 (2010): 471–90 (474).

29. Ibid.

30. Walt Whitman, *Song of Myself*, ed. Stephen Mitchell (Boston: Shambhala, 1993), 145.

31. John Fowles, *The Aristos* (London: Triad/Granada, 1982), 24.

32. Jeffrey J. Cohen, "The Sea Above," in *Elemental Ecocriticism*, ed. Jeffrey Cohen and Lowell Duckert (Minneapolis: University of Minnesota Press, 2015), 105–33 (107).

33. Rasula, *This Compost*, 1.

34. Ibid., 8.

35. On this point see Stacy Alaimo's reference to Ladelle McWhorter's ruminations about growing and eating a tomato and thus about "dirt and flesh." *Bodily Natures: Science, Environment, and the Material Self* (Bloomington: Indiana University Press, 2010), 12.

36. Reinhard Böhme, "Hygenic Safety in Organic Waste Management," *Proceedings of the 10th International Conference of the Ramiran Network on Recycling of Agricultural, Municipal, and Industrial Residues in Agriculture* (Slovak Republic: University of Veterinary Medicine Research Institute, 2002): 17–30 (18). According to Böhme, "Recycling of biological waste from agricultural, municipal or industrial sources with or without biotechnological treatment . . . is a necessity in order to protect the environment and to save limited natural resources like phosphorus" (17). See http://www.ramiran.net/DOC/A1.pdf.

37. Barad, *Meeting the Universe Halfway*, 392.

38. Levi Bryant, "Knots: For an Interactivist Ontology," *Larval Subjects* (blog), May 27, 2015, https://larvalsubjects.wordpress.com/2015/05/25/knots-for-an-interactivist-ontology/.

39. Schwartz, "Introduction," xvii.

40. Stanley Cavell, "Companionable Thinking," in *Philosophy and Animal Life*, ed. Stanley Cavell et al. (New York: Columbia University Press, 2008), 91–126.

41. Barad, *Meeting the Universe Halfway*, 381.

42. *Wordsworth and Coleridge: Lyrical Ballads and Other Poems*, Wordsworth Editions Limited (Hertfordshire: Wordsworth Poetry Library, 2003), 131.

43. David Abram, *Becoming Animal: An Earthly Cosmology* (New York: Vintage, 2011), 143.

44. Sagan, *Cosmic Apprentice,* 166.

45. Charles Olson, "The Kingfishers," in *The Norton Anthology of American Literature,* vol. 2 (New York: W. W. Norton, 1985), 2283–88 (2286).

46. Walt Whitman, "This Compost," in *Leaves of Grass* (New York: Bantam, 2004), 307–9 (308).

47. Ibid., 309.

48. Ibid.

49. Felix Guattari, *Chaosmosis: An Ethico-aesthetic Problem,* trans. Paul Bains and Julian Pefanis (Bloomington: Indiana University Press, 1995), 20.

50. Whitman, "This Compost," 308.

51. Clayton Eshleman, "Narration Hanging from the Cusp of the Eighties," in *The Difficulties* (Kent, Ohio: Viscerally Press, 1980), n.p.

Attune

TIMOTHY MORTON

Les non-dupes errent.

—JACQUES LACAN

Since a thing cannot be known directly or totally, one can only attune to it, with greater or lesser degrees of intimacy. This is not a "merely" aesthetic approach to a basically blank extensional substance. Since appearance can't be peeled decisively from the reality of a thing, attunement is a living, dynamic relation with another being.

The ecological space of attunement is a space of veering, because rigid differences between active and passive, straight and curved, become impossible to maintain. Consider, for example, the phenomenon of adaptation, a complex and curious event. An evolving species is adapting to another evolving species, since what we call rather glibly "the environment" (that which veers around) is nothing but other lifeforms and what one Darwinist calls their "extended phenotypes," the results of their DNA mutations and that of their symbionts, such as spiders' webs and beavers' dams.[1] Moving targets adapt to moving targets, in a constantly morphing adaptation space. This process simply cannot be "perfect," because *perfect* means that motion stops. Adaptation just is movement in adaptation space, and perfection would mean the end of adaptation. So when we talk about how lifeform x is "perfectly adapted" to the swirl of phenotypes— including those that are "its own," such as its also-constantly-evolving bacterial microbiome—we are trying to *contain* or *stop* the veering of attunements of lifeforms to one another, if only in thought. Teleology is the gasoline of "perfect adaptation," and teleology, namely Aristotelian concepts of species, is precisely what Darwinism liquidates.[2]

We find a special and revealing adaptation mode in the syndrome we call *camouflage*. An octopus takes on the palette of the surface on which she is resting. A stick insect disappears into the foliage, to avoid predators. And at a basic level, to be alive is to adapt, without disappearing completely—to be protected by one's attunement, but not to the point of dissolving altogether. These brief glimpses of how what appears "only" aesthetic to our eyes suggests that attunement is not a case of having a blank, block-like substantial being whose superficial qualities are tuned while the substance remains the same—like what we think happens when we tune a violin, a topic not irrelevant to the notion of *ecotone* (*tone,* tension). The strings and the wood and the curvature of the violin form a unit such that tuning the strings by turning the pegs at the top of the instrument is not like arranging the apps on a smart phone, because the "platform" is being altered by the tightening or loosening of the strings.

A lifeform is like that must-have eighteenth-century equivalent of the iPod and Bose speakers, the Aeolian harp, a string instrument placed in an open windowsill that resonates to the breezes that veer around the house. The haunting, harmonic-rich, phasing sound this attunement system produces is strangely contemporary, as if Jane Austen characters were listening to a drone piece by Sonic Youth while they sipped their tea and played cards and wondered about the intentions of Mr. Bingley. But sipping tea and playing cards are also attunement systems, exemplifying in this case the upper-class mode of consumer performance, in which establishing and maintaining a certain sense of "comfort" is the basic tone to which the system is tuning: everyone must feel at ease, disturbance to the status quo must be minimized. All aristocratic attunement is about drones, sustained tones that waver as little as possible. If converted into sound, the space of polite interaction would indeed resemble a Sonic Youth drone piece.

In every single-celled organism there is a chemical representation, more or less accurate, of the realm in which it is floating. A perfect match—exactly the same chemicals—would equal death, which in a sense is a term for when a thing actually and wholly becomes its surroundings. Freud's *Beyond the Pleasure Principle* is a startling consideration of this fact.[3] Copying, mimicry, influencing and being-influenced by, being tuned and tuning—something causal is happening when these happen, which is why we think "primitive" (not-us) people imagine that photographs are stealing their soul.

Or perhaps photographs do steal your soul. Or perhaps photographs show you that your soul isn't yours in the first place, and that it certainly isn't inside you like a vapor in a bottle. The realm of attunement is thus like the mesmeric realm of "animal magnetism." It is a force simultaneously discovered and repressed at the inception of modernity. When in the film *Dark City* the protagonist finds out that he can "tune," what this means is that he can perform telekinesis.[4]

While modernity allowed agricultural logistics to destroy Earth even more successfully than before, it also unleashed, ironically and unwittingly, the nonagricultural ("Paleolithic") idea of an interconnective, causal–perceptual aesthetic force. We have recently rediscovered attunement, which is unsurprising, as we have recently rediscovered nonhuman beings outside the flattening, reifying concept *nature,* which almost seems to have been designed to dampen our awareness of attunement space, perhaps just as the "well-tempered" keyboard is designed to reduce the spectral harmonics that haunt a sound owing to its necessary physical embodiment: there is no sound as such, no pure tones, only the sound of a string, the sound of a sine wave generator. Objects thus have what is called *timbre,* and this is not an optional extra. Appearance is like that: appearance is better thought not in an ocularcentric manner as candy decorating a cupcake, but rather as an object's timbre.

We have rediscovered the veering brotherhood and sisterhood of nonhuman beings, smoothed and packaged as Nature and indeed as "the environment." Kinship, as in sisterhood, as in *humankind,* precisely has to do with an uncanny intimacy, which is why the replicant Roy shouts that word ("Kinship!") as with one arm he lifts his bleeding enemy up onto the roof of the tall building at the end of *Blade Runner,* his own kind (unbeknownst to both) in the form of the agent Deckard. A weird ruse of history has caused the flight from veering, which is the flight from our material embodiment, the timbre that haunts us with our affinities to chimpanzees, fish, and leaves trembling on the tree outside the window, to have resulted in a return to veering. The Owl of Minerva didn't just fly at dusk as Hegel famously said. She veered straight out of a dream into the dreams of sleepers convinced they had woken up from every last trace of the so-called primitive. When we study attunement, we study something that has always been there: ecological intimacy, which is to say intimacy between humans and nonhumans, violently repressed with violent results.

To begin to track this flight, then, is to veer toward a veering. And the first question that we might ask, in the context of an essay collection on veering and ecology, is whether nouns really are uninteresting until we make them more like verbs, because nouns denote things, and things are static things that underlie (sub-stance) appearances, which I have been arguing is fundamentally motion. Consider this noun: *future*. The future, or what Derrida called decisively *l'avenir*—the radically open future that is a possibility condition of the predictable future: is this term *future* unmoving? What happens at the end of this sentence? Does its meaning arrive? Arrive fully? This sentence means something, but you don't quite know what yet, as if the meaning, which is to say the tone to which it is tuning, were lying just off of its end, elephant, seaweed, gamma-ray burst. Is the future a thing? What is a thing? Haven't we already smuggled in a basic, default ontology before we start to think or talk about thinking, if we say *thing* is a noun and *noun* is static, must be put in motion to be worthy of inclusion in a collection of essays on veering? All the objects in the world must be rounded up and forced to march and march until they drop, because that's the kind of work that makes them free?

Underlying this, isn't there a Neolithic binary between *moving* and *staying still*, one that underlies most default (and incorrect) mechanical causality theories? A binary, moreover, that is part of the built, social space of Neolithic agriculture that eventually required carbon emissions to reproduce itself?

Object-oriented ontology (OOO) wants to reconfuse us, much like deconstruction, about the status what we take for granted. Language doesn't want to stay put, even if our version of staying put (fixing meaning) is making everything move around. Why? OOO argues that this is because things in general won't stay put, even when they are to all intents and purposes utterly still.

So I am going to resist saying *attune* rather than *attunement*, as if *attune* were a verb (and worthy) and attunement, a gerund (technically a

noun made from a verb) were not (and unworthy). I resist this direc-
tive especially in the context of a grammatical novelty not unlike the
tactics of the world of contemporary products: a painkiller called *Alleve,*
a high-end home furnishing magazine called *Dwell,* an English kitchen
store called *Cook,* a restaurant called *Eat.* In the case of a perfume called
Givenchy Play Intense, the placement of the adjective after *Play* seems to
make *Intense* into a vernacular adverb and *Play* into a verb, a noun intensified before its adverbial qualification by having been transmogrified
into an imperative, like what you see on the side of a Coca-Cola bottle
(Enjoy).[5] Play hard! Work hard! It makes you free!

Debt scoops open the future. Capitalism is a "verbal" economic form:
all that is solid melts into air. Static objects are what it despises. Are we
sure that using verbs and not nouns is truly subversive?

Hesitation? Is it a verb or a noun? Is it movement, or stillness? Veering hesitates: in a way, hesitation is a quantum of veering. So we will start
this veering process by seeming to veer off course altogether, hesitating
to begin. And you may find yourself examining the protocols of the discipline we now call sociology, and you may say to yourself, how did we
get here? And will this rhetorical veer, this "assay of bias," as Polonius
puts it, this curveball of an essay topic, catch the slippery "carp of truth,"
as Polonius also puts it?[6] If it does catch something, would it be more
like *seizing* (thrusting your hands into the cold water, grasping a fish) or
being seized? In that latter sense, I *catch* the flu—but I am also mesmer-
ized by the little fish, I track its darting movements with my eyes that
begin to dart, fish-like, in their sockets. When I catch a fish perhaps I
need to have been caught by it.

Max Weber pioneered the discipline of sociology over a century ago,
but sociology's structuring principle excludes one of its foundational
concepts: charisma. Weber argued that charisma-based societies give
way to "disenchanted," bureaucratic societies. But sociology does not
see its task as related to exploring disenchantment. Sociology acts just
like the bureaucratic society that Weber argues is its birthplace; sociology is part of the logistics of what Weber called "the disenchantment of
the world."

Its founder's concept was a little scary at the time too, since charisma
has to do with forces that many described as supernatural or paranormal. One might regard modernity—world history since the later 1700s—

as a profoundly awkward dance of including and excluding the paranormal. Freud, for instance, developed his theories as a way to bowdlerize the theory of hypnosis, which was in turn a bowdlerization of animal magnetism, a hypothetical force discovered by Anton Mesmer later in the eighteenth century. Animal magnetism is practically identical with The Force of *Star Wars* fame; it is, as Obi Wan Kenobi observes, an "energy field" that "surrounds" and "penetrates" us, and we can interact with it, with healing and destructive consequences.[7] Marx argues that capital makes tables *compute* value as if they were even weirder than the dancing tables of the quasi-religion of spiritualism.[8] And so on—examples of this secret, almost completely untold history of modernity are everywhere once you start to look.[9]

The paranormal is what religion was already excluding, religion being the way Neolithic society—otherwise known as "Axial Age" or "agricultural" society, agricultural according to models such as that established in the Fertile Crescent 12,500 years ago—monopolized what Weber calls "charisma," restricting it to the king who has the direct line to the god whom he hears ringing in his ears, telling him to tell the people what to do, *what to do* never being "dismantle agricultural society, which has created patriarchy and tyranny in the name of sheer survival, and return to hunter-gathering and a less violent, less hierarchical coexistence with nonhuman beings." Heaven forbid we stop the logistical functioning of the world of agriculture, which eventually gave rise to global warming, which was precisely and ironically what it was set up to evade in 10,000 BCE. Heaven forbid we drop the anthropocentric equal temperament by which everything else becomes keyed to our teleological reference tone.

Destructuring this logistics, which elsewhere I've called *agrilogistics,* is usually considered out of bounds, because it implies accepting a non-"modern" view.[10] The modern view was established on (although it thinks itself as a further disenchantment of) now ancient and obviously violent monotheisms, which in turn find their origin in the privatization of enchantment in the Neolithic with its "civilization." We are all still Mesopotamians. We are Neolithic humans confronting the catastrophe wrought by the Neolithic fantasy of smoothly functioning agricultural logistics, and we want to hold on to the philosophical underpinnings of those logistics for dear life, because otherwise . . .

The logic underpinning Neolithic logistics is very obviously (when you study it) riddled with unsustainable paradoxes that result in cognitive, let alone social violence (in the conventional sense, between humans) and ecological violence (in the conventional sense, regarding nonhumans). Equal temperament is riddled with awkwardly cramped and fudged frequencies, precisely to eliminate "beating," the production of rhythmical pulses between tones, because the human manipulator of the instrument should be in charge of beating it according to what the human telos of the tune happens to be. Beating is also revealingly called a *wolf tone*. It is biologically true that we aren't totally Neolithic—we have three-million-year-old bodies infused with Neanderthal DNA, and so on—but it is also philosophically and politically true. Because it is never true that there can be a perfect adaptation to one's phenotype, such that the search for perfection, now visible in seeds genetically engineered for tolerance to pesticides, must be destructive on numerous levels. Just as agricultural patriarchy restricts the (female) body to a limited range of sexual organs, so equal temperament dampens the haunting harmonics of an instrument's timbre, monoculture dampens biodiversity, logocentrism dampens the play of the signifier . . . and the dream of "ecological" society as immense efficiency (the fantasy of perfect attunement) dampens the uneasy coexistence of lifeforms. We think we don't like veering—until an electric guitarist bends a note.

It is *too easy* to dismantle the philosophical basis of our "world" (aka "civilization"). Without this basis, that world would collapse. The only thing inhibiting us is our habitual investment in that world, visible in the resistance to wind farms—we like our energy invisible, underground in pipes, so that we can enjoy the view. The very mention of changing our energy throughput raises the specter of the constructedness of our so-called Nature. Think of the birds the turbines will kill! (Think of the entire species wiped out by *not* having the turbines and so forth.) (Are birds "perfectly adapted" to oil pipes?) Think of the dreams we will be disturbing! We spent all this time tuning the world to anthropocentric tones, then delimiting attunement space. We might have to teach birds to tune to wind turbines, and this will be a drag. We want to be comfy in our unwavering, thanatological world. Death is comfy, as Freud observed: the tension between a thing and its environing veering beings is lowered to zero.

Dismantling the underpinnings of agricultural logistics involves dismantling the "metaphysics of presence," the idea that to exist is to be constantly present: to exist is to be a lump of extended stuff underlying appearances. Reality is a plastic, unformatted surface waiting for us (humans) to write what we want on it: "Where Do You Want to Go Today?" (the 1990s Windows ad); "Just Do It" (Nike); "I'm the Decider" (George Bush); "We create realities" (Iraq War press conference, 2005). There is the regular flavor of this metaphysics, basic default substance theories. We scholars all think we are superior to them, but they shape our physical life, which we happily reproduce, and we retweet them in the cooler flavored upgrades, which speculative realism calls correlationism, which is the Kantian (and post-Kantian) idea that a thing isn't real until it has been formatted by the Subject/History/human economic relations/Will to Power/Dasein . . . In a way it's a worse (in the sense of more ecologically destructive) version of the regular substance ontology flavor. Now there aren't even blue whales—there are only blue whales when we say there are. And lo and behold, it came to pass—there were no longer any blue whales . . .

Happily, that particular extinction didn't occur. It didn't occur because people became enchanted by recordings of whale sounds in the mid-1970s. Enchanted. What does it mean? In terms of charisma, it means some of us submitted to an energy field emitted by the sounds of the whales. The fact that this is a wholly unacceptable, beyond the pale way of describing what happened is a painful and delicious irony.

Just as attunement is the fuel of veering, so charisma is the fuel of attunement. Charisma makes us hesitate, wavering in its force field.

What if charisma were *actual*? What would the emission of such an energy field imply? It would imply, for a start, that art isn't just decorative candy. It would imply what "civilized" philosophy from Plato on has been afraid of, the fact that (shock horror) art has an effect on me over which I am not in control. Art is *demonic*: it emanates from some unseen (or even unseeable) beyond in the sense that I am not in charge of it and can't quite perceive it directly, in front of me, constantly present. A dangerous causative flickering: magic. Magic is taboo cause and effect, or unthinkable cause and effect: either ridiculous or dangerous or impossible, or some weird borrowed-kettle combination of all three. (How can something be impossible *and* dangerous?) What we are talking about is what Einstein called *spooky action at a distance*, by which he meant quantum entanglement, but which also means what happens if you visualize the Rothko Chapel even if you aren't there, even if you have never seen the Rothko Chapel per se, perhaps even if you have never actually seen a Rothko painting, or a postcard of a Rothko painting.

We might conventionally argue that the charisma of the Rothko painting is bestowed on it by humans: this would be the acceptable Hegelian way of putting it. We make the king be the king by investing in him. Investing what? Psychic energy—which if you recall, is a bowdlerization of the Force-like animal magnetism. What if this attitude were not only masochistic in the extreme, but also—incorrect? After all, as Schrödinger already argued, the one thing you can rely on is that at the very least two tiny things (an electron, a photon—but now physicists are experimenting on scales trillions of times bigger than this, with positive results) can be "entangled" such that you can do something to one of them (such as polarizing it, changing its spin), and the other will, for instance, polarize in a complementary way instantly—which is to say, faster than light. And this complementary behavior happens at arbitrary distances. You can now observe two particles separated by kilometers behaving this way: one is on the other side of town, one is on board a satellite, and so on—*arbitrary* means "even if that particle is in another galaxy." And there are to date no loopholes: there is not some underlying substance that means the two particles are really one, for instance.[11] Causality just is magic. But magic is precisely what we have been trying desperately to delete.

Magic implies causality and illusion, and the intertwining of causality and illusion, otherwise known in Norse-derived languages as *weirdness*. *Weird* means strange of appearance, and it also means having to do with fate.[12] Neolithic ontology wants reality not to be weird. Eventually weirdness is confined to Tarot cards and vague remarks about synchronicity. What does it mean though, to entangle illusion and causality? What it means is that how a thing appears isn't just an accidental decorative candy on an extension lump. Appearance as such is where causation lives. Appearance is welded inextricably to what things are, to their essence— even "welded" is wrong. Appearance and essence are like two different "sides" of a Möbius strip, which are also the "same" side. A twisted loop is exactly what *weird* refers to, etymologically speaking. The minimal topology of a thing is the Möbius strip, a surface that veers all over, where a twist is everywhere. This is because the appearance of a thing is different from what it is—yet the appearance is inextricable from it. There is no obvious dotted line between what a thing is, and thing data. Attuning is like studying a Möbius strip.

Unfortunately for the scientistic ideology that dominates our world and the neoliberalism that forces us to behave in scientistic ways to ourselves, one another, and other lifeforms, *the idea that appearance is where causality lives is also just straightforward modern science.* David Hume's argument was precisely that when you examine things, what you can't see directly is cause and effect. All you have are data, and cause and effect are correlations of those data: things that are given, aka appearances. And Immanuel Kant underwrote this devastating insight. All we have are data not because there is nothing, but because there *are things,* but these things are withdrawn from how we grasp them. Kant's example: raindrops fall on your head, they are wet, cold, spherical. This is raindrop data, not the actual raindrops. But there are raindrops, not gumdrops. And they are raindroppy: their appearance is entangled with exactly what they are.[13]

What art gives us, argues Kant, is the *feel* of data, the data-ness of data, otherwise known as givenness (*datum,* Latin for *what is given*). This data-feel is, he argues, an attunement space, the one place in the whole universe where mesmerizing hesitation can happen—a very important mesmerizing hesitation, because it underwrites the existence of a priori synthetic judgment, because in this experience, I get a magical taste of something beyond my (graspable) experience, a transcendental beyond-ness that Kant wants to restrict to the transcendental subject's capacity to mathematize. But Kant's analogy—he was afraid of analogies because they might veer away from his conscious control—is the raindrop, and the raindrop's mathematizable properties such as size and velocity are also, he states, on the side of appearance.

The aesthetic dimension is a necessary danger, a tiny bowdlerized zone of mesmeric attunement without which we couldn't know that there is a weird gap between what things are and how they appear, which is why we know we should treat the beings we call *people* as ends and not as means, because your uses of me never exhaust me in principle. Through this attunement, I get to discover that my inner space is infinite—but it's equally likely, according to the implicit logic of Kant's *Critique of Judgement,* if not its explicit argument, that the beautiful thing is also bigger on the inside, as Doctor Who says of his dimension-transcending spacecraft, the TARDIS. For the beauty experience is precisely that phenomenon in which I find it impossible to tell who started it: was it me or was it the thing? Yet Kant concludes that this secretly means *we (the subject) started it.* A small piece of mesmeric, magical dynamite is embedded at the crucial point in modernity's architecture: a tone rich with harmonics "disgustingly" not quite keyed to human teleological reference frames. The self-transcending subject is underwritten by a mysterious power emanating from the nonsubject (the "object"). I may be the one who gets to decide whether the light is on in the refrigerator or not (correlationism); but there needs to be a refrigerator in the first place, and for some reason I find myself drawn to it. Why are the Sami people of northern Scandinavia reluctant to cast counterspells to those woven by global corporations? Because that would involve their culture with corporate culture in a mutual attunement space: their culture would be distorted by the attempt.[14]

Things are exactly what they are, yet never as they seem, and this means that they are virtually indistinguishable from the beings we call *people.* A person is a being that veers in just this way. Humans are more like nonhumans, and nonhumans are more like humans, than we like to think—and those two phrases do not quite add up. It is radically undecidable whether we are reducible to nonsentient, nonconscious, nonperson status—or whether things that aren't us, such as foxes or teacups, are reducible upward to conventional personhood. I might be an android—this android might be a person: that's the best we can do. Deleting the hesitation by reducing either one to the other is what is called *violence.* I am playing a tune called *myself* to which you are attuning, but which is itself attuned to you, so that we have an asymmetrical chiasmus between *myself* and me, between me and you.[15]

We live in a world of tricksters. How we conduct ourselves in this world, the ethics of the trickster world, has to do with respecting that subjunctive, hesitant, might-be quality. It has to do with attunement, which is a dance between completely becoming a thing, the absolute camouflage of pure dissolution (one kind of death), and perpetually warding off that thing (another kind of death), the mechanical repetition that establishes walls, such as cell walls. Between *I am that* and *me me me*. What is called *life* is more like an undead quivering between two types of death, a deviance that is intrinsic to how a thing maintains itself, like how a circle is how a line deviates from itself at every point, thanks to the seductive force of a number existing in a dimension perpendicular to that of the rational numbers (pi).

The Rothko Chapel causes a certain kind of viewer, invested in a certain kind of hermeneutic of suspicion dependent on a certain Neolithic ontology, to want to contain the experience, which is of being *tuned* by a painting—we attune to the gate-like rectangles of aubergine space, because they are already tuning to us, waiting, beckoning. A Rothko Chapel painting is a portal: just what might come through? Such a painting is a doorway for what Derrida calls *l'arrivant* (verb or noun?), the *future future*, the irreducible, unpredictable one. Philosophy, which is wonderment (hence horror, or eroticism, or anger, or laughter) in conceptual form, is an attunement to the way a thing is a portal for the future future. The love of wisdom implies that wisdom isn't fully here, at least not yet. Perhaps if it ever succeeded in teleporting down perfectly, it would cease to be philosophy. Thank heavens philosophy *isn't* wisdom. If it is, I want nothing to do with philosophy.

We might want to contain the aesthetic experience by framing it as "art" in some predictable, preformatted sense. Going further, we might think art is a reflex of the commodity form, which would really help us to keep our suspicious distance: heaven forbid we be seduced by anything. Heaven forbid we fall in love—the beloved might be a shill for the Coca-Cola corporation. Art shows us how a disturbingly ambiguous pretense is woven into aesthetic experience: wonderment is based on the capacity to be deceived. The more we are okay with being lied to, the wiser we might become. "Ever get the feeling you've been cheated?" (Johnny Rotten, a pop singer, at a pop concert at which he was singing—was he cheating at the time, or was this part of the art, or . . . ?). So perhaps we could dismiss a Rothko painting, as Brian O'Doherty does in his famous essay on the commodification of art space, the dreaded "white cube" of the contemporary gallery, now replicated in a million minimalist townhome interiors.[16]

Attunement is the feeling of an object's power over me—I am being dragged by its tractor beam into its orbit. We are not to be manipulated.

We write essays such as *Inside the White Cube* about how white cube spaces inevitably seduce us all—except for me, the narrator of the white cube essay, and you, the sophisticated reader whom the essay is interpellating, rising above it all, exiting the poor beastly body and the abject world of objects, like the Neoplatonic soul transcending the body. "Obey Your Thirst" (Sprite ad, 1990s) has no effect on us. Everyone gets conned by objectification, except for me, the one who writes the sentence *Everyone gets conned by objectification*. All sentences are ideological, except for the sentence *All sentences are ideological*.

Critique mode is the mode of the pleasure of no-pleasure, the sadistic purity of washing your hands of the crime of being seduced, as if *detuning* (discord, dissonance) were about exiting attunement space rather than what really happens, which is only *retuning*. The opposite of attunement isn't detuning or discord: I love being seduced by Schoenberg. The opposite of attunement is the ignoring, straightening out or otherwise destruction of attunement space, whether in thought, word, or deed. The worst thing that could happen in critique mode is that you could make or enjoy kitsch. Happily children have never heard of such things. My son Simon tells me that if you cross your eyes and stare at a Rothko painting just so, the red lines will start to vibrate and float towards you, and you will feel nauseous and giddy—and that these are exciting, oddly pleasant sensations, like spinning in a swivel chair. Apparently the paintings aren't just commodities sitting primly in a shop window. Apparently they even exceed their human-keyed "use-value." For O'Doherty, the best kind of art, which he calls postmodern, is an endless conversation between (human) subjects about what good art might be, as if tuning up were not part of the orchestral performance—a myth rapidly dispelled by the first few seconds of "Sergeant Pepper's Lonely Hearts Club Band."[17] Actually letting yourself enjoy a thing is pleasurably avoided. For a six-year-old child, it's obvious that Rothko is trying to blow your mind.

Art is a realm of passion for no reason: I just like this particular shade of blue, I want you to feel the weight of this metal toe, come in to this installation, look, peer through the curtain. The time of novels is the time of lust—the first novels were necessarily pornography (consider the works of Aretino). So when we talk about art, we are talking in the region of love and desire, those unsteady, uneasy, wavering partners.

Let us widen our gaze from the artwork to a more general description of this region. Love is not straight, because reality is not straight. Everywhere, there are curves and bends, things veer.

Per-ver-sion. En-vir-onment. These terms come from the verb *to veer*. To veer, to swerve toward: Am I choosing to do it? Or am I being pulled? Free will is overrated. I do not make decisions outside the universe and then plunge in, like an Olympic diver. I am already in. I am like a mermaid, constantly pulled and pulling, pushed and pushing, flicked and flicking, turned and opened, moving with the current, pushing away with the force I can muster. An environment is not a neutral empty box but an ocean filled with currents and surges. It environs. It veers around, making me giddy. An aesthetic wormhole, bending the terrestrial and ecological into the cosmological. The torsion of deep space, beaming into the cold water of this stream like bent light, the stream I was caught by the fish I was catching a few pages ago.

Space-time as such is a bending, a curvature. It isn't correct to say that space-time is first flat, then distorted by objects. Objects directly *are* the distortion of space-time: space-time is the distortive force field that emanates from them. Curvature, lumps and bumps, a strange plenitude everywhere, no dead air. Space-time isn't a flat blank sheet that gets disturbed. Space-time is disturbance. Disturbing lens of matter-energy, we see as much as we can see, always less than all, through the convex kaleidoscope of space-time. A thing is dappled with time. But not a lump coated with time, improved by the makeup of motion. Better: a thing *is* this temporal dappling.

John Ruskin argued that the modern tendency to want to clean old buildings, very much in effect today, was a sacrilegious erasure of what he liked to call *the stain of time*.[18] In a sense Ruskin was aiming at something like an ontological redescription of things: to remove the time stain is to harm the actual thing, because a thing *actually is* this temporal staining. To want to cleanse a building of what is taken to be a supplementary stain is to assume that a thing underlies its appearance, the old default substance ontology. To allow things to get dirty is to allow that things are not at war with time. There are not things plus time, like eggs dropped into water. The "dirty" Sistine Chapel ceiling is exactly how it would have been seen in flickering candlelight.

Newton's world is a realm of straight love, instant beams of gravity that are God's love, everywhere, all at once, outside time, omnipresent force of an omniscient being acting on static extensional lumps, exciting them, pushing and pulling them around like cattle.

We do not live in Newton's world.

Einstein's world is a realm of perverse desire, invisible ripples of gravity waves that make up space-time, the invisible ocean in which the stars float submerged. We love the dead. We love fantasies. Do they love us back? We are pulled toward them and as this happens, time expands and shrinks like a polymer. No God could be omniscient in such a world, where time is an irreducible property of things, part of the liquid that jets out of a thing, undulating. There are parts of the universe that an observer will never be able to check. They are real. Things happen there. But some observers will never know *where* they are happening or *when* they are happening. Some people in the universe will never know you are reading this, because they never *can* know. And you won't be able to know them.[19]

In a universe governed by the speed of light, parts are hidden, withdrawn, obscure. The dark Dantean forest of the universe, an underwater forest of rippling weeds. You should find this idea extremely comforting. It means that you cannot be omnipresent or omniscient. It means that you cannot look down on the poor suffering beings of the universe from a position outside time and smile sadistically at their pain, a smile we often call *pity*.

Each entity in Einstein's universe is like the veering turbulence in a stream, a *world tube* or vortex that cannot know all. There is a darkness that cannot be dispelled.

Consider now the even stranger, and even more accurate, description of things we call *quantum theory*. In quantum theory the binary between moving and staying still—between a certain concept of *verb* and *noun,* or between a certain concept of *object* and *quality*—becomes impossible to sustain. Objects isolated as much as possible from other objects still vibrate without being pushed—that is to say, without being subject to mechanical causation.[20]

And this isn't surprising, because "traditional" agrilogistics ends up as our current version, so that there is a line from the notion of the guiding weight of tradition to the play of infinite (human) freedom and "choice."

The aesthetic dimension is commonly imagined as a special glue that sticks these two poles together (as in Schiller), by allowing humans to impose the proper form, to adapt their world perfectly to their requirements. But this is not how it works. We have seen that this dimension is deeply entwined with things as such, not with (human) formatting. There is a certain courage in letting yourself fall asleep (as Lingis observes) and allowing dreams to come, which resembles the courage of allowing art to affect you.[21] Hallucinatory phantasms are a condition of possibility for seeing anything at all. Hearing is a chiasmic crisscross between sounds emitted by my ear and pressure waves perturbing the ear's liquids from the outside. The not-me beckons, making me hesitate. Come.

Notes

1. Richard Dawkins, *The Extended Phenotype: The Long Reach of the Gene* (Oxford: Oxford University Press, 1999).

2. Gillian Beer, introduction to Charles Darwin, *The Origin of Species* (Oxford: Oxford University Press, 1998), xxvii–xviii.

3. Sigmund Freud, *Beyond the Pleasure Principle* (New York: W. W. Norton, 1989).

4. *Dark City,* directed by Alex Proyas (New Line Cinema, 1998).

5. Many verbal perfumes exist: Gardez-Moi (Bertrand Duchaufour), J'Adore L'Or (Christian Dior), Quand Vient la Pluie (Guerlain), En Passant (Editions de Parfums Frederic Malle).

6. William Shakespeare, "Hamlet," in *The Norton Shakespeare Based on the Oxford Edition,* 2nd ed., ed. Stephen Greenblatt et al. (New York: W. W. Norton, 2008), 1683–1784.

7. *Star Wars,* directed by George Lucas (Twentieth-Century Fox, 1977).

8. Karl Marx, *Capital,* vol. 1, trans. Ben Fowkes (Harmondsworth: Penguin, 1990), 1.163.

9. Jeffrey Kripal, *Authors of the Impossible: The Paranormal and the Sacred* (Chicago: University of Chicago Press, 2010).

10. Timothy Morton, *Dark Ecology: For a Logic of Future Coexistence* (New York: Columbia University Press, 2016).

11. The recent work of Anton Zeilinger has been devoted to eliminating loopholes in nonlocality theory—in other words, maintaining the paradox of two entities tuning to one another simultaneously.

12. *Oxford English Dictionary,* s.v. weird, adj.1. *Oxford English Dictionary,* s.v. weird, adj.3.

13. Immanuel Kant, *Critique of Pure Reason,* trans. Norman Kemp Smith (Boston: Bedford, 1965).

14. I am grateful to Tanya Busse for discussing this with me.

15. Alan M. Turing, "Computing Machinery and Intelligence," in *The Philosophy of Artificial Intelligence,* ed. Margaret A. Boden (Oxford: Oxford University Press, 1990), 40–66; René Descartes, *Meditations and Other Metaphysical Writings,* trans. Desmond M. Clarke (London: Penguin, 2000).

16. Brian O'Doherty, *Inside the White Cube: The Ideology of the Gallery Space* (Berkeley: University of California Press, 1986).

17. The Beatles, *Sergeant Pepper's Lonely Hearts Club Band* (EMI, 1967).

18. John Ruskin, *The Seven Lamps of Architecture* (1849; Project Gutenberg, 2011), http://www.gutenberg.org/files/35898/35898-h/35898-h.htm.

19. This is a deep implication of Herman Minkowski's geometrical proof of relativity theory.

20. Aaron O'Connell, "Making Sense of a Visible Quantum Object," TED Talk, March 2011, http://www.ted.com/talks/aaron_o_connell_making_sense_of_a_ visible_quantum_object.html.

21. Alphonso Lingis, *The Imperative* (Bloomington: Indiana University Press, 1998), 11, 14, 22, 42.

Sediment

STEPHANIE LEMENAGER

"All the sediments I have met with were amorphous," writes Edmund Lloyd Birkett in *Bird's Urinary Deposits* (1857).[1] The *Oxford English Dictionary* records this medical usage—poetic and oblique in its decontextualized solitude—for posterity. Birkett's gloss on what urine yields as sediment also invokes a type scene for the nominal meaning of *sediment* as "matter composed of particles which fall by gravitation to the bottom of a liquid." This scene of the urology lab spurs imagining perhaps too vivid to represent common usage. I envision the experimenter as white coated, precise in his rendering. Is he affronted by the amorphousness of the samples? As noun, *sediment* doesn't typically summon the amorphous. Rather, *sediment* suggests ungainly—dirty—form, "earthy or detrital matter deposited by aqueous agency." *Sediment* also suggests *dregs,* which can be a social category. *Dregs* indicates the poor, those humans without the socioeconomic capacity for mobility and for the sharpening of personality that Georg Simmel urges is necessary to be recognizable (grievable?) in the bustling nineteenth-century city. "The last sediment of the human stew that had been boiling there all day, was straining off," Charles Dickens writes of such people in *A Tale of Two Cities* (1859). Dickens uses *sediment* to imply the "lowly," who are the foundational human layer in sentimental fictions. Sentimental fictions, in turn, posit the fixity of these human "dregs" only to perform the literary miracle of individualizing and lifting some one or two of them toward the reader's sympathetic attention. *Sediment,* as a verb, is always already falling into noun, as in settling out into dregs or deposits. My choice to begin thinking with

168

sediment from a quandary of imprecision in the laboratory reflects my desire to reanimate the noun form of *sediment* by returning to a primal scene in which *sediment* (n.) does appear "amorphous" and open. As will become clear, my larger project is to reanimate *sediment* in order to rethink settling, settler colonialism, and the kinds of economic disenfranchisement that historically have marked some persons and nonhuman life as lowly dregs.

To sediment (v.), infinitive. Or, the imperative: *sediment!* There are several ways to conceive the veering of this noun-bound verb toward animacy, ecological resonance, and social justice. One is an essentially green gloss on the *longue durée,* what we might call *the rivering of time,* wherein we imagine temporality as a settling out of matter, century upon century, layer upon layer, by a hybrid agency both aqueous and cultural, material and semiotic. As Jeffrey J. Cohen has written of the great North American river, the Mississippi, "The river composes with ice, stone, potent flows of water, heterogeneous biosystems, and tumbling sediment." Cohen describes the river more broadly as an "earth artist."[2] Here, *to sediment* (v.) is to act as time, while *sediment* (n.) records such continuous action.[3] *Sediment* (n.) forms a cultural–geologic record and book open to rereading and therefore to perpetual reanimation, re-verbing, if you can forgive the pun. "The sediments are a sort of epic poem of the earth," as Rachel Carson writes in *The Sea around Us* (1950).[4] Sedimentary layers provoke questions of relations among objects, vital assemblage. What better example of sediment as the making of time, storied time, lithic *and* aqueous, solid yet interpretable, than our contemporary discussion of the Anthropocene, when the culture-making gesture of epochal naming collides with the empiricist desire to hitch culture to the stratigraphic record?

The Anthropocene Working Group associated most readily with its chairman, Jan Zalasiewicz, favors the mid-twentieth century as the location of the Anthropocene's "golden spike," due in part to the high proportion of radioactive isotopes discernible in rock layers from the atomic years.[5] Scholars and pundits have argued about what is the best spike year to mark the start of the Anthropocene: Could it be 1945, or 1950, or 1964? The atomic age, with its links to the Great Acceleration in production/consumption and in population growth in the Global North, might well stand as the age when humans ineluctably ruined the planet. The atomic age also speaks to a well-known history of ecology, the scientific movement meant to introduce interspecies ethics (broadly conceived) in the age of the bomb.[6] Ecology kidnapped "science" from the military-industrial complex, suggests the historian Donald Worster. Through the publication of *Silent Spring* (1962), Carson spoke for ecology to North Americans, reminding us that science might be a public and even pastoral project—for example, in the peacetime citizen science that she invokes by sampling everyday Americans' accounts of the diminishment of wild bird populations in U.S. suburbs. As a golden spike year, 1964 in particular connotes the stirrings of North American environmentalism, a culturally specific—largely white, middle-class—ethical response to world destruction on heretofore unimaginable scales. The bomb. The ban on DDT set in motion by John F. Kennedy and mobilized—initially—by the publication of Carson in the *New Yorker*. Shall we sediment here?

The Dutch atmospheric chemist Paul Crutzen, who coined the Anthropocene concept with Eugene Stoermer, an ecologist often forgotten in the legacy of this lithic idea, prefers the late eighteenth century for the golden spike. This time frame commemorates James Watt's design for the steam engine, which ushered in the industrial age in Britain and North America. More sensitive to the imperialist dimension of Anthropocene dating than most Northern geologists, Simon Lewis, of the University of Leeds, and Mark Maslin, of University College London, elect 1610 as the year of the golden spike. The year 1610 shows evidence of vast reforestation in the Americas, which is symptomatic of the collapse of Indigenous agriculture and thus of the massive genocide of Indigenous Americans provoked by European voyages of discovery and trade across the Atlantic.[7] Building genocide and implicitly chattel slavery into the stratigraphic record—recall that Bartolomé de las Casas endorsed African slavery in response to the die-off of Amerindians in the time of Columbus—opens the *longue durée,* the slow rivering of time, to the repetitive and haunted time of racist and colonial trauma. *Sedimentamos aquí?* My Spanglish intends to put a mestizaje pinprick in the Northern-centeredness of the Anthropocenic balloon, which only occasionally attaches to sedimentary layers beyond the cultural parameters of the triangle trade.

Why not drive our spike into the sediments marking intensified Chinese coal usage in the twelfth century[8] or—as some scholars have suggested—into the birth of agriculture on the central Asian steppes? The

Anthropocene will not decolonize our minds, fellow Northerners. It intends to hail those in the wealthier world into the feeling of species-being that historian Dipesh Chakrabarty worries might erase the violent intentionality of settler colonialism.[9] Yet that hailing into species-consciousness, too, can encourage environmental justice, if only to humble a global elite associated with the speeded up time of fossil fuel extraction and the forgetting of intergenerational memory that is a primary attribute of modernity. *To sediment* is to remember the material histories of modernity, reconstitute our own materiality as animals, and make of memory something durable: matter.

Où sedimentez-vous? Whether 1964, 1610, the 1780s—only the dates associated with Watts's steam engine mark a residual of human techne that we might imagine as triumphant, evidence of a desire to sediment, to be remembered in stone *here,* and *for this.* The Great Humiliation, as Michael Pollan prefers to name our epoch,[10] implies that the ecological importance as global elites falling into self-recognition as a species pertains not to deliberate acts of making. The unromantic externalities of mundane and habitual practices, such as driving our cars to the supermarket, will be the legacy of the global rich. To sediment in such circumstances is to become a Pompeian fossil, caught in the act of the everyday, caught half-naked, poignant in prosperity and unpreparedness. "The challenge of the Anthropocene is to learn to see ourselves not at the open end of Earth's timeline but within its bounds, as fossils in the making," writes journalist Michelle Nijhuis, translating the scientists' geologic imaginary into (elite) popular culture.[11] But the rivering of time remains an open process still, even if we—humans, global Northerners—are not at its leading edge. Let us not drive our ecological significance into lithic certainty precipitously, without thinking about what conceiving ourselves as fossil does, ethically. Do not sediment too readily. *Not when to sediment* means to give in to a kind of fossilization that implies extinction and the consolation of Apocalypse, sparing *Homo sapiens ("sapiens northernensis")* the trouble of inventing a practice of collective continuance through the climate disruptions that probably will not end the world or kill all of us off.

Sedimentation (the noun implying the verb-as-process) can lead to unreasonable expectations. It is the process of settling into sediment that, over millennia, has baked marine life into the material of modernity's easy energy, conventional fossil fuel. Barbara Freese, the assistant attorney general of Minnesota, inspired by her knowledge of that state's dirty electricity to write the germinal Anglo-American history of coal *(Coal: A Human History)*, reminds us that coal is a testament to the efficiency of plant life in storing energy and multiplying its concentrated force over millennia. Energy essentially is the *longue durée,* time concentrated into greasy rocks that once were ferny plants and fantastic trees, such as the *Lepidodendron,* named for the Greek *lepid,* meaning *scale* and *possessed of a lizard-skin bark.* "For millions of years," Freese writes of the U.S. coal region of Pennsylvania, "the shoreline periodically swept back and forth . . . as distant glaciers formed and melted. . . . When the rising seas inundated the dense jungles, they buried them in marine sediment, leaving behind multiple layers of coal—geologic souvenirs."[12] Such geologic souvenirs recall past climate changes and, implicitly, long skeins of time forgotten in the quick-burn psychology and place-defying infrastructures of modern energy. The geographer Kathryn Yusoff has described coal as a co-agent and underwriter of modern human subjectivity. In so doing she restores the most abject of fossil fuels—hated because of its connections to visible pollution and back-breaking labor—to its rightful position in an assemblage of actors (human and nonhuman) who make up the modern everyday, with its portable climate and light.[13] Lowell Duckert, writing as both a literary theorist and a resident of the coal state of West Virginia, tells us that "coal is a reminder of our ongoing shaping of and being shaped by the earth, of living within the world rather than upon its outer surface."[14] Coal is the greasy stuff of ourselves, with the "we" here indicating the beneficiaries of coal heat and light even more profoundly than those who labor to get the stuff out of the ground. Climate makers and destroyers, coal beneficiaries typically live far from the scene of extraction. As far as the beautiful, ecoconscious Puget Sound from Wyoming's Powder River Basin, for example.

Coal and oil and natural gas, buried deep in sedimentary layers of shale in boom regions like North Dakota or Venezuela. These sources of energy are, in cultural terms, substances set loose from the gold standard of geologic time, ideologically a paper currency. How many of us associate "energy"—charismatic, characterological—with their ungainly matter? How many times a day does any one of us, do I, for instance, think about how the light switch in my office connects, from contemporary infrastructure through extraction zones through deep history, to the Carboniferous forest? Extractivism means, among other things, abstraction, which is time-saving and aesthetically clean (no sedimentary muck), just as forgetting is time-saving and aesthetically clean. For Paul Connerton, modernity *is* forgetting in an explicitly disembodied way, so that we no longer have the physical habits of our ancestors, their sensory knowledges.[15] Modern energy is released from its ties to sedimentation and to slow histories, to enduring intergenerational habits, to rivering time, oceanic time. Because of this disconnection—ideological and infrastructural insofar as energy is often extracted and produced far from where it is delivered—"energy" acts as a profoundly asocial and unsettling concept. In contrast, to sediment is to deposit as sediment, essentially to settle down, and to settle in. (This is to think a kind of settling in which power is redistributed justly across human and nonhuman bodies, a model to supplant settler colonialism.) As the historian Richard White reminds us in his history of hydropower generation on the Columbia River, power has been the ability to command energy. To command energy rarely implies a respect for the places and bodies that make it, over time. To sediment must be to act against the antimaterialism of power.

When the poet and labor activist Mark Nowak writes of energy in the poetry and photography collection *Coal Mountain Elementary*, co-created with photographer Ian Teh, he juxtaposes the cultural value of laboring bodies—bodies conventionally trivialized or demoralized, like miners in West Virginia and southern China—against the value of coal itself. Both coal and coal miners are abject in the sense that Julia Kristeva has written of the abject—extruded matter, dead matter, the corpse-in-the-making. They become "energy" (charismatic and light, part of the national plan) only by way of metaphor. The witness testimony that makes up Nowak's found poetry from West Virginia tells us that dead miners trapped in a pit must be reported in company parlance as "items," not persons. Supposedly this rhetorical sleight is meant to avoid alarming the living who wait anxiously to hear about their loved ones in the mines—who will nonetheless, of course, discover that their loved ones are dead and grieve them as persons.[16] Meanwhile the American Coal Association's Lesson Plans for elementary school children, created in 1981 for public school usage, encourage metaphoric complicity with the dehumanization and denaturalization of coal production. Children are instructed to practice cost/benefit analyses in the process of "cookie mining," using toothpicks and paper clips (more costly tools, more precise) to remove chocolate chips from cookies. They then efficiently restore their cookie properties to their "original" condition by containing the leftover crumbs within circles drawn around the cookies before they were "mined." This process of piling the crumbs into the circles is called, in the lessons, "reclamation."

This "reclamation"
should also be timed
(no more than three minutes)
and students may only use
their tools, not fingers.[17]

Could mountain-top removal, with its massive redistribution of rubble that was mountain peaks into once-fertile West Virginian valleys—"accelerating by hundreds of millions of years erosion processes," as Freese reminds us—be submitted to such simplistic calculation?[18] The mountains and valleys, like the miners, are violently elided in the story of cheap energy and power. Again, extractivism means, among other things, abstraction.

The beauty of Nowak's serial poem "PROCEDURE (cont'd):," which wittily cribs from the American Coal Association's Lesson Plans, is its quantitative certainty. Laid out on the page, the graphic logic of the poem tends toward the ever-more simple and compact, visually dwindling down to the truism "profit/or loss."

After time is up,
collect additional
reclamation costs
($1) for each square covered
outside the original outline.

. . . .

Have students use
the Cookie Mining Worksheet
to calculate

their profit
or loss.[19]

There is no sedimenting here. There is no materiality to the time spent, calculated, won, or lost. "Mountain" crumbles either can or cannot be contained by paper circles. They either do or do not cost students by spilling outside the lines, onto the grid. Student workers either do or do not mine their cookies in the time allotted, which is figured as a number of minutes equal to a monetary value. This is not time, material. This is not memory, with the social resonance and sensory thickness that memory implies. This is math. It is money. It is "energy" as an idea removed from laboring bodies, both human and nonhuman, and from material histories reaching back millennia, to the *Lepidodendron,* that Carboniferous tree. The logic of extractivism is an emphatic "Do not sediment!"—fail to become embodied, refuse to matter—to be matter, in time. Under the sign of extraction, *sediment* can only be an abstract noun, the standing reserve that must be dug out of the ground. It is what will be called forth for ordering by clumsy toothpicks and paperclips and bulldozers, regardless of what the river valleys expect of the mountaintops or what a miner's children might hope to see of him tonight, and tomorrow.

Let us sediment, as a deliberate act. Imperative: Sediment! Remember mattering . . . conceive your future . . . as . . . mattering. Let this be a movement, Occupy with a nod to lithic temporalities, making a commons of deep time. What kind of fossil are you? What's your residue? Aspire to sediment in a way that remembers the future. And consider—if you have not, or are not—Indigenous scholars and writers, who have thought, written, taught, and acted often from the front lines of climate destruction and extractivist violence. For example: The continuous oil spillage in the Niger Delta that helped spur Ken Saro-Wiwa's leadership of MOSOP (Movement in Support of the Ogoni People). This is the Indigenous, antiextraction movement that Naomi Klein recognizes as originary for Keystone XL activists in the Global North. As the Indigenous philosopher Kyle Powis Whyte argues, Indigenous knowledges aren't supplemental to nonindigenous "science," something to add to a settler-colonialist toolkit that has already proven unsustainable. Traditional ecological knowledges (TEKs) affect "an Indigenous community's capacity to be adaptive in ways sufficient for the livelihoods of its members to flourish into the future."[20] TEKs are primarily for Indigenous peoples—not for settlers, not for me, even as I look to these knowledges (what can be made public, that is) to summon possibilities of governance and futurity. To sediment is to recommit to living in place—if you have lived with the worldview of a settler colonialist—to refuse extractivism and the erasures of indigeneity (of peoples, places, living in place) undergirding colonialist practice.

∼

Repeat. *To sediment* (v.) To settle into place over time, in ways that might transform the relations of violence bound up in settler colonialism. To transform the meaning of settling with a nod to the new materialism and to anti-extractivism, so as to respect all bodies and the dignity of matter at multiple scales. Thinking both molecularly and planetarily is necessary in this time of Anthropocene hubris and despair. The First Nations scholar Leanne Simpson writes that the alternative to extraction and abstraction, to coal subjectivity, is "respect . . . relationship . . . responsibility."[21] When the oil sands companies in Alberta speak of the peaty substance known as *muskeg* that overlays bitumen deposits and supports the boreal forest, they (extractively, abstractly) call this multiform, multipersoned life "overburden." As in: that which burdens us. That which hangs over our treasure. That which must be removed, by our giant, lurching CATs.[22] What could be a more concentrated call on us *to sediment* (here the verb summons an ethics and even a social movement) than this derogatory erasure of the liveliness and multiplicity of living, material time, set down over centuries, as forest?

Rita Wong, the Canadian Chinese poet who has walked among those protesting the oil sands and worked alongside the Indigenous movement Idle No More, writes of the metaphor "overburden":

"overburden removal" leaves poisonous polycyclic aromatic
     hydrocarbons, pah
the PAHs stink—swallow them and die a slow cancerous death[23]

Wong plays with death in these short lines, punning it: *pah*—small case—the noise of exhalation, of the poet's irritation, even disgust. PAH, a killing toxin. When I visited Fort McMurray, Alberta, in the summer of

2011, the friendly spokesperson on Suncor's energy bus tour told us that the company is virtually "farming" PAHs, that they have a way of drying them out and planting them in earth. Suncor, as its name implies, aspires to be something other than an extractive industry, aspires to primal, solar force, to be a player—reciprocal, responsible—in the ecological *longue durée*. Their metaphors, those linguistic helpmates in the culture of extraction, carry us across cognitive categories—from "toxins" to "humus," from "killing" to "farming." Wong calls this noxious playfulness and childishness, this envy of the slow rivering of time that intends to mimic its dignity and its grandeur a "blunt ambition, hydrocarbon hubris." She writes:

> hydrocarbon hubris,
> . . . .
> banal tyranny, drunk-on-god-trick, devastating heists from
>   the banks of the river scarring the earth
> scaring the unborn who hover, wondering what meets them[24]

If the Athabaska River takes in more of the pollutants housed in the oil sands industry's infamous tailing ponds, what will it become? The unborn might be the children of Fort Chipewayan, a First Nations community already devastated by a rare bile duct cancer. Or the unborn might be fish, plant life, energy-producing microbiota in the great river—all of which are dying or morphing into strange, ungainly shapes that indicate their degeneration.

Thomas King, the Native novelist and historian living and writing in Canada, imagines the tailing ponds bursting into the Athabaska, and the toxicity flowing north—as the Athabaska is a northern-flowing river—into the McKenzie, and then into the Arctic Sea.[25] Finally the Arctic melts in its toxic burden. And what, then, of the great ocean currents that regulate Earth's climate, let alone that of the low-lying coastal cities (Manhattan, Mumbai) sure to be inundated by sea level rise? These aren't science-fiction scenarios, even if they are still, we hope, understandable as worst cases. Whyte mocks the Anthropocene concept as a settler-colonialist realization of world-scale destruction—a scale of injury already lived, and survived, by Native communities.[26] Anthroposcenes can be stagy, even narcissistic indulgences, erasing intergenerational and transcultural memories. To sediment, in turn, is to respect what rivers know of time, the cumulative knowledges that shift and settle into place, over time. Wong laments that "miles of living medicines made by rivers over millennia/are unceremoniously eradicated, annihilated, wasted."[27] The situation of rivers in Canada, most famously the Athabaska, is mirrored to an extent throughout the United States and Canadian Northwest, where both great rivers (the Columbia) and lesser-known ones are affected by rising mercury levels as a result of coal trains speeding through the region. Each train loses between five hundred pounds and one ton of coal dust en route.[28] Sediment—the greasy coal rock sifting from the tumbling coal cars—ironically alters processes of sedimentation, the rivers' slow creation of living medicines.

In Wong's collection *undercurrent,* sediment stands for both ancestors and oil, for intergenerational memory and for the mineral treasure that incites us to ignore it. In "A Magical Dictionary from Bitumen to Sunlight," Wong offers keywords with multiple and conflicted definitions. For example,

> bitumen: buried ancestors, unearthed & burned to expand the ocean
>   : pitched sacrifice zone wherever it bubbles up, hellishly excavated[29]

Buried ancestors, marine fossils, unearthed to "expand the ocean"— meaning what? Sea level rise makes oceans larger, diminishing human habitat, the habitat of animals and plants on coastal lands. "Expand the ocean" also calls to mind, if perversely, the shortening of ocean distances, the demystification of oceans due to transoceanic trade made possible by fossil-fuel-burning engines, fiery boats loaded with oil sands crude for Asia and returning, perhaps, with petrochemical plastic toys for the children of North America. If bitumen is both the ancestor and excuse for creating more sacrifice zones, destroyed places, then ancestors are, in Wong's dictionary, both cellular memory and petrochemical plastic, fitted to a culture of speed. She writes:

> ancestors: holding my body up through cellular memory, anonymous
>   : condensed over eons into mineral wealth
>   : material in the headlights, reconfigured as a vintage car more
>   retro than we know, heavy metals millennia old[30]

Wong refuses to segregate a "good" and "bad" sediment, the heavy metals millennia old that configure a culture of speed versus the cellular memory in her own body, the body's liquid sea. Ancestors are what we burn, destroying ourselves and other life. Ancestors are what sustain, carried within us as healing medicine, adaptive knowledges. This duality means that to sediment is never a single thing, never an act of purity. To sediment is to be alive in a timely fashion, where time isn't abstract, mathematical, monetary, or linear, but always embodied, coalescing, "eddying"[31] into a story about itself.

As coda rather than conclusion, let me propose one last re-verberation for *sediment*. *To sediment* might mean to sleep wakefully. The ambition to become more our embodied selves, material and animal selves. To sleep until the eyes collect silt, borne from the body's inner sea. To sleep knowing that sleep steals time back from speed, from contemporary capitalism as a mode of extractive hyperproductivity where the need to sleep can only be seen as a problem, a theft of time from mathematics, efficiency, and money. The body, ungainly in its repose. I am paraphrasing and expanding on the art historian Jonathan Crary, who writes that "in the context of our own present, sleep can stand for the durability of the social."[32] But I am thinking of Walt Whitman, too, and of his poem "The Sleepers," in which the night figures as the *sediment* (n.) of the day, a repository of the day's dynamic time—with sleep itself a sedimenting—in which all human types come together, as equals under the sign of the body's animal need. Sleep becomes a playful metaphor of alliance across class, race, and differential access to power. Whitman writes of all persons whom his hovering persona observes in sleep:

> How solemn they look there, stretch'd and still,
> How quiet they breathe, the little children in their cradles.
> . . . .
> The wretched features of ennuyes, the white features of corpses, the
> livid faces of drunkards, the sick-gray faces of onanists . . .
> . . . .
> The night pervades them and infolds them.

Whitman recalled himself from the din of his own voice—from ego, from a prospective fame he sensed—by settling into matter, into realizations of embodiment. Elsewhere, in "As I Ebb'd with the Ocean of Life," it is sediment itself that recalls him to a sense of the human as interdependent, responsible, humble, ecological. Walking the beach swale on Long Island Sound, he observes "the rim, the sediment that stands for all the water and all the land of the globe." This sediment recalls him from "this electric self out of the pride of which I utter poems." Sediment is a silent way of speech—"chaff, straw, splinters of wood, weeds, and the sea-gluten"—a piling up of animate and inanimate knowledges before capture, before science or philosophy. *To sediment* might mean, finally, to invite ourselves, as aspiring humans, especially we settlers, more often to quietness. To listen and to practice what it means to be social with other human and nonhuman persons, to be alive beyond pride and its signal, speech.

Notes

1. *OED Online*, s.v. *sediment*, n., December 2015, http://www.oed.com/.

2. Jeffrey Jerome Cohen, introduction to *Prismatic Ecology: Ecotheory beyond Green* (Minneapolis: University of Minnesota Press, 2013), xix. See also Cohen's longer, lyrical gloss on the river as composer of geologic time in the section "Messenger," xxvi–xxviii.

3. In one of his many rich glosses on a draft of this chapter, Jeffrey J. Cohen reminds me—and us—that "in unconformities within sedimental strata James Hutton read what he called 'annals of a former world' and realized the depths of geological time." Jeffrey J. Cohen, personal communication, March 15, 2016.

4. Thanks to Lowell Duckert for generously recalling me to this reference. Rachel Carson, *The Sea around Us* (New York: Oxford, 1989), 76.

5. Jan Zalasiewicz et al., "Epochs: Disputed Start Dates for Anthropocene," *Nature*, April 22, 2015, http://dx.doi.org/10.1038/520436b.

6. Donald Worster, *Nature's Economy: A History of Ecological Ideas* (Cambridge: Cambridge University Press, 1994), 342–88.

7. Simon L. Lewis, and Mark A. Maslin, "Defining the Anthropocene," *Nature*, March 26, 2014, http://www.nature.com/nature/journal/v519/n7542/abs/nature14258.html. See also Mark A. Maslin and Simon L. Lewis, "Anthropocene: Earth System, Geological, Philosophical, and Political Paradigm Shifts," *Anthropocene Review* 2, no. 2 (2015): 108–16. Here Maslin and Lewis both explain the geologic time scale and the conventions of epochal dating in geology and argue for the value of "the more fluid and broader use of the Anthropocene concept" in the humanities and social sciences (114).

8. Barbara Freese, *Coal: A Human History* (New York: Penguin Books, 2003), 205.

9. Dipesh Chakrabarty, "The Climate of History: Four Theses," *Critical Inquiry* 35, no. 2 (Winter 2009), 197–222.

10. Michael Pollan, "The Intelligent Plant," *New Yorker,* December 23 and 30, 2013. http://www.newyorker.com/magazine/2013/12/23/the-intelligent-plant.

11. Michelle Nijhuis, "When Did the Human Epoch Begin?" *New Yorker,* March 11, 2015. http://www.newyorker.com/tech/elements/holocene-anthropo cene-human-epoch.

12. Freese, *Coal,* 111.

13. Kathryn Yusoff, "Geologic Life: Prehistory, Climate, Futures in the Anthropocene," *Environment and Planning D: Society and Space* 31 (October 2013), 779–95.

14. Lowell Duckert, "Earth's Prospects," in *Elemental Ecocriticism: Thinking with Earth, Air, Water, and Fire,* ed. Jeffrey Jerome Cohen and Lowell Duckert (Minneapolis: University of Minnesota, 2015) 239.

15. Paul Connerton, *How Modernity Forgets* (Cambridge: Cambridge University Press, 2009).

16. Mark Nowak, *Coal Mountain Elementary* (Minneapolis: Coffee House Press, 2009), 150.

17. Ibid., 118.

18. Freese, *Coal,* 180.

19. Nowak, *Coal Mountain Elementary,* 118.

20. Kyle Powys Whyte, "What Do Indigenous Knowledges Do for Indigenous Peoples?" in *Keepers of the Green World: Traditional Ecological Knowledge and Sustainability,* ed. Melissa K. Nelson and Dan Shilling (Cambridge: Cambridge University Press, forthcoming).

21. Naomi Klein, "Dancing the World into Being: A Conversation with Idle No More's Leanne Simpson," *Yes,* March 5, 2013, http://www.yesmagazine.org/ peace-justice/dancing-the-world-into-being-a-conversation-with-idle-no-more -leanne-simpson.

22. See also Duckert on the cruel rhetoric of "valley-fills" in mountain top removal in "Earth's Prospects," 251.

23. Rita Wong, "Fresh Ancient Ground," in *undercurrent* (Gibsons: Harbour, 2015), 17.

24. Wong, "too long a sacrifice," in *undercurrent,* 25.

25. This scene appears in King's novel *The Back of the Turtle,* but his argument that it is not science fiction can be found in "Thomas King's Water Treatment," *Quill and Quire,* n.d., http://www.quillandquire.com/authors/thomas -kings-water-treatment/.

26. Kyle Whyte, "Our Ancestors' Dystopia Now: Indigenous Conservation and the Anthropocene," in Ursula K. Heise, Jon Christensen, and Michelle Niemann, *The Routledge Companion to Environmental Humanities* (New York: Routledge, 2017).

27. Wong, "Fresh Ancient Ground," 17.

28. "Key Facts," Coal Train Facts, http://www.coaltrainfacts.org/key-facts.

29. Wong, "A Magical Dictionary from Bitumen to Sunlight," in *undercurrent*, 29.

30. Ibid., 28.

31. This apt verb is Jeffrey J. Cohen's, and a keen image for thinking of the contiguities of past and present—a circling back or remixing of time as opposed to linear trajectories from past to present—visually present in sedimentary formations. Personal communication, March 15, 2016.

32. Jonathan Crary, *24/7: Late Capitalism and the Ends of Sleep* (New York: Verso, 2013).

Environ

VIN NARDIZZI

At the end of the twentieth century, the consensus would seem to have been that the term and concept *environment* were no longer (and may never have been) critically productive. According to Wendell Berry in "Conservation Is Good Work," "The idea that we live in something called 'the environment' . . . is utterly preposterous." The prime position accorded to this noun in "the language we are using to talk about our connection to the world" signals, for Berry, both the anthropocentrism and the "inadequacy" of twentieth-century ecodiscourse. Berry aims to correct for this paucity in language by itemizing concrete nouns that have been abstracted into "the environment": "The real names of the environment are the names of rivers and river valleys; creeks, ridges, and mountains; towns and cities; lakes, woodlands, lanes, roads, creatures, and people."[1] Around the same time, Michel Serres also described in *The Natural Contract* a conceptual shortcoming of the noun *environment* as he observed it to be "commonly used." As had Berry, Serres designated anthropocentrism as the problem and also proposed that an ecodiscourse would do better without the term: we need to "forget the word *environment* in this context" and correct for its hubristic and projective distortions by reconfiguring relations on Earth so that "things [are] in the center and us at the periphery, or better still, things all around and us within them like parasites."[2]

But the noun *environment* has had tremendous sticking power in ecocritical circles.[3] Lawrence Buell, for instance, calls his important manifesto for the field *The Future of Environmental Criticism* because, for him, "'environmental' approximates better than 'eco' the hybridity of the subject at issue—all 'environments' in practice involving fusions of 'natural' and 'constructed' elements."⁴ Even though, more than ten years after Buell predicted the future for environmental criticism, *ecology* as an organizing principle would now seem to have more theoretical purchase and marketability in the broad field of ecocriticism, *environment* and *environmental* persist as powerful keywords and conceptual anchors among the field's practitioners. Numerous contributors to this volume belong to ASLE (the Association for the Study of Literature and Environment), and many of us would say that we conduct research in the environmental humanities and teach environmental art in our classrooms.

Our editors, too, wittily embrace—rather than jettison—environment. Indeed, inspired by *The Natural Contract,* where, in Jeffrey Jerome Cohen and Lowell Duckert's words, Serres critiques the anthropocentrism of "merely environing," our editors nonetheless encourage us to literalize— so as not to forget—the etymology of *environment.* They observe that the collection's title plays upon the root verb within *environment*: French *virer* means "to turn." In "tak[ing] the ecological turn quite literally" and, in so doing, "acknowledge[ing] a world full of inhuman forces, dynamic matter, and story-filled life that inevitably go off course,"⁵ Cohen and Duckert imagine *Veer Ecology* as a suite of ecoverbs for the future that theorizes the spirals, vortices, and whirls that organized matter and metaphor (*matterphor,* or the "tropic-material coil, word and substance together transported: of language but not reducible to linguistic terms, agentic and thick") in *Elemental Ecocriticism.*⁶ The ecoverb at the heart of this chapter, *environ,* is a close relative of the verb animating this collection's veerings, but I would be hard-pressed to detail and endorse unequivocally its future utility for ecocriticism, as other contributors in this volume do for their verbs. Instead, I aim to elaborate the history of an *ecomatterphor* embedded in *environment* and crystallized in *environ* that articulates an anthropocentrism unlike the kind that Berry and Serres would have us forget. In this story, anthropocentrism proves oppressive, overwhelming, and even deadly, since acts of environing were not always (and in some cases are still not always) desirable or pleasant for those who are so surrounded and thus, in Tim Ingold's terms, formed as human figures.⁷ The historical meanings of the ecoverb *environ,* then, may help account for why it is that colleagues might desire environmental criticism to veer in the first place.

In an effort to bring the story of this ecoverb to light, I focus on the moment in premodern literary culture just before the noun *environment* entered print circulation in England. In pursuing this enviro-philology, I attend to what came before *environment* and so what actions and other grammatical positions helped to create it. When we unseat this noun from its pride of place in ecodiscourse, we glimpse, in its absence, the brutal and terrifying movements—acts of siege, both physical and spiritual—vehiculated through the verb *environ* in popular texts composed in English before the early seventeenth century. We also see that *environ* was commonly and surprisingly used as an adverb, noun, and preposition before *environment* first appeared in print. The grammatical multifunctionality of *environ* thus prompts me, at chapter's end, to veer somewhat suddenly and unpredictably off course toward a compound adverb that characterizes greenwashed ecodiscourses in the twentieth- and twenty-first centuries—*ecofriendly.* Such veering is a methodologically queer motion, for, as Laurie Shannon observes, "whatever is queer stands (or moves) at an angle to something else; it encodes a relationality that, as such, precipitates a perspectival dilemma between two moving points."[8] It is a movement that is "utterly preposterous" but in a queerer way than Berry likely imagined.[9]

Environ

The noun *environment* first appeared in print in England in 1603. According to the online *Oxford English Dictionary (OED)*, it debuts in Philemon Holland's translation of Plutarch's *Morals* in a section called "Natvrall Qvestions" and specifically in response to the query, *Why doth the Polyp change his colour?*[10] I have begun to tell the story of this neologism and its tantalizing emergence in Holland's translation (and of its just-as-sudden disappearance from the print record) elsewhere.[11] Searches of the *Middle English Dictionary (MED)* and the *OED* reveal that, prior to Holland's introduction of this noun form, writers employed the word *environ* as an adverb, a noun, a preposition, and a verb in a range of texts and genres. In my accounting of the *OED* database entries, I observed this information for the period from 1350 to 1670:

1. There are four witnesses of *environ* as a preposition before its use as this part of speech vanishes in 1500.

2. There are eight witnesses of *environ* as an adverb before this grammatical function becomes obsolete in 1600.

3. There are fifteen witnesses of *environ, environs, environment, in viron,* and *viron* as nouns, most of which occur in prepositional phrases that serve as adverbial markers of place and are also likely translations of the Latin noun *circuitus (circuit)* and only two of which—*environs,* which John Evelyn used twice after 1640 to name areas beset by military incursion and the plague—might be more recognizable to contemporary readers.[12]

4. There are fifty-eight witnesses of *environ* and *viron* employed as verbs.

The *MED* generally corroborates this accounting of grammatical usage for the late-fourteenth and fifteenth centuries, although it does not record examples of *environ* and its variants as nouns; instead, it categorizes a prepositional phrase *(in viroun)* as an adverb. On the basis of such quantitative evidence about *environ,* we could hypothesize, as the Renaissance ecocritic Ken Hiltner does in the context of the vogue for seventeenth-century country-house poetry in England, a relation between urban expansion around London and the conceptualization of the countryside as an "environment."[13] We can also see that, before such developments (in both senses of the word), the verb *environ* predominated in English literary culture.

What sorts of entities were environed and what or who did the environing in the English literary imagination before Philemon Holland apparently introduced *environment* to the print record? In preparing answers to these questions, I decided that I would not rehearse here in an exhaustive way the witnesses gathered in the *OED,* although I carefully reviewed its entry for the verb *environ.* In the spirit of veer ecology, I instead turned to the texts that were immediately around me: the writings of Christopher Marlowe, William Shakespeare, Sir Philip Sidney, and Mary Sidney, countess of Pembroke, all of which I had just been teaching in my courses during the fall 2015 semester. Surprisingly, especially with regard to Shakespeare, the *OED* does not enlist these authors to exemplify meanings for the verb *environ.* Even so, as my more informal survey demonstrates, these literary figures employed *environ* in ways that are fully consonant with the writers who are catalogued in the *OED.* Collectively, they offer a snapshot of environing during the late 1580s and

1590s as that action appears in popular drama and in elite manuscript culture, both of which registers of literary production were highly influential in English letters. In this sampling of texts, *environ* and its verbal variants track with interlinked discourses of physical protection, militarism, and spiritual peril. Thus associated with *environ,* such matters are literally embedded in—and perhaps haunt—more contemporary discourses of environment and environmentalism.

In Marlowe's *Tamburlaine the Great,* part 1, men encircle other men in acts of environing during peacetime and in war. After having successfully defeated all the foes he will confront in part 1, Tamburlaine, who has otherwise been an anarchic social force in the play, establishes civil order. He commands loyal supporters: "Cast off your armour, put on scarlet robes, / Mount up your royal palaces of estate, / Environèd with troops of noble men, / And there make laws to rule your provinces" (5.1.524–27).[14] Although Tamburlaine imagines as framed by male courtiers these newly installed petty rulers, who have exchanged their metalwear for softer fabrics, they nonetheless remain environed—surrounded and, if need be in a time of political transition, defended—by "troops." Two earlier uses of *environ* in the play more explicitly blend matters of security and militarism. In advance of his encounter with Tamburlaine at Damascus, the sultan of Egypt flatters his regiment: "Methinks we march as Meleager did, / Environèd with brave Argolian knights, / To chase the savage Calydonian boar" (4.3.1–3). The shield of "brave . . . knights" to which the sultan likens his men, charmed by the aura of classical allusion, fails to protect against Tamburlaine's army, which the sultan terms "a monster of five hundred thousand heads, / Compact of rapine, piracy, and spoil" (4.3.7–8). The sultan here seems to recall with some measure of accuracy the report of a messenger, who, in an earlier scene, informs him that Tamburlaine leads "three hundred thousand men" on horseback and "five hundred thousand footmen threat'ning shot, / Shaking their swords, their spears, and iron bills, / Environing their standard round, that stood / As bristle-pointed as a thorny wood" (4.1.21, 24–27). The weaponized circle gathered around Tamburlaine's war banner (and possibly the man bearing it) is as impenetrable and harmful—*atypical ecowords in contemporary environmental discourse*—as is a dense, unwelcoming wilderness, and the sultan's own wall of environing soldiers will prove no match for it.

Marlowe's *Jew of Malta* further links the act of environing with military might, but here the verb draws a circle around an island. Indeed, the play employs the verb *environ* to describe a military tactic, the siege. The vice-admiral of Spain, Martin del Bosco, recalls how the "hideous force" of the Turkish army once "environed Rhodes" and so beset the "Christian isle" that "not a man survived / To bring the hapless news to Christendom" (2.2.31, 48, 50–51). Later in the play, having "viewed the city, seen the sack, / And caused the ruins to be new repaired," Calymath, the Turkish leader, expresses disbelief that his army accomplished a similar feat in Malta (5.3.1–2), particularly in light of "the situation / And how secure this conquered island stands" (5.3.5–6). Malta is, as Calymath observes after having "walk[ed] about" it (5.2.18),

> Environed with the Mediterranean Sea,
> Strong countermured with other petty isles,
> And, toward Calabria, backed by Sicily
> (Where Syracusian Dionysius reigned),
> Two lofty turrets that command the town–
> I wonder how it could be conquered thus. (5.3.7–12)

In this reckoning, architectural ("lofty turrets") and topographical ("the Mediterranean Sea") fixtures exist to safeguard Malta from attack. They also conjoin forces, at least metaphorically, to repel invaders: the "other petty isles" function as if they were Malta's outer walls and first line of defense. One act of militaristic environing in *The Jew of Malta*—the Turkish "sack"—thus overwhelms the insulation afforded by other forms of protective environing—the sea, neighboring islands, and battlements. No amount or kind of security in Marlowe's Malta can fend off the incursion of an environing "military formation."[15]

1 Henry VI is a Shakespearean collaboration that bears the impress of Marlowe's dramatic style.[16] It is also stages and reports on multiple military sieges. But the play does not explicitly designate these tactics as acts of environing. Instead, it employs a series of synonyms for the verb *environ* in its martial sense: the legendary English soldier Talbot is "round incompassed and set upon" by a huge company of French soldiers (1.1.114); the English troops are also reported to be "enclosed . . . with their enemies"

(1.1.136); and Talbot is later imagined as "girdled with a waste of iron / And hemmed about with grim destruction" (4.3.20–21). In its final act, when the play invokes the verb that shadows these enwrapping terms *(encompass, enclose, girdle,* and *hem), environ* veers off in a new direction. Joan Puzel, the play's caricature of Joan of Arc, after having been captured by the English and before she is to be escorted offstage, where she will be burned at the stake for witchcraft, bestows upon her enemies a "curse":

> May never glorious sun reflex his beams
> Upon the country where you make abode,
> But darkness and the gloomy shade of death
> Environ you, till mischief and despair
> Drive you to break your necks, or hang yourselves. (5.3.86–91)

Joan, whom Talbot had earlier called the "Foul fiend of France and hag of all despite, / Encompassed with . . . lustful paramours" (3.2.51–52), confirms in the articulation of this curse Talbot's chauvinism as she turns its rhetoric against all of England (his "country"). In this speech act, *environ* switches registers, as its associations with a menacing militarism (the "hideous force" that "environed Rhodes" in *The Jew of Malta*) transform into a solar nightmare that feels as if it were inspired by the dark moods of Senecan revenge tragedy.[17]

Such imagery seems to have appealed to Shakespeare and his collaborators in the early 1590s because both *Richard III* and *Titus Andronicus* feature set pieces inflected by it. In the former play, Clarence recalls a terrifying dream in which "a legion of foul fiends / Environed" him "and howled in [his] ears / Such hideous cries, that with the very noise" he was jolted awake, unable to determine if he "was in hell" (1.4.58–60, 62),[18] while in the latter play the eponymous character, after seeing his daughter Lavinia's mutilated body, exclaims,

> For now I stand as one upon a rock
> Environed with a wilderness of sea,
> Who marks the waxing tide grow wave by wave,
> Expecting ever when some envious surge
> Will in his brinish bowels swallow him. (3.1.93–97)[19]

The sea does not overwhelm Titus, since it is only invoked in a simile that he uses to describe his emotional state, but Clarence's dream does presage his fatal immersion in a vat of booze, or a "malmsey butt" (1.4.269). The three early plays in the Shakespeare canon, then, link the verb *environ* with sensory deprivation (the threatened darkening of all light and life in *1 Henry VI*) and with stimulus overload ("ears" crammed with "hideous cries" in *Richard III* and a body so imaginatively inundated by water in *Titus Andronicus* that the sea itself turns into a gigantic alimentary system that "swallow[s]" Titus up). Being environed in these plays is a harrowing affective and physical state in which death and its otherworldly agents press upon—and sometimes into—the human figure.

Both the spiritual peril encoded in Joan's curse and in Clarence's nightmare and the faith-based wars depicted in Marlowe's plays suggest that vernacular religion in England may have been as foundational in elaborating sixteenth-century meanings for *environ* as was imagery associated with the classical underworld or period military tactics. In the eleven different definitions provided for the verb *environ* in the *OED,* religious texts, including a mid-fourteenth-century prose psalter and the Wycliffite Bible, are cited in six definitions as witnesses to earliest usage. The *OED* also marks nearly all six of these witnesses as translations of two Latin verbs, *circuire (to ensnare* and *to orbit)* and *circumdare (to surround* and *to enclose).* Composed by Sir Philip Sidney and completed, after his death, by his sister, Mary Sidney, countess of Pembroke, the Psalms afford a glimpse into how ardent Protestants updated spiritual environing in the latter quarter of the sixteenth century. In metrical translations that circulated in manuscript and that shaped the religious poetry of John Donne and George Herbert,[20] the Sidneys employ the verb *environ* in moments during which standard English translations of the Psalms (Geneva Bible and Authorized Version) tend to use *compass.*[21] For instance, in Psalm 27 *(Dominus illuminatio)* Sidney's speaker reports that "wicked folk" and "my foes, to utmost of their pow'r / With raging jaws environ me, / My very flesh for to devour," and yet "they stumble so, / That down they go" (lines 7–12). In Psalm 118 *(Confitemini Domino)* the speaker elaborates on this scenario. He recounts how, on three occasions, "enemies all sorts that be, / On every part environed me" (lines 29–30). After the third attack, these "enemies" "swarming fast like bees they flew" (line 38). The speaker ultimately proves resilient in Psalm 118, but "enemies" also populate other psalms, suggesting that they (or their kin) are still out there, environing, regrouping, and swarming.

There is one moment in the Sidney Psalms when men will environ God, thus reversing the more typical pattern in which wickedness and evil surround a beleaguered speaker. In Psalm 7 *(Domine, Deus meus)*, the speaker wishes for a time when God "in wrath . . . up set / Against such rage of foes" who torment him (lines 16–17); after God executes "high doom" on them, the speaker imagines this universal congratulation: "So shall all men with lauds environ thee" (lines 18–19). On the basis of the phrase's future tense and the poem's general pleading tone, we can infer that such celebratory environing marks the culmination of an unfulfilled apocalyptic fantasy.

Unfriendly Environings

Incomplete though it is, the enviro-philology that I outline confirms the sense that the history of environing is fully and belligerently anthropocentric, but it approaches this conclusion from a different angle than do Berry and Serres, for whom anthropocentrism signals humanity's supreme pride and misunderstanding in the world of things. From Marlowe to the Sidneys, single human figures (the sultan in *Tamburlaine* and the speaker of the Psalms) and entire human settlements (Rhodes and Malta) are located at the exact center of constrictive or insufficiently protective acts of environing. In these texts, difference, which is mainly articulated in religious terms, besets the figure who (or which) has been fully environed. Environings thus organize relations that put the health of individuals and populations at risk and endanger life. I would venture that this is not the most common definition for what environments (can) do in our contemporary moment.

In making this point about a particular set of perils associated with anthropocentrism in the sixteenth century, I claim that, then as now, though in different registers, acts of environing (and so the creation of environs and environments) can persist as hazardous enterprises for the human figures so environed. For more contemporary contexts, Stacy Alaimo, Mel Y. Chen, and Rob Nixon have explored the idea that an environment can negatively affect people's well being just as much (and no doubt because) more privileged groups of humans have behaved, historically, in ways that are *environmentally unfriendly*.[22] An *environment,*

which term Serres would have us "forget" for its anthropocentric chauvinism and which we are at political, personal, and even theological pains
not to despoil any further, is an ambience that can also make us feel stifled
and overwhelmed. We might know this act of environing, if we experience its effects at all, as allergy, sensitivity, and environmental illness.

Environmentally unfriendly: these adverbs propel this chapter's conclusion off course. In ecodiscourse, adverbs can modify verbs, turning
their meaning toward this or that way. According to a proverb recorded
by the early seventeenth-century dramatist John Ford, they also prove
the measure by which God weighs human actions ("This man not only
liues, but liues well, remembring alwayes the old adage; that God is the
rewarder of Aduerbes not of Nownes").[23] This adverbial pair aptly captures the multiple and overlapping unkindnesses that inhere in the acts of
environing that I've been describing: the harm done by humans to their
immediate and more far-flung surroundings, the danger humans pose
to others (humans and nonhumans) through acts of environing, and the
lethal threat posed by environments to their human and nonhuman inhabitants. Historically in the West, in contexts sometimes more officially
secular than Ford's England, we have not lived well. But we try.

These conjoined adverbs thus also press against the popular logic
of living well or in a manner that proves ecofriendly. According to a
handy online definition, "Eco-friendly literally means earth-friendly or
not harmful to the environment. This term most commonly refers
to products that contribute to green living or practices that help conserve resources like water and energy. Eco-friendly products also prevent contributions to air, water and land pollution. You can engage in
eco-friendly habits or practices by being more conscious of how you use
your resources."[24] We can determine how ecofriendly an item's manufacture or use is by reading the information included on a product's eco-
label, but such labeling can be—and has been—coopted by companies
that greenwash. Most cynically, we could regard endeavors to live eco-
friendly as (poor) compensation for the damage that human beings,
especially those of us settled in the West, have inflicted on the planet; in
making these consumer "decisions," of course, we also pay a duty for
assuaging guilt and for wanting not to make matters any worse. Being
ecofriendly thus allows us to buy comfort for our collective and personal

and enduring entanglement with anthropogenic ecological change on a global scale.

In this way, the marketing of the ecofriendly (an adverb turned into a noun) may well be yet another late twentieth- and twenty-first-century symptom of the anthropocentrism that has come to epitomize the Anthropocene, which arguably has its origins in the "ecological globalization" founded by European colonialism, militarism, and overseas trade during the early seventeenth century—that is, to say, around the moment that the literary environings I've explored were imagined.[25] Those of us who try to be ecofriendly, among whom I am most certainly included, thus take individual and group measures—any measure that we can take within our current economic framework—to *clean up, conserve, preserve, save,* and *sustain* environmental health and planetary biodiversity. Tellingly, readers of *Veer Ecology* will find that contributors have turned in no entry for any of these ecoverbs for the heterofuture. I offer *environ* as an ecoverb of the past and, in some measure and for some humans and nonhumans alike, of the present. I do so because environmental criticism would do well to remember the long and devastating history of one of its key terms in an effort to imagine more amicable futures.

Notes

I completed this research under the generous auspices of the Peter Wall Institute for Advanced Studies at the University of British Columbia. I thank my editors as well as my readers (Ignacio Adriasola, Phanuel Antwi, Kyle Frackman, Gregory Mackie, and Mo Pareles) for their feedback and critique.

1. Wendell Berry, *Sex, Economy, Freedom, and Community* (New York: Pantheon Books, 1993), 34, 33, 35.

2. Michel Serres, *The Natural Contract,* trans. Elizabeth MacArthur and William Paulson (Ann Arbor: University of Michigan Press, 2003), 33.

3. Even so, more recent ecological theory persists in highlighting the inadequacy of *environment* as an umbrella term. See Timothy Morton, *Hyperobjects: Philosophy and Ecology after the End of the World* (Minneapolis: University of Minnesota Press, 2013), 128–29. Morton's assessment of how the term and "the idea of nature [are] getting in the way of properly ecological forms of culture, philosophy, politics, and art" also informs my thinking. Timothy Morton, *Ecology without Nature: Rethinking Environmental Aesthetics* (Cambridge, Mass.: Harvard University Press, 2007), 1.

4. Lawrence Buell, *The Future of Environmental Criticism: Environmental Crisis and Literary Imagination* (Malden, Mass.: Blackwell, 2005), viii.

5. Jeffrey Jerome Cohen and Lowell Duckert, "Introduction: Welcome to the Whirled," in this volume.

6. Jeffrey Jerome Cohen and Lowell Duckert, ed., *Elemental Ecocriticism: Thinking with Earth, Air, Water, and Fire* (Minneapolis: University of Minnesota Press, 2015), 11.

7. Tim Ingold, "Whirl," in this volume.

8. Laurie Shannon, "Lear's Queer Cosmos," in *Shakesqueer: A Queer Companion to the Complete Works of William Shakespeare,* ed. Madhavi Menon (Durham: Duke University Press, 2011), 176.

9. On the availability of the preposterous to queer reading practices, see Jonathan Goldberg, *Sodometries: Renaissance Texts, Modern Sexualities* (Stanford: Stanford University Press), 180–81; Patricia Parker, "Preposterous Events," *Shakespeare Quarterly* 43, no. 2 (1992): 186–213.

10. I consulted the online *Oxford English Dictionary*'s first definition for *environment* on December 28, 2015. The word appears in Plutarch, *The Philosophie, Commonly Called, the Morals,* trans. Philemon Holland (London, 1603), 1009, emphasis in original.

11. Vin Nardizzi, "Remembering Premodern Environs," in *Object Oriented Environs,* ed. Jeffrey Jerome Cohen and Julian Yates (New York: Punctum Books, 2016), 179–83. Outside the scope of this chapter is a full elaboration of the transition from *environment* as a hostile force encircling human beings to its status as an imperilled thing that humans must save for themselves because they have been hostile to it.

12. For more details about these uses, see *The Diary of John Evelyn,* vol. 2, ed. E. S. De Beer (Oxford: Oxford University Press, 1955), 38; *Diary and Correspondence of John Evelyn,* vol. 1, ed. William Bray (London: George Bell and Sons, 1889), 419.

13. Ken Hiltner, *What Else Is Pastoral? Renaissance Literature and the Environment* (Ithaca: Cornell University Press, 2011), 49–66.

14. I cite all references to Marlowe's play from *Christopher Marlowe: The Complete Plays,* ed. Frank Romany and Robert Lindsey (London: Penguin Books, 2003). I note all references to them parenthetically.

15. Throughout my discussion of Marlowe, I am indebted to Patricia A. Cahill, *Martial Formations, Historical Trauma, and the Early Modern Stage* (Oxford: Oxford University Press, 2008), 24–70, esp. 49–52, where she examines moments in the *Tamburlaine* plays in which military leaders detail step-by-step procedures for enacting a siege.

16. On the play's collaborative authorship, see the introduction to William Shakespeare, *King Henry VI, Part 1,* ed. Edward Burns (London: Bloomsbury Arden Shakespeare, 2014), 73–84. I note all citations to this edition parenthetically. Burns's footnotes alert readers to moments in the play that echo or share subject matter with Marlowe's dramas.

17. I have in mind Thomas Kyd, *The Spanish Tragedy,* ed. J. R. Mulryne (London: New Mermaids, 1990). The frame of this popular drama features the descent of a murdered soul (Don Andrea) into the classical underworld where he meets Revenge, the allegorical figure who will preside over the play's multiple and confusing retribution narratives.

18. I note parenthetically references to William Shakespeare, *King Richard III,* ed. James R. Siemon (London: Arden Shakespeare, 2009).

19. I note parenthetically references to *Titus Andronicus* from *The Norton Shakespeare: Early Plays and Poems,* ed. Stephen Greenblatt (New York: W. W. Norton, 2016).

20. On the availability and influence of the manuscripts containing the Sidney Psalms, see the introduction to *The Sidney Psalter: The Psalms of Sir Philip and Mary Sidney,* ed. Hannibal Hamlin et al. (Oxford: Oxford University Press, 2009), xv–xvi. I note all references to these poems parenthetically.

21. In tracking this pattern, I consulted the online Geneva Bible (www.gene vabible.org) and the King James Bible (London, 1611).

22. Stacy Alaimo, "MSC Matters: Material Agency in the Science and Practices of Environmental Illness," *TOPIA: Canadian Journal of Cultural Studies* 21 (2009): 9–27; Mel Y. Chen, *Animacies: Biopolitics, Racial Mattering, and Queer Affect* (Durham: Duke University Press, 2012), 189–221; Rob Nixon, *Slow Violence and the Environmentalism of the Poor* (Cambridge, Mass.: Harvard University Press, 2011), 1–44.

23. John Ford, *A Line of Life* (London, 1620), 92.

24. "What Does Eco-Friendly Mean?" Home Guides, http://homeguides. sfgate.com/ecofriendly-mean-78718.html.

25. The phrase is from Steve Mentz, "Enter Anthropocene, c. 1610," http:// arcade.stanford.edu/blogs/enter-anthropocene-c1610. Mentz adds literary texture to the findings of Simon L. Lewis and Mark A. Maslin, "Defining the Anthropocene," *Nature* (March 12, 2015): 171–80. These findings, of course, are not without controversy, and it is reported that the Working Group on the Anthropocene (WGA) will date the epoch's start in the mid-twentieth century.

Shade

BRIAN THILL

The rich tradition of linking ecological thought to spectacle—to the practices of bearing witness, on film and through documentary photography, to visible instances of environmental degradation and destruction—is no longer sufficient for confronting the existential threats posed by contemporary ecological crises. Carbon levels, species extinction, the collapsing ice shelf, rising seas, and other indicators of humankind's impacts on the environment exceed our capacity to witness and document the true scope of the damage directly. Because it grants special weight to ecological spectacle and tableau that evoke strong feelings in us, the ecological image can only offer us an extremely limited portrait of what is taking place every day. As a result, the privileging of the visual register of our ecological impacts has distorted our ecological thinking and our collective action, directing our attention (if not always our political energies) in many cases toward those displays of ecological catastrophe that are much more camera friendly or camera ready than others: photogenic polar bears rather than insect colony collapse or extinct worms; consumption's grimy residues rather than our relatively clean and slick sites of consumption; smog, calving ice, and trash heaps rather than cold data on temperature spikes. Against the tyranny of the spectacle, ecological thought might benefit from a blinkered ecology, one structured on the nonvisual, the invisible, the unfilmable and the unfilmed—on those equally real catastrophes that are clouded and obscured from view, if not from planetary experience. In order to do this we must shade our ecological thinking, blotting out some of the most familiar and well-lit spots so that we can work with what's left in the darkness. By throwing a veil over our praxis in this way, we can give attention to those elements that are not spotlighted, those objects and practices and modes of ecological thought that all too often only appear as phantoms to us, when they appear at all. What this process entails is less a turning away from the traditional forms of ecological spectacle than an active process of obscuration. In other words, under current visual regimes, we may not find ourselves in a position to combat them directly with an alternative ecological spectacle (nor would we want or need to). Therefore, obscuration of existing spectacles serves as a better method of *curating* the global visual archive as it is imposed on us and as it besieges us with its familiar and outmoded appeals to empathy by way of the visual sense. This entails a deliberate and systematic process of shielding ourselves from the image, since the image itself is not (or is no longer) politically efficacious.

"On behalf of the land and everything living on it," Lucy Lippard argues in the closing page of *Undermining*, "new image wars must be waged." Photography, she argues, plays a major role in communicating land-use issues, and conveys firsthand experience like no other medium.[1] But then she goes on to note that ecological insight can be difficult to communicate and that dramatic landscape imagery, placed often in the service of larger ecological imperatives, can, despite its good intentions, be rendered banal and apolitical—that it can fail, as the great nature photographer Robert Adams had recognized time and again. So if there are new image wars to be waged in the battle for our planet, let us choose pacifism. The argument I offer is an argument against the ecological spectacle and its politics of visuality, and it is at the same time a call for the resurrection of mere language, data, or introspection to assume correspondingly larger roles in the battle against planetary devastation. Turn the cameras away from the drowning polar bear, dim the lights we consistently shine on the enormous vistas of waste we funnel into our air, our earth, and our water; and you are still left with those planetary crises and more. What has changed is that we no longer have the affective crutch of *the image* of suffering and decimation.

We should also abandon this method of ecological thinking and documentation because, like many other large-scale human-made crises and atrocities, the devastation of our planet is by its very nature not amenable to being meaningfully captured in images. Any effort to comprehend the totality of nature, which is part of the ambition of any ecology worthy of the name, must necessarily fall short, for there are deep pockets of darkness that no theory and no ecological spectacle can adequately illuminate. The future depends on reading a totality that is on some level illegible; confronted with it, we are plagued by what Tim Morton in *The Ecological Thought* describes as "dark thoughts," wherein the book of nature is not unified or unifiable and whose contours refuse to provide us with a linear or syntactically clear image of itself. We will not only need to take responsibility for what we cannot see, as Morton notes;[2] we will also need to pursue the goal of not seeing it in order to fulfill that responsibility. By definition, large-scale environmental transformations taking place over vast distances and long periods of time are resistant to visualization and spectacle. Our images, powerful and immediate as they are, capture the tail end of long historical processes or provide a literal snapshot of a moment in that immense parade of destruction and annihilation; and the time has come to think the whole parade instead. At the same time that we move further into the era of total surveillance, outfitting our atmosphere, streets, and bodies with ever more sophisticated and pervasive equipment for image capture, we also move further into the era where many of the most significant global crises are to a large degree not surveillable. Millions of handheld cameras, millions of surveillance technologies (cameras, recorders, drones, police body cams), the relentless visual mapping of everyday life, and yet the long inexorable processes of ecological plunder and exploitation continue apace.

Furthermore, to the extent that something as catastrophic and immense as global warming can be captured or communicated in some fashion through visual images, those images are not only grossly insufficient but also, I would argue, unnecessary and perhaps even counter-productive at this stage in history. The opposition to the struggle for our ecological future has become a master manipulator of images too: the climate deniers, of course, but also the ad men, the market hucksters, those who sell visions of nature to us in predictable and banal ways. To throw a veil over the whole ecospectacular project would be to short-circuit at least some of these oppositional antiecologies. "Beauty" as a feature of the imperiled natural world is a market notion, so we should have done with it and cast it into the shadows. Save the mountaintops from removal not because mountains are majestic but because we have the sense to know that they should not be decapitated and obliterated. Don't save the eco-systems because the animals there are adorable or magnificent; save them because they simply need them to survive, and for us to survive in turn. The saturated visual field bleaches the brightly lit worlds captured in its images. When even progressive political gestures have become part of the omnipresent visual field, it is clear that the deaths of rare birds or the collapsing of a majestic glacier are part of the saturated market of images competing for our attention. They have in large part been reduced to an indistinguishable facet of the entire visual field to which we are routinely subjected, endeavoring nobly and often unsuccessfully to reach us in the way they seem to want to reach us, if only they could.

At the same time, the technological innovations that have allowed for the efflorescence of these visual archives of ecological spectacle—new and cheaper and higher-quality modes of film or digital production, global communications networks on which to broadcast those spectacles, the elaborate system of intercontinental travel and transportation, the ornate global webs of financing that underwrite these noble projects—are, like everything else we do, themselves contributors to the very calamities they so lovingly document in their well-meaning fashion, just as a university press book about rethinking ecology is printed on harvested trees, with inks from distant sources, shipped out to hungry readers thousands of miles away. This is not to issue a wholesale indictment of the ecospectacle industry for its hypocrisy or any such thing, since part of the modern condition is in fact to have been born into and participated actively in this complicity and hypocrisy (we're all on the hook for that), but it is to suggest that the nature of nature photography has begun to ossify, its powerful and important images desiccated by the endlessly streaming spectacle-culture of which it is a part. Without meaning to, ecological spectacle finds itself in the unenviable position of documenting crises that those tragic documents will do little to address.

~

Do something! we think the image cries; but in truth the most it usually says to us is, *Feel something, here and now, as you gaze on this sad tableau.* As the vast hordes of image makers and manipulators have always known, using images to make someone feel something is reliable and easy. They also know it is ephemeral. And the fate of the earth and everything on it is far too much for ephemerality to handle. For all their virtues, images have a hard time doing sustainable ideological or political work. Even the most powerful and resonant image has a very difficult time embodying the idea, value, or belief that is required to sustain political action and direct engagement, particularly in the service of deep ecology. What is distinctive about these particular kinds of spectacle is that, unlike spectacles of direct or slow motion violence against other humans (torture, war, refugee crises), the self that watches these environmentally conscious spectacles can find in them no creature that can see itself through the eyes of others. The imperiled animals within them are clearly not ignorant of our direct and indirect presence, and not even quite indifferent, but perhaps at least *indisposed*: we are here to look on what we are doing to them and their habitats and their lifeways, but they can only go on trying to endure while our film rolls; they can offer no other testimony or interest beyond their own existential imperative. At this stage, how much more of this footage could we possibly need? "What I see not, I better see," Emily Dickinson wrote, manifestly *not* watching a bird eat a worm this time, and giving us a glimpse of a darker, sharper ecology instead: "For frequent, all my / sense obscured / I equally behold / As someone held a light / unto / The Features so beloved." Dickinson's formulation offers us the precise tableau to describe the shading process. It is not that we look away from the object as much as it is that, aware of its location and presence as possible spectacle for our visual consumption, we choose instead to seize on the mental image instead, the residuum of our visual blindness. The mind that can grasp or sense what we allow our visual sense to produce in us can function just as effectively, if not more effectively, once the limits of our crude vision are willfully darkened. Choose not to look, she suggests in at least two senses: because we do not need to look in order to apprehend, and because the things we are endeavoring to understand could not be manifest to our mere vision to begin with.

The privileged and familiar spaces for ecological spectacle are the shore, the deep sea, the tundra, the steppes, the desert, the forest, the jungle; all the places where humankind's catastrophic footprint is *felt* more than seen, as a ghostly shade or revenant among the natural spaces where all other species endeavor to survive in our wake. Mostly we venture out beyond our cities, to the relatively untrammeled hinterlands where a vast array of plants, animals, and geological features wage battle against their erasure. "Nature" in well-meaning ecological spectacles is presented as the alternative space to humanity and modernity, one intimately and symbiotically connected to our lives, of course, but one that posits a vaguely Archimedean distance from which to contemplate the inferno we've created. Neil Smith was among those who recognized that nature (or the putatively nonhuman world), far from being the antithesis to human productive activity, was in fact an integral part of that activity. As Marx reminds us, Smith notes, "animals and plants, which we are accustomed to consider as products of nature, are in their present form, not only products of, say last year's labour, but the result of a gradual transformation, continued through many generations, under man's superintendence, and by means of his labour. . . . In the great majority of cases, instruments of labour show even to the most superficial observer, traces of the labour of past ages."[3] In *The German Ideology*, Marx spoke in related terms, noting that the "nature" that preceded human history effectively no longer exists anywhere, save for those rare nooks and crannies where human productive activity has not yet gained access—a small part of the globe in his time, to be sure; and a far smaller part now.

So what we are looking at when we gaze on photographs and films of dying animals and beleaguered ecosystems and shrinking glaciers is, of course, ourselves. In this sense, the planet is (to draw on Marx once more) a prolongation of our body, not in some quasi-mystical sense but in a starkly materialist one. The living things and the landscapes alike appear here not in and for themselves, for us to "see" and comprehend, but instead as dark reflections, as discrete images that document moments in the long historical process of our transformation of "nature" into a second nature that spans the globe, now including the tiniest creature and the remotest reef and the farthest reaches of our thin atmosphere. Meanwhile, ecological imagery continues to establish its familiar and reductive oppositions, playing off the tensions between natural beauty and human-made destruction, between epic grandeur and human terror; in short, between imperiled "nature" and destructive "man." The visually excluded middle in this formulation, though, is always the actual transformative activities of humankind on the natural world. We do not directly see our individual car's exhaust or our own home's air conditioning unit melt a specific glacier forever. At best we can, with no small degree of effort, *comprehend* the complicated but real causal relationship between those two things (or refuse to, as the world burns), at which point we all too often look to the spectacle to gaze at the *effect zones* for appropriately damning footage that catches a glimpse of some tiny facet of this long and elaborate interrelationship. But the spatial and temporal boundedness of the image cannot truly accommodate this complex form of planetary causality, because it is causality without either proximity or immediacy; and it is only the proximate and the immediate that an image can truly capture. Therefore its capacity to structure ecological thinking and praxis is minimal at best, and is now best left behind altogether.

"Why look at animals?" John Berger asked. Animals and humans, he says, look at each other from across an abyss of noncomprehension.[4] To Berger's question we might add an ecological update: Why look at crumbling glaciers, slow extinctions? In the present context, we don't need to understand them visually; we only need to stop destroying them. Empathy, identification, attention, cuteness, respect, comprehension, and the like: we don't need the bright light of the image for these anymore. It is instructive in this regard to consider the relationship posited between the image of ecological spectacle and its viewer, for it would appear to be something more or something other than merely bearing witness to calamity. The overriding emotional appeal invested in such images is that they are a goad to empathy, and therefore (it is hoped?) to action. An additional problem is that the very capacity for empathy has been diminished, or at the very least radically fragmented, in an age where we are subjected to a seemingly endless array of spectacles of every kind: advertisements, films and programming, video clips, GIFs, memes, all in an endless artificial stream, wherein the horrifying and consequential clamors for attention alongside the whimsical and the banal. The "society of the spectacle" is now at least several generations old, but the pervasiveness and polymorphousness of the spectacle continues to writhe, contort, and expand in all directions. Ours is the perspective of the spectator or the voyeur, which is of course a position that first begins by establishing a break or difference between the viewer and the thing he beholds. Part of the argument in favor of ecological witness is that we are all, to one degree or another, complicit in the slow-motion atrocities being shown to us. This is part and parcel of what Rob Nixon describes as the "slow violence" of our environmental catastrophes: not the nightmarish spectacles of the Deepwater Horizon oil spill or Hurricane Katrina, but the leaky, invisible, diffuse destruction of lifeforms, human and non-human alike. In this way the nonhuman world is subjected to precisely the same globalization of indifference and destruction visited on "the poor," the teeming masses least responsible but most directly impacted by the slow and inexorable processes of environmental destruction.[5] This is indisputably true; and yet it is in the nature of film to always keep that identification partial or incomplete. The time of its witnessing, the selection of shots and edits, the narrative or auditory accompaniments, its slow unfolding at a consistent pace not fully dictated by the viewer: all of these work against the impulse to full enmeshment and identification with the subjects' pain or erasure. So while our moral sympathies are being called on, the actual grammar of film always works as a kind of crosswind or counterforce. Even as we have now been trained to fix our gaze on the flickering image and to accord it a special kind of weight, its prevalence and its structure push back against any overt actions that might arise spontaneously in response to the cruelty or injustice that the spectacle is ostensibly designed to reveal to us.

⌒

Traditional forms of ecospectacle also impose a particularly unhelpful relationship to *time* on their viewers. In *Madness, Rack, and Honey,* Mary Ruefle compares the competing views of Henri Bergson and Ezra Pound on the question of what power images might or might not possess. For Bergson, she notes, the image in its isolation cannot seize the intuitive sense, but a broad array of diverse ones can perhaps approach that possibility; whereas for Pound, the image is that which presents an intellectual and emotional complex in an instant of time.[6] But Bergson's and Pound's time is not quite ours anymore. Where the image of ecological catastrophe is concerned, neither the comprehensive sentiment expressed by Bergson nor the immediacy gestured toward in Pound can offer much aid. The endless expansion of the visual archive of planetary catastrophe no more approaches the possibility of seizing our intuition than a narrower field of images ever did; and while the image may yet present an intellectual and emotional complex, it is that nagging problem of its *presentness* that is unsuitable for our deep ecological purposes. We're not trying to save that particular bear, or not *only* that; we are trying mightily to find some way to stop the atrocities, of which that one bear's fate is just a single instance among countless others. Better to shade the lamp, draw the blinds, kill the lights; stop looking and beholding and watching the images in hopes of some grand Bergsonian epiphany or Poundian puncture. To shade our ecology would be to work toward our impossibly challenging future in the dark. Without either the easy consolations of the empathetic ecological spectacle or its tendency to individualize or fragment global crises into digestible images, we are freed to do the work of ecological transformation without the need for witnessing individual acts of violence and cruelty, while also beginning to see more clearly that such images, for all that they accomplish, actively work against exactly the kind of systemic thinking required to address them with sufficient impact.

The *longue durée* of global warming suggests that ecocritical thinking might benefit from an approach grounded in seeing ecological thought as part of "event history," given the colossal timescales and geographies of our subjects of study and action. In this sense, the value of the image and the spectacle rapidly dwindles. The time signature of the visual image, then, poses a serious challenge to the time signature of justice. There is an inescapable belatedness to the ecological image. This problem of belatedness, rather than a mere critique of the image as such, is a too-often-neglected component of our immersion in spectacle. It is often said that the image is powerful precisely because of its presentness and its immediacy, but to say this is to miss something fundamental about the slow-motion atrocity footage that fills our feeds. It misses the fact that things are already so bad we've got footage of the horror. And once the horrors of global warming can be captured on film, we must ask ourselves how much time is actually left to address the things the picture has shown us. The historical trajectory of our comprehensive abuses of the planet is such that we can expect to see more frequent (and more "telling" or ominous) images communicating some small or large display of violence that stands in for the totality of the ecological crisis we have created in general. But this in itself is not something that is likely to transform collective responses and collective actions, because as we know from so many other historical instances, humankind's capacity for absorbing and processing (but also, consequently, managing and deflecting) information will also increase; and so catastrophe after catastrophe will unfold in a long procession of increasingly perilous images that will not land with the politically mobilizing force required of them and expected of them.

We will watch the death of the last black rhino; we will watch the corrosion of the coral reef; we will watch the sea engulf the low-lying communities. We will have a fertile visual archive of all the things we have allowed to be destroyed.

There are additional problems with shining more light on ecological crises, if we imagine that the goal of these images is to spur us to action, policy changes, social transformations, reconfiguration of political and economic institutions, and more. While it is clear to anyone paying attention that the biosphere (of which we are an integral part) depends for its survival on diversity in all its forms, the visual rhetoric of ecological peril still grants an inordinate amount of attention to mammals, correspondingly less to birds, then reptiles and such, and in some cases making its way down (and there is a clear hierarchical supposition here) to vermin, insects, mosses, and the like, in a Great Chain of Seeing. The rapidly expanding genre of environmentally conscious "nature film" seems obsessed with maintaining this lopsided consideration of the biosphere, and its steady stream of the (very real!) crises of polar bears, elephants, whales, and so forth indicates that the hierarchies expressed by those who see man as lord of all creatures are still very much at work in their supposed antithesis. There are many reasons that mammals tend to "count" for more than other classes of living things, but our aesthetic definitions of beauty and sympathy would seem to have a lot to do with it.

This forces us to ask: What is the field of the "camera friendly" or "camera ready"? Who weeps for the obscure mite or millipede? The idea that the beauty of nature should be employed as a method for moving forward the political and economic imperatives of actually *saving* it has become such a bit of environmentalist commonsense that we have not considered it might be better to dispense with that method entirely. Part of the reason for this is that there is still an enormous amount of high-Romantic aesthetic associated with images of capital-N "Nature." Glaciers and meadows and gamboling fauna are perhaps more lovely to look on than a struggling insect or fungus, I suppose; but the point is: who cares? To shade our ecological thinking, then, would be to do more than to simply turn away from the visual regime and the spectacle of ecological crisis; it would be to turn away from the politically retrograde matters of beauty and equally retrograde Romantic version of visual aesthetics entirely and see feelingly instead.

Guy Debord noted that the spectacle is not a collection of images but rather a social relationship between people that is mediated by images.[7] "Since the spectacle's job is to cause a world that is no longer directly perceptible to be *seen* via different specialized mediations," he writes, "it is inevitable that it should elevate the human sense of sight to the special place once occupied by touch; the most abstract of the senses, and the most easily deceived, sight is naturally the most readily adaptable to present-day society's general abstraction." But ecology is a guide to political action, not merely philosophical abstraction. Commodities, Debord suggests, are now all there is to see.[8] Whatever else a drowning elk is, then, it is first and foremost the cinematic document of an imperiled commodity. By zooming in closely on the rare foxes and panthers, we trade our geographical and temporal distance for what Debord identifies as a "spectacular separation."[9] To shade our ecological thinking would be to endeavor to heal that separation by refusing the simple consolations of an endlessly commodified landscape and an endlessly commodified natural world. Furthermore, under the guise of "natural beauty," we can all too easily slide toward notions of beauty and ugliness that indicate that the field of ecological mindfulness may be fraught with some of the same perils as any other arena of life in which appearance and aesthetic judgment is granted a special place. To shade our thinking would, among other things, helps to remove certain unhelpful aesthetic considerations from ecology (beauty, cuteness, grandeur, majesty, sublimity, and other counterproductive categories) and begins to replace them with an unromantic aesthetics, one grounded more deeply in an affective relationship that does not have its foundation in vision. To veil it in this way is to attempt to stop looking at the planet and start *thinking* it. The ecological witness, even of the sympathetic kind, can too often fall into the mode of seeing described by Michel Serres in *Les Cinq Sens* as the "bearer of the look," he who does not move, but sits down to gaze on the thing in a condition of rest.[10] The shaded aesthetic that would be far more impactful at the present moment would more closely resemble what he describes as the *sensible* rather than the merely visual. It is that condition of apprehension whose object of contemplation is always in movement, capable of shifting directions on us even as we seek to pin it down with the limited visual sense. Beyond the merely visual we must venture into the visitation, an itinerant vision that replaces the spectacle and the tableau we are used to witnessing with a deflected and shaded sensory experience.

~

Elaine Scarry was interested in exploring the political consequences of the inexpressibility of *human* pain. For Scarry, it was precisely the incommunicability of one's pain that contributed to the problem, and, most importantly, prevented "recognition," the very legitimacy of another human being's existence.[11] Here the barrier is that between the suffering and the full expression of that suffering in language. In our present case, however, where the landscapes and creatures and ecosystems in question do not possess any language we can recognize, the barrier is between their suffering and the full visual display of that suffering. Ecological display then comes to stand in for the narrower dream of communicability that Scarry had documented in her attention to interhuman interactions. Since the jaguar and the coral reef and the river cannot communicate their suffering or damage directly to us but can only endure it as long as they can, we employ the ecospectacle to communicate those injuries to ourselves, in a visual grammar we invented and which only we understand.

And yet we find in Scarry a relevant concept beyond recognition and direct human communicability. "The only state that is as anomalous as pain," Scarry says, "is the imagination." Pain and imagination, she adds, are each other's missing intentional counterpart.[12] It is my heartfelt belief that *imagination* is the antithesis to spectacle. To shade our ecological thinking is to effect a *détournement* of our traditional habits of witness, document, and gaze. It may provide us with some measure of emancipation from the regime and age of the image. To shade is to work in the dark, to act without requiring the objects of spectacle, without making the environmental calamity available to the rudimentary senses, which are always first and foremost placed in the service of survival and comprehension in the present. Imagination describes more than just the profound ability to "see" what is not amenable to mere vision, because of time, nature, or circumstance; imagination is also the name for the process occasioned by a willful act of shading, of choosing to depend on aesthetic senses available to us when our eyes remain shut, and the objects of our interest have been cloaked in darkness. In other words, to shade is avert one's eyes, and to *imagine* instead. To imagine is to see only with the mind's eye, and this is the most productive way we ought to see the planet and its condition of crisis now and see a path toward doing something meaningful about it.

In *The Sixth Extinction,* Elizabeth Kolbert describes a journey to the tiny rock-island outcropping of Eldey, a short jaunt from Iceland's Reykjanes Peninsula, which, after the destruction by volcano of the Geirfuglasker, or great auk skerry, served in 1844 as the site of the auk's last sizable colony before they were finally hunted to extinction by gentlemen for their skins and eggs. The last of the auks now long dead, this jutting rock is now blanketed with tens of thousands of gannets. Beneath a modern pyramid-shaped structure atop the island rests a platform built by an Icelandic environmentalist agency, where their webcam, set up to watch and record the animals, is now completely buried in guano. "The birds do not like this camera," her guide says.[13] The shitpile of Eldey can communicate something important to us about the endgame of ecological spectacle. We shit on species, and ourselves, by driving them to extinction; and later, in putatively more enlightened times, we set up our cameras to document their plight. But the birds, in their simple imperative to endure, can only shit in turn on the cameras that were set up to document their plight. To shit on the camera that documents your plight is another (nonhuman) way to shade, to cover in excrement the instruments of ecological inquiry that follow in the wake of ecological devastation. What we have yet to learn, the birds, in their incessant activity toward life rather than annihilation, intuitively understand. The guano-covered camera means you can no longer see what this little imperiled pocket of nature is up to; but at this point in the crisis, you can probably imagine.

Notes

1. Lucy Lippard, *Undermining: A Wild Ride through Land Use, Politics, and Art in the Changing West* (New York: New Press, 2014), 167, 190.

2. Timothy Morton, *The Ecological Thought* (Cambridge, Mass.: Harvard University Press), 99.

3. Neil Smith, *Uneven Development: Nature, Capital, and the Production of Space* (Cambridge: Basil Blackwell, 1990), 181.

4. John Berger, *About Looking* (New York: Vintage Books, 1991), 4.

5. Rob Nixon, *Slow Violence and the Environmentalism of the Poor* (Cambridge, Mass.: Harvard University Press, 2011), 5.

6. Mary Ruefle, *Madness, Rack, and Honey* (Seattle: Wave Books, 2012), 46.

7. Guy Debord, *The Society of the Spectacle* (New York: Zone Books, 1994), 12.

8. Ibid.

9. Ibid., 120.

10. Michel Serres, *Le Cinq Sens* (New York: Bloomsbury, 2009), 405.

11. Elaine Scarry, *The Body in Pain* (Oxford: Oxford University Press, 1985), 39.

12. Ibid., 162, 169.

13. Elizabeth Kolbert, *The Sixth Extinction: An Unnatural History* (New York: Henry Holt, 2014), 63–64.

Try

LOWELL DUCKERT

Uncap
Empty
Watertight jugs
Set free
Entombed
Lingering droplets
Rescue
Clouds from bottles.

—NORMAN JORDAN, "A Go Green Flash Dream"

I start with a twisting trial of strength. Off I go; with turns of the wrist, this essay attempts to encapsulate two years' worth of teaching, tries to condense time's course into instructive droplets. Distilled, one lesson is that something lingering always escapes a teacher's endeavor, gets set free by the jugfuls across multiple-month journeys. A pedagogical paradox in Jordanian blue: rescuing is an effort to uncap, never to contain. To en-flash, to dream. *Try it with me.*

I began teaching at West Virginia University in the autumn of 2012, and I have been grateful for the opportunities to engage my research interests—environmental criticism, ecotheory, and early modern literature—every semester since. To help orient my students on the first day, I typically read aloud learning objectives on the syllabus like these: "Our goals are (1) to inquire what environmental criticism *is* and *does,* and (2) to discover the complications and joys that arrive when we theorize how physical place affects the imagination." And then to destabilize the present time for us, I tend to recite a statement such as this: "How can past works of literature not only *resemble* the present, but *influence* it, and consequently *bring about* livable futures for as many human and nonhuman beings as possible?" In general, each course I teach stretches environmental history from the Middle Ages to the twenty-first century, establishing a transhistorical continuum that eschews periodization, elongates deep ecological time and its concerns, and views historicism and presentism as allies rather than enemies.[1] At least they *try*; my ethical and methodological motivations are always outlined on page one: "How might early modern works of art speak to the present, but also create new ecologies, desirable futures? . . . Overall, you will understand how literature theorizes ecology as much as ecological theory informs literature, thereby revealing literature's potential to reshape, and even redress, contemporary issues of environmental health and justice." This investigation, I tell them, is where they will *try*—and that means taking risks—in their assignments and discussions, both within and without the classroom. I am *trying* too, I assure them; we are not here to pinpoint, but to ponder, even to *unknow*, and to leave each class with more questions than answers. When I talk casually about our next four months together, I stress ecology's complications; I note that the course will scrutinize complex issues of race, gender, class, sexual orientation—incorporating feminism, queer studies—that will not necessarily "fix" injustices but perhaps further perplex them. I underline the "experiential learning component" (more on that later). I warn them that thinking of nature as "outside" reinstates harmful divides. I call attention to our breath, for example: air from the coal plant a half mile away; seismicity, the new fracking site three miles south and across-river from the Morgantown Water Treatment Plant; and catastrophe, whatever local/global disaster that unfortunately happened over the preceding break. Without one whiff of Ulrich Beck's work, they all know what living in a "risk society" looks, smells, and tastes like already.[2] To engage this idea of shifting epistemologies as well as a heightened attention to place, my students usually write weekly "ecojournals" responding to our readings but also to a guided prompt. Undergraduates in The Hydrological Turn of fall 2015, for instance, thought about their bodies in relation to a specific body of water: from a kitchen glass (home), to the rain (atmosphere), to the Monongahela River (region). Since most students hail from West Virginia and various socioeconomic backgrounds, I invite them to amplify their lived experiences, engage multiple disciplines, and add personal stories. I thank them, finally, for signing up, for *trying* a class that employs an unusual pedagogical style, that considers classes and pages as experimental spaces for *trying-out*. For reasons undisclosed, a handful frequently drops, to be seen or heard from nevermore. Turning entrenched risk into a site of potentiality, I tell myself, is never easy.

It is different writing this essay now, I confess, with pedagogical hindsight a little clearer (yet never perfect), being a bit more confident though no less tenured in the spring of 2016. As such, this essay faces futures even as it tries to remember the past: specifically, the semesters of fall 2013 and spring 2014 and two unsteady classes, a graduate Seminar in Renaissance Studies, 1550–1660 I named Elemental Shakespeare and a trial run of what could become a permanent undergraduate Environmental Criticism course if rated "successful." I had been preparing to revise my dissertation into a book manuscript on water,[3] contemplating how premodern descriptions of ecological ills, whether or not written by authors we would now deem "writer-activists,"[4] might actually invigorate (or "re-activate") current environmental movements, particularly local ones. These early courses taught me that the abovementioned ethical thought experiments were (and still are) often unsettling for my students. I began to realize that my greatest challenge was not working against students' "apathy" but being able to listen to their frustrations without supplying a ready answer; to adapt somehow to the everyday probability of "staying with the trouble" for fifty or seventy-five minutes in a classroom;[5] to navigate an exploratory space where opinions are freely given, some more passionately than others, which promised a stimulating discussion oftentimes respectful (although never guaranteed). Lynne Bruckner's article "Teaching Shakespeare in the Ecotone" spoke to me during these personal *trials*; in her words, an ecocritical effort "requires something new from us—a deliberate heterodoxy, a willingness to take risks and break rules, a commitment not only to examining our own historical, material, political selves as we really live in the world, but also asking our students to do the same."[6] (A requirement, to be sure, that does not just apply to Shakespeare.[7]) *Trying* the "unorthodox" proves stressful, and necessarily so; a trial of "breaks" may mark as well as make the tension inherent to any ecotone (from the Greek *tonos*, "tension, tone"). Traversing archives "historical" and contemporary reveals "material, political selves" who might ask for our "commitment," their strident voices discordant, disequilibrial, striving. I worried when Greg Garrard announced in a recent (2012) collection dedicated to teaching ecocriticism that "the point of ecocritical pedagogy is to make its existing environmentality explicit and, above all, sustainable."[8] I discovered in a single half year that one of my more pronounced and perennial teaching tensions is critiquing (not achieving) sustainability—the "Anthropocene" is a close second—by choosing to craft intra-catastrophic alternatives instead. I was not alone in this endeavor, either, since my students were only eager to describe the discursive failure of "sustainability": from their "wild and wonderful" setting of Morgantown flunking air quality standards set by the United States Environmental Protection Agency (EPA), to having to withstand harangues from environmentalist groups *as well as* antienvironmental organizations, being blamed for their unwillingness to simply "*act!*" or, oppositely, being ignorant of their surroundings and what is "really" happening. They, too, wanted to know how to live, negotiate, navigate with/in cataclysmic events that characterize the present as "postsustainable," these *trying* times that routinely occur in this area of Appalachia.

Like many places, West Virginia is a precarious ecotone, one that confronts sustainability's promise of eventual, not-too-distant equilibrium. In the fall of 2013 I had been calling the state my home for barely one year, and getting to know it better—then and now—has been an ongoing challenge: it is the second-poorest state in the country with little economic diversification beyond coal mining; chronically divided over (supposedly) inexhaustible sources of mineral wealth, its residents embroiled in endless arguments regarding energy production and consumption (coal keeps the lights on, but coal also kills);[9] red politically but blue psychologically; predominantly white (95 percent of the 1.85 million people) with racism directed toward other whites (discarded as "trash") as well as persons of color; glorified historically as a "sacrifice zone" for the nation's industries but a resource colony in reality;[10] whose workers' demands for better treatment led to the only time in history that the U.S. military has bombed its own citizens (the Battle of Blair Mountain in 1921).[11] Every week the amount of explosives used in the surface-mining technique known as mountaintop removal (MTR) equals the amount of force that leveled Hiroshima, its pollutants and powers of displacement unparalleled.[12] Rampant addiction to prescription painkillers and heroin—and the highest rate of drug overdose deaths in the country—is wrongfully attributed to indolence, pure and simple, not to the widespread joblessness and pervasive medical mistreatment at its core. A chemical spill in the capital city of Charleston on January 9, 2014, left three hundred thousand people without water. Due to Freedom Industries' negligence, ten thousand gallons of a toxin used for cleaning coal (MCHM) leaked into the Elk River; later that summer a containment trench at the same site overflowed. Besides the popular mining and engineering program within the College of Engineering and Mineral Resources, my university has suspicious ties to the industry, supported by funds from environmentally harmful companies and interests, such as the Koch Foundation, who financed (in part) the Center for Free Enterprise within the College of Business and Economics. The billionaire brothers also created a conservative "think tank," The Public Policy Foundation of West Virginia, which has funded faculty positions within the school and, arguably, their publications.[13] And yet, despite this litany of loss, my students are often the first in their family to attend college, and they bring with them an enthusiasm for change *because* of the hardships I just indicated: some of my students volunteer with agencies like Friends of Deckers Creek, while the Reed College of Media's "Stream Lab" employs innovative sensor-style journalism to monitor water quality and the anti-MTR film *Blood on the Mountain* (2016), introduced by three widely known activists from the area, opened the spring 2016 on-campus Human Rights Festival. The region is astoundingly beautiful, containing some of the oldest mountains and rivers on earth; a deep-rooted and multicultural Appalachian identity, celebrated by youth movements like the Rise Collective, have compelled many to stay and improve the lives of both human and nonhuman residents, together.[14]

Try Out

This is only a test: an essay that aims to understand how the ve(e)ry word *try* enacts a ceaseless process of ecological experimentation. I have been purposefully inflecting the word thus far in an effort to convey the task of "try": not just to try and imagine alternative futures but to examine how trying actively helps bring them about, opens up modalities upon possibilities, tries out trajectories, engages in world making. Seeing the world's fullness as well as its depletion, the verb *try* troubles images of a pessimistically entombed biosphere; far from simple idealization, it hovers within (not above) reality, moving between "clouds [and] bottles," fleeing the tightness of ecotones once believed to be bounded. Allow me to outline a few of the things I will try just a little bit harder to accomplish here: (1) I chose the word *try* for its sift-like character, from the Old French *trier* ("sift") of unknown origin. *To try,* in my estimation, is to test etymological and epistemological predestinations, allowing for more errancy than the related noun *trial* affords with its concrete sense of right and wrong, a truth solved once *sieved* correctly. *Try* in fact tests the frameworks of environmental justice rather than submits to predetermined outcomes, asks whom our current laws serve, interrogates their human and nonhuman interests, spotlights whom we put on trial (or not) and why. Not an abandonment of justice but a test of its scales: *to try* means breaking old rules in order to recast them, if unresolved, and yet anew. (2) Students dread what a *trial* means; the noun form expresses the trials of analytical reading and writing, the anxieties of oral and written exams, late-semester defenses in argumentative form that professors then "test the quality of" by assigning a grade. But as Nicholas Royle reminds us, any essay—like this one—is the outpouring outcome of an embarked-on trial, a testing of the fluid waters of "analysis" (Greek for "unloose"), the true liquidity of learning: "a practice of critical thinking and writing newly attuned to the strangeness of literature (including its relation to law and democracy); the animal that you are; spectrality; and the environment."[15] Environmental trials summon the veering potential of *virer* ("to turn, veer").[16] As vertiginous producer, a try can un/learn information as quickly as it settles it into lore. Instead of debating whether trials turn, we should be advocates for *giving turns a try.* (3) The word's relation to *assaying* uncomfortably underscores the commodification of West Virginia and other endangered zones facing the "logic of extraction"[17] and, since both are kinds of labor, pedagogical anxieties about assigning, grading, eco/tripping, and evaluations, testy times when instructors are bureaucratically put on *trial.* But assaying also bears witness; it is feeling the stress and tiredness of trying, of retiring with the world's distress, sadness, and injustices, especially against racialized bodies that tackle these tribulations differently (as we will see). (4) I want to give, lastly, hope a try.[18] Royle adds that "the essay is untimely, anachronistic, never on time," and yet it arrives, and it writes, not of *a* moment but of moments that irrupt into your own, slowly, polyvocally.[19] As the Appalachian and African American ("Affrilachian") poet Norman Jordan wrote on August 8, 1970:

> There is no break
> Between
> Yesterday and today
> Mother and child
> Air and earth
> All are a part
> Of the other
> Like
> With this typewriter
> I am connected
> With these words
> And these words
> With this paper
> And this paper with you.[20]

There (and here) is no "break," only delays in the observance of being "connected" across times, all our "yesterday[s]" and "today[s]." Jordan's is a prominent voice around which "this paper," and now "this paper with you," turns. A keyword in redefinition, typed out: the *try* is a temporary traveler on its aleatory way to other trajectories, spiraling trials, further tests. In my own ecojournal known as "Try," I will try to detail my experiences for you—for me, steps on a way to becoming a better pedagogue if not *trier*—risking, perhaps, that something may be learned. I will try "the trouble" through in- and outdoor classrooms of collaborative collision. All of this amounts to an uneasy commitment to trying, to turning the language of teacherly pinpointing into multidirectional pointing "with this paper," and it comes via two tries. It is wearisome sifting through a bottomless unknown wrought by linguistic nonorigin. *But won't you try?*

First Try

In September 2013 I took my thirteen Elemental Shakespeare graduate students to Kayford Mountain, an active MTR site just south of Charleston and about a three-hour drive from Morgantown. We read *King Lear* (1605) and selections from the May 2012 *PMLA* "Sustainability" cluster to think about how (and why) to carry on in an ecocatastrophic world.[21] In the short paper assignment called "Is this the promised end?"—quoting Kent's horror in the final act (5.3.262)[22]—I invited them to "formulate an argument that speaks to the play's engagement with a fluctuating, unpredictable world: Is Shakespeare advocating a kind of post/ sustainable way of life? Or does he expose the difficulties of initiating political change in the face of mountains reduced to 'nothing'?" They could incorporate informal dialogue or data from suggested websites if they wanted, but, "far from simply transposing current ecological thought onto an early modern play, I want[ed] [them] to think about how all of these authors are mutually informative." Standing on the brink together with our guide, Elise Keaton, an environmental lawyer and an organizer for the Keeper of the Mountains Foundation, we observed the destruction firsthand. She gave us a tour of the property: bullet-holed trailers riddled by pro-coal sympathizers; the anti-MTR activist Larry Gibson's family home; and the crumbling contours of Stanley Heirs Park, where children avoid mountain-top removal missiles while playing. Being a Sunday afternoon, the massive machines were silent and, to my surprise, so were my students. But then I thought that maybe they, like me, felt overwhelmed by the enormity of the place, trying to comprehend Elise's narrative of failed accountability, of how laws could be violated in broad daylight. (Companies simply file for bankruptcy and reopen under a different name, after which they are not responsible for the clean-up.) Reflecting on my own grief, I checked in with students individually as we walked to the "reclamation" site. We sat and listened to Elise talk about how an area with some of the highest levels of biodiversity in the world is literally downgraded to just a few plant species and spray-on grass; how the companies' power seems illimitable regardless of statistical evidence such as extremely rare forms of brain cancer occurring in neighboring communities; how they manipulate this information to dodge indictment, so when stray boulders roll over houses outside worksites they are pronounced "acts of God"; how West Virginia has been a supply state since its inception, divided from Confederate Virginia in 1863 *not* because of abolitionist fervor but in order to preserve the railroad for the Union and its armies during the Civil War. Retaining this information took a toll; the sun came out to reveal a stick of land carved by descending bulldozers—a family cemetery whose bodies are occasionally exposed—looming like a sea cliff in a losing battle with erosion. One of my students volunteered his own experience growing up in coal-replete Wyoming County, then all went quiet. Dwelling on the parameters of the assignment, and in an attempt to provoke discussion, I asked Elise how she could find *any* hope when facing this affront, what she tells not just her tour groups but herself when observing the "end" of mountains. She stated three morals that surely represented the Keeper's motto—be a critical consumer, know your representatives, educate—but she also discussed the top-down subpoena (starting with the presidency) as a way to rethink political ecology at home. And perhaps more powerfully, she stressed Gibson's legacy (who had died just over a year before): how he straightforwardly asked people, even those who harassed him, to have a seat and talk, who believed that nonviolent engagement was a means to find common ground, even if that mutual recognition of shared suffering is the cruelty of coal companies breaking up communities like toxins breaking apart bodies and mountains.

I cannot remember if I said this to myself or aloud before reentering the van: What if we have never been postlapsarian, coming *after* the biblical "Fall," but, like Gloucester believes himself to be, exist in a suspended free fall off Dover Cliff instead, within a constant downturn of catastrophe, living without a prelapsarian paradise to which we may return, where the promise of an immediate (or ever) "Up—so" (4.6.65) of aid is unforeseeable? The ride back was hushed once more, the scenery rolling by, situated pleasantly (deliberately, we now knew) along the interstate corridor. I gave them a week off to write their papers; during our next class we would discuss their work and the daytrip in its entirety. That meeting was productively tense: we talked about post/colonial space (Lear's territorialization and obsession with the map in 1.1, mountains being wiped from them); problems of scale (Edgar's dizzying view in 4.6, our vertigo on the mountainside); disability studies (the play's, and our, emphasis on blinded sight); genre (the implicit pastoralism within a title like "Keepers" and how the play likewise vacillates between pastoral and tragedy, Lear *"fantastically dressed with wild flowers"* at 4.6.81); class (vagabonds like the "poor Tom" of 2.3 who ventured early seventeenth-century highways, the two camouflaged men on ATVs who visibly intimidated my students even after they were revealed to be "friendly"). Their papers, some of them outstanding, reflected our onerous, and obligatory, conversation. Thus I was astonished to learn over the remaining months of the semester that the visit did not have the impact I intended, that my attempt at affective pedagogy did not pay off, and that it had created, in fact, a mutually trying in-class environment as opposed to the community of shared risk taking I had hoped to construct with them. Many (but not all) of my students were outspoken in their frustration with the experiential learning requirement of the course, stating that they felt "forced" to complete an assignment they did not wish to do, that I was pushing them to become "activists," or both. I was downtrodden, to say the least, to the point where I honestly reconsidered having any environmental engagement in future courses. But I talked through it with others; I gleaned (what little I could) from my course evaluations; I revisited some of my favorite sections of the state; and I added details from the trip to "Earth's Prospects," my coaly contribution to *Elemental Ecocriticism: Thinking with Earth, Air, Water, and Fire*, the volume Jeffrey Jerome Cohen and I were coediting.[23] I kept trying, that is, to try. To quote *Pericles*'s (1607–8) bewildered Thaisa who provided the prompt for the seminar's final paper assignment on Shakespeare's elementality and a relevant environmental issue—"What world is this" (13.103)?—I *tried* to imagine how experimental approaches to learning could usefully expand the ecocritical encounters that we debated in the classroom, augmenting everyday experiences in order to alter the worlds to come, the future (of the past) of environmental criticism.

Second Try

Over winter break I readied myself to try and teach again: an introduction to ecology and literature for undergraduates titled Environmental Criticism. *Prismatic Ecology: Ecotheory beyond Green* had just been released in December 2013; on the syllabus, I paired each primary text with a style and a color of ecotheory: think of Mary Shelley's *Frankenstein* (1831) with a queer-ecological bent in quivering "Pink" (Robert

McRuer) and "White" (Bernd Herzogenrath). Because all four of my students were native West Virginians, for their first short paper assignment I selected colors one would commonly associate with the state ("Blue," "Black," and "Grey," a palette of toil), to be viewed through the prism of local author Denise Giardina's diptych of historical fiction—*Storming Heaven* (1987) and *The Unquiet Earth* (1992)—two tales that painfully narrate multigenerational struggles to survive in the twentieth-century coalfields. I tried my tested exhortation once more, this time encouraging them to "formulate an argument that speaks to the novel's engagement with a precarious, 'multihued' world. As [Giardina] writes in *The Unquiet Earth*, 'Nothing is permanent, not even death.' Is [she] advocating a kind of post/equilibrial way of living in the coalfields? Or does she expose the difficulties of initiating political change in the face of ecosystems reduced to 'nothing'?" Eileen A. Joy's "Blue" granted us the optimism of empathy, of cohabiting another's melancholia as a way forward; Levi R. Bryant's "Black" the denigrated color of coal and of skin, but also a presence that speaks of inextricable absorption within a "wilderness" of relations; Jeffrey Jerome Cohen's "Grey" the ashen bodies of abject laborers, zombie miners, the objects of dehumanization. Their papers, too, proved to be extraordinary; but when I asked them beforehand if they had any qualms about the assignment's anomalies, they returned only quizzical looks. I took a chance; remembering the evaluative component of pedagogical "tries," I confided to them my trying times with the former group—how I misread them, the failures that ensued—and used the opportunity to ask outright if they had any reservations about experiential learning. They were altogether assuring and excited; one of my students, Andrew Munn, was (and still is) an activist in the southern coalfields, and with his help I arranged a daytrip for us to the abandoned mining community of Nuttallburg along the scenic New River.[24] Nuttallburg was like many boom-and-bust coal towns: founded in 1870, its citizens manufactured coke (a high-carbon fuel made from coal) until diminishing demand closed the mine once and for all in 1958. But what makes Nuttallburg more unique, and more troubling, is the fact that it was racially segregated. I had designed the course to ask precisely this question about which beings are allowed into the *oikos* of ecology, and what is at stake when the commons is de/limited: Why are *certain* human and nonhuman voices unquiet and others are quieted? Why are some heard and others ignored? To help us dwell on this (unapparent or pur-posefully forgotten) aspect of the river gorge's history, we stopped at the African American Heritage Family Tree Museum in Ansted and met its cofounder, Norman Jordan.[25] Not only was Carter G. Woodson the father of Black History Month, he told us, but he was also a West Virgin-ian who mined Fayette County as a young man. Before we left, Norman read us a few of his own poems, after which I asked him why he thinks many people (myself included) are unaware of the area's and even the state's African American ties. He replied through his art: "Poetry," he said, "is about telling stories."

My students and I, shuttled by the writer and radio producer Cathe-rine Venable Moore, thought about Norman's storied response as we stood next to the foundations of Nuttallburg's "Black School" and "Black Church." It was a beautiful early spring day; the wind was the only audi-ble noise, the rusted compound solemn. We talked about the interpretive sign's disturbing language: African American miners had a tough life, but "slavery was worse." When we crossed Short Creek, the trickling line

Figure 1. Black School.

that separated the camps, it felt ridiculous to us that something so small could divide races, and yet we were aware of the real danger that the small stream would have posed. The surrounding area is now known primarily for its tourism, particularly whitewater rafting and the breathtaking scenery visible from the New River Gorge Bridge (a state emblem, and one of its most recognizable features). Being nine hundred feet above the water at the Canyon Rim Visitor Center earlier in the day had reminded us that vistas can be occluded as well, that traumatic histories are often whitewashed in the name of sport, aesthetics, industry. Or they can try; Carolyn Finney's thoughts on "the complexities and contradictions" of African American environmental history seemed apposite: "We have collectively come to understand/see/envision the environmental debate as shaped and inhabited primarily by white people," she asserts, "and our ability to imagine others is *colored* by the narratives, images, and meanings we've come to hold as truths in relation to the environment."[26] Crossing Short Creek did not catch us unaware of our own race (we all self-identified with "white") and ignorant of our privilege; rather, we were compelled—by a deteriorated foundation, a dried-up creek bed—to critique the *perceived* absence of black bodies in *presumably* white spaces. Places like Nuttallburg were not initially con- and preserved for whites' use but segregated for two races to live separately and yet somehow work equally. Black lives were used for the benefit of primarily white capitalism (John Nuttall); in beholding their spectrality, the materiality of race could be tried and justifiably interrogated. We passed the coke ovens and imagined what the heat and smoke must have felt and tasted like and how fragile, how illusory, a placard's romanticized notion that "men went into the mines different colors, but they all came out the same" truly was. Illness. Transcorporeality across choked bodies flaked in white and black.[27] The "Appalachian Ghost" that haunts Jordan's museum and the Hawks Nest Tunnel disaster near Gauley Bridge was close, just about ten miles from the gorge. In 1931 it was discovered that the predominantly African American workforce of five thousand who had been carving the nearly four-mile tunnel were dying from silicosis and being buried in mass graves. Estimates range from 109 (admitted) deaths to the thousands. "As a boy," Jordan recalls, "I remember men / Coming up the mountain . . . covered with white silicon dust . . . carrying in their lungs / Industrial diseases / Slowly and painfully / Killing them and their dreams."[28] The coal tipple led our eyes hundreds of feet up the hill to the mine. It was a scrambling, wending way. About an hour later we stood, sweating, next to warped and rusted pieces of metal with leafy offshoots, vegetable–mineral machines of no more use. Pieces of coal littered the ground. We peered into the mine's mouth as far as we could, trying to look beyond the boulders that were deliberately placed there to block our entry. It was cold. And although together, it felt lonely. I thought of Joy's blue note to "get *fated* and *outcast* together . . . as the crafting of a more heightened sense of the co-melancholic implication of pretty much everything."[29] Of Jordan's exclamatory "Kuumba," Swahili for "creativity":

> Life is
> Creative force
> In motion
> MOVE!!![30]

I began crafting a story unawares at this height, "Try," of my trials. On the way down the hill, the New River looked as scenic as ever, and yet I could tell that our relationships with the riverscape had gained unexpected, and necessary, complexity. Later that evening the mood lightened as we enjoyed dinner together in the nearby town of Fayetteville. The trip set the tone for the rest of the semester—one of my favorite classes so far—for it helped us think about the stories that are told and are yet to be told, their potential to intervene in our lives and our policies, and who cares, or is willing, to listen. It was a long, though companionate, two-hour drive back to Morgantown.

Third Try

One more try: may we all offer something new from ourselves, in every tone and in every form imaginable, in semesters and in seasons. I would "set free" this "dream" in conclusion rather than supply its rubric. Any reliable measure for determining whether a try is worth giving would be superfluous since trial conditions are always contingent—ranging from the amount of departmental support (if any), to the discovery of a degraded site nearby, to an accidental meetup with student activists—and because the mode with which tests are tried out is unavoidably singular and thereby subject to individual circumstances. Even without this

Figure 2. New River.

unwavering standard for approving or authenticating a given try, the eco-logical act of trying operates via an essential desire to deviate, perforce, with others. We do justice to texts' infinite variability, then, when trying to counter the world's injustices, by experimenting on and with poems in a "flash." An essay's previously decided designs are important as much as they are unpredictable, forever liable to change by un/foreseeable f/actors: be it a segregated coal camp (speaking to specters of race), for exam-ple, or a mountaintop removal ridgeline (confronting the catastrophes of capital). Determinants, therefore, threaten to overdetermine try's uncer-tain trajectory. The trier teleologically hazards an objective (intention) and outcome (result) but also recognizes that the end goal may never be reached, that the targets might surprisingly "MOVE!!!" and change course. Arresting aim's linear functionality requires performing its diver-gence in addition to repurposing its paths. Trials are ethically ambitious precisely for this risky reason: we cannot conceive of a world without tries because tries are what enable us to conceive a different world. A testament to trying implores us to keep trying, to count trials of countless numbers. I originally wrote most of this essay in the final weeks of August 2014. While revising in the spring of 2016, I discovered that Norman Jordan had died the summer before. I turned to his poem "August 22":

> Liquid creation
> flows
> in and around us
> like
> the tip of a stream
> of water
> running down a
> dry mountain side.[31]

I wanted to thank him, again, for helping us try that day, today.

Notes

I am beholden to Brucella Jordan, curator of the African American Heritage Family Tree Museum, for permission to reprint her late husband's poems; to Lynne Bruckner for her work and wisdom; to my students, previous and present, who inspire me to take the pedagogical risks presented here; and to my coeditor,

Figure 3. Dedicated to Norman Jordan (1938–2015).

Jeffrey Jerome Cohen, who initially invited me to write about my teaching for *In the Middle* (blog) ("Teaching Literature in the West Virginian Ecotone," published on August 30, 2014, http://www.inthemedievalmiddle.com/2014/08/) and who encouraged me to elaborate on it for this collection. This essay acknowledges anyone who teaches across disciplines and ecotones, and it advocates for all those unable to *try* since they lack the occupational security or the support—departmental, collegial, institutional—they deserve.

1. An example I like to circulate (even if the course is not listed as "early modern") is Sharon O'Dair's "Is It Shakespearean Ecocriticism if It Isn't Presentist?" in *Ecocritical Shakespeare,* ed. Dan Brayton and Lynne Bruckner (Farnham, UK: Ashgate, 2011), 71–85.

2. See Ulrich Beck, *Risk Society: Towards a New Modernity,* trans. Mark Ritter (Thousand Oaks, Calif.: Sage, 1992).

3. Lowell Duckert, *For All Waters: Finding Ourselves in Early Modern Wetscapes* (Minneapolis: University of Minnesota Press, 2017).

4. Rob Nixon, *Slow Violence and the Environmentalism of the Poor* (Cambridge, Mass.: Harvard University Press, 2011), 22–30.

5. Donna J. Haraway's uncomfortable phrase in *Staying with the Trouble: Making Kin in the Chthulucene* (Durham: Duke University Press, 2016).

6. Lynne Bruckner, in Brayton and Bruckner, *Ecocritical Shakespeare,* 224. For her, an ecotonal Shakespeare necessitates "letting the archival and the presenist collide, even compete, to achieve something that matters. Unless we invite such collisions and collaborations, we are unlikely to find a meaningful way of doing ecocritical work with Shakespeare in the classroom" (236–37). Early modern ecostudies has been remarkably attuned to pedagogy: not only does the collection in which Bruckner's essay appears dedicate three (of its thirteen) essays to teaching, but a brand-new edition is even more generous: *Ecological Approaches to Early Modern English Texts: A Field Guide to Reading and Teaching,* ed. Jennifer Munroe, Edward J. Geisweidt, and Lynne Bruckner (Farnham, UK: Ashgate, 2015), devotes five (of seventeen) essays. If ecopedagogy is not period-specific, it is certainly pronounced in some areas more so than others. *In the Middle,* of course, is a significant example of an all-inclusive space; see Dan Kline's posts from January–February 2008, "Pitched Past Pitch of Grief," for instance: http://www.inthemedievalmiddle.com/2016/01/sometimes-i-have-to-be-reminded-what.html.

7. I do not want to give a comprehensive list here; see Greg Garrard's introduction to *Teaching Ecocriticism and Green Cultural Studies,* ed. Greg Garrard (New York: Palgrave Macmillan, 2012), 1–10, for a brief bibliography of ecopedagogical writing, including, to Garrard's credit, its historical roots in ecofeminism. He also, and admirably, calls for an expansion beyond North American contexts, a tendency that began, arguably, with Frederick O. Waage's *Teaching*

Environmental Literature: Materials, Methods, Resources (New York: Modern Language Association of America, 1985) and perpetuated by *Teaching North American Environmental Literature,* ed. Laird Christensen, Mark C. Long, and Fred Waage (New York: Modern Language Association, 2008).

8. Garrard, introduction, in Garrard, *Teaching Ecocriticism,* 9.

9. Almost eleven thousand jobs have been lost since 2013, two thousand since January 2016. Chris Lawrence, *WV Metro News,* January 26, 2016, http://wvmetronews.com/2016/01/26/coal-industry-reeling-from-newest-layoff-. The U.S. Energy Information Administration corroborates coal's decline: http://www.eia.gov/todayinenergy/detail.cfm?id=26012.

10. Rebecca R. Scott defines the "sacrifice zone" as "a place that is written off for environmental destruction in the name of a higher purpose, such as the national interest" (31). See *Removing Mountains: Extracting Nature and Identity in the Appalachian Coalfields* (Minneapolis: University of Minnesota Press, 2010), particularly chapter 1. The fact that almost a third of West Virginia's coal is shipped overseas (as of 2013), coupled with the lack of local economic prosperity, suggests how unpropitious these sacrifices really are: http://www.eia.gov/state/?sid=WV.

11. Union membership has fallen from 30 to 12 percent nationwide in fifty years, their numbers further reduced by black lung disease, which claims a thousand lives annually (according to the United Mine Workers of America). Republicans in the West Virginia legislature passed a "right-to-work" law in February 2016—granting workers the option to refuse payment to the unions who must still legally represent them—a bill Democratic Governor Earl Ray Tomblin vetoed in vain.

12. MTR's extensive environmental ravages are listed at http://ilovemountains.org and http://appvoices.org/end-mountaintop-removal.

13. Paul J. Nyden at *The Charleston Gazette-Mail* broke the story on February 25, 2014: http://www.wvgazettemail.com/News/201402250167. Then-professor Russell Sobel's *Unleashing Capitalism: Why Prosperity Stops at the West Virginia Border and How to Fix It* (2007), for example, argues for the impracticality of mine safety laws.

14. Also see the STAY (Stay Together Appalachian Youth) Project and Y'ALL (Young Appalachian Leaders and Learners): http://www.thestayproject.com; http://appalachianstudies.org/members/committees/yall.php.

15. Nicholas Royle, *Veering: A Theory of Literature* (Edinburgh: Edinburgh University Press, 2011), 62.

16. "Nowhere is the need for rigour and inventiveness more urgently demanded than in the experimentations of 'environmental writing.'" Ibid., 63.

17. See Scott's introduction to *Removing Mountains,* esp. 6–20.

18. Teresa Shewry writes keenly of hope as a creative and critical engagement: "Hope . . . is a relationship with the future that involves attunement to

environmental change . . . nonhuman beings . . . people, and deep, irreversible loss." *Hope at Sea: Possible Ecologies in Oceanic Literature* (Minneapolis: University of Minnesota Press, 2015), 2.

19. Royle, *Veering*, 62. "It is not a matter of simply following the thinking of another writer. When it comes to the theory and practice of the essay, veering is the art of the singular and perverse" (63).

20. Norman Jordan, *Where Do People in Dreams Come From? And Other Poems* (Ansted, W.V.: Museum Press, 2009), 31. Frank X. Walker coined the term Affrilachian in 1990 to "def[y] the persistent stereotype of a racially homogenized rural region." See http://www.theaffrilachianpoets.com.

21. Sustainability was the "theories and methodologies" subject of *PMLA* 127, no. 3 (2012).

22. All quotations from Shakespeare refer to *The Norton Shakespeare*, 2nd ed., ed. Stephen Greenblatt et al. (New York: W. W. Norton, 2008).

23. "Earth's Prospects," in *Elemental Ecocriticism: Thinking with Earth, Air, Water, and Fire*, ed. Jeffrey Jerome Cohen and Lowell Duckert (Minneapolis: University of Minnesota Press, 2015), 237–67.

24. I retrieved most of my information from the National Park Service's website: https://www.nps.gov/neri/learn/historyculture/nuttallburg.htm.

25. Learn more at http://aaheritagefamilytreemuseum.org.

26. Carolyn Finney, *Black Faces, White Spaces: Reimagining the Relationship of African Americans to the Great Outdoors* (Chapel Hill: University of North Carolina Press, 2014), xii. Her work "On Living Color" is crucial for "not only critiqu[ing] the historical absence of African Americans from mainstream environmental narratives, [but] also focus[ing] on foregrounding the narratives of African Americans who have shown leadership, creativity, and commitment to engaging environmental concerns within and in relation to their communities" (6).

27. Stacy Alaimo's chapter "Eros and X-rays: Bodies, Class, and 'Environmental Justice'" in *Bodily Natures: Science, Environment, and the Material Self* (Bloomington: Indiana University Press, 2010) remains ever-pertinent.

28. Jordan, *Dreams*, 69.

29. Eileen A. Joy, "Blue," in *Prismatic Ecology*, ed. Jeffrey Jerome Cohen (Minneapolis: University of Minnesota Press, 2013), 218.

30. Jordan, *Dreams*, 30.

31. Norman Jordan, "August 22," in *Above Maya* (Cleveland: Jordan Press, 1971), 31.

Rain

MICK SMITH

Rain, midnight rain, nothing but the wild rain
On this bleak hut, and solitude, and me
Remembering again that I shall die
And neither hear the rain nor give it thanks
For washing me cleaner than I have been
Since I was born into this solitude.

—EDWARD THOMAS, "Rain"

What is man? A transition, a direction, a storm sweeping over our planet,
a recurrence or a vexation for the gods?

—MARTIN HEIDEGGER, *The Fundamental Concepts of
Metaphysics: World, Finitude, Solitude*

The summer storm inundates the city, washing air clean of particulates, reviving and refreshing a world exhausted and choked by
constant traffic. Breath comes more freely, colors appear more vibrant.
Watered earth gently steams through turf and last year's leaves. To stay
without shelter under cloud-wracked sky, soaked through, is to receive a
rare gift, albeit one often spurned. After all, we have so much to busy
ourselves with and so ignore or demean such worldly providence, a gen-
erosity that would recall us, however momentarily, to our elemental, but
never isolated, composure, our being-there-in-the-world together—our
"we(a)ther-ing." To feel our exposure to rains falling and not forget to give

thanks is to partake in events precipitating about us, to be ecologically open to life, despite our eventual mortality.

Rain conjoins sky and earth, life and (when its nocturnal and intemperate falling occasions darker thoughts) life's passing. Edward Thomas's poem "Rain" recalls an earlier event walking the ancient Icknield Way; a feeling of being encompassed at night by the "black rain falling straight from invisible, dark sky to invisible, dark earth" (280).[1] "The summer is gone, and never can it return" (281). "Even so," he muses, "will the rain fall darkly upon the grass over the grave when my ears can hear it no more" (ibid.). Rain-narrowed horizons presage the drawing in of life. Rain's incessancy enfolds the phenomenal opening that *is* his own worldly existence, foretelling a future extinction that leaves only a senseless body, flesh (re)turning to earth, humus, buried and decaying under wet sods.

For Martin Heidegger, such intimations of death are not morbid but indicative of coming to a self-understanding of the finitude of one's being-there *(Dasein),* an acknowledgement of one's life and mortality, on the earth, under the sky. Here too, in midnight's dark rain, as in the cleansing summer storms, one's awareness of this (solitary) worldly finitude contrasts with self-involved bus(i/y)ness—our being caught up in a hectic economy that cages rather than engages any understanding of life's worldly possibilities.

To meditate on finitude and solitude, whether in Thomas's poem or Heidegger's philosophy, is to consider life and its limits, not to reiterate the platitude that, after all is said and done, we all die alone. On the contrary, death, unlike standing out in falling rain, is *never* an *event for us,* it is not something we live through. Death is not aloneness, it is no-"thing," nonexistence, while solitude, even being-alone, is actually only possible in the midst of life. Our individuation, the uniqueness of *our* being-there and the realization of our individual finitude, is only possible in and through our worldly relations. Our partaking in and appropriating the world, but also our ek-static being, our standing out into the world as locus of apparently solitary experience and potentially reflective and creative ("poietic")[2] understanding, should therefore, Heidegger suggests, be appreciated as, and through, a mode of staying (dwelling) *with things.* For to be mortal is not just a matter of being able to think about the cessation of our individuated consciousness, it too involves a broader understanding of the world:

> To know oneself to be mortal is not (merely) to know that one will oneself
> die; it is to know that all one knows and most cares about—*everything*;
> every thing—is contingent upon a constellation of circumstances that will
> someday no longer hold together. To acknowledge one's mortality is to
> acknowledge that abyss over which everything precariously juts.[3]

We might say then that to *appreciate* the gift of life and its precariousness requires ways of thinking *and* thanking those things that serve to
creatively gather together this world, to make manifest the world's worlding. Such mindfulness about the precariousness of how things "hold
together" could, perhaps, be termed *ecological*.

Be that as it may, to forget to give thanks for such things as compose
the world constitutes a "lack of appreciation," the consequences of which
go far beyond any mere failure of politeness. Such forgetfulness, Heidegger suggests, is indeed characteristic of that mode of existence which
has fallen under the sway of "technology," a situation marked by the universal predominance of a thoughtless appropriation of every "thing" as
just another potentially transformable *object,* a resource passively awaiting human instrumental purposes. In this now dominant worldview,
rain is reduced to just a source of water for agricultural production, a
supply of potable liquid, a matter of environmental security, a coolant,
solvent, and, of course, in terms of Heidegger's own example, a source
of hydroelectric power.[4] Such a technological enframing of the world
has, as several environmental philosophers have recognized, very serious ecological repercussions,[5] some of which affect the seasonality of the
falling rain itself as it is climatologically altered, globally warmed, acidi-
fied, and so on.

But as Thomas's poem and summer storms attest, rainfall resists being
reduced to a passive object. Indeed, given certain *appropriate* conditions,
rain precipitates experiences, poetic understandings, and forms of appreciation that take us outside of any narrowly bounded self-interests and
also bring *things* home to us in unexpected and sometimes uncomfortable ways. That is to say, rainfall can initiate a different mode of *appropriation,* it is an *event (Ereignis)* that has the potential to call us, however
momentarily, out of technology and capital's narcissistic economy. Raining *happens* to allow things to come into view as things and not as objects
under the domination of a barren instrumentalism, not under the guise

of commodities, or even under any predetermined theoretical *concept* but rather, as Heidegger says, "inceptively," where inception is "an enduring origin that opens up a whole realm of events and meanings."[6] The inception *(Anfang)* might be thought of as a cascading enactment of thoughtfulness rather than the repetitive reproducible reification of concepts applied routinely to emergent contexts.

So if rain is a "thing" then it is so in Heidegger's more fundamental sense, not as an object, but an environing *gathering* together of aspects of a precarious world. Here, in this emphasis on "gathering," Heidegger draws on an etymological association of *thing*, in the sense of the old German *dinc*, with gathering or assembly, as expressed, for example, in the ancient Icelandic "parliament," the *Althing*. From this perspective then, rain, as Heidegger says of the simple water-containing jug, "is a thing in neither the sense of the Roman *res*, nor in the sense of the medieval *ens*, let alone in the modern sense of object"; it "is a thing insofar as it things"[7]—that is, insofar as it serves to gather and make manifest aspects of the world together. And as, in Heidegger's gnomic terms, the "thing things," so we might more easily admit that rain rains. Just as for Heidegger the question of being is inescapably temporal, a matter of being *and* time, and so fundamentally concerned with being as verb, as exis*ting* and not primarily with existents or specific beings, then the *thing*, which we commonly assume to be definitively or archetypically noun-like, should also be understood as a temporal and temporary holding together, a staying of, but also holding a potential to stay with (in the sense of living with and appreciating), the world's transitory nature. Rain occasions the gathering together of earth and sky and mortals, even as it washes away technologically produced particulates.

Indeed, in the course of attempting to speak of the "things" that compose the world and of the "event"—*Ereignis*, the appearance and realization (in both the sense of the unfolding awareness and the appearing and appropriated reality) of what is happening—Heidegger also recognizes a fourth aspect. The thing not only gathers earth, sky, mortals, but also, he claims, "divinities." Just what Heidegger means by divinities is a matter of contention, given its mystical overtones and the ethical and political dangers that might be associated with such mysticism given Heidegger's own track record. However, Heidegger argues, the world is appropriately understood as world, indeed as worlding, only through an event that realizes and makes apparent the gathering together of this fourfold in things. The fourfold consists in a "*primal* oneness",[8] the conjoining of earth, sky, divinities, and mortals, each of which can only be appropriately understood in relation to the other three.

"Earth," "sky," "mortals," and "divinities" as employed by Heidegger thus take on subtly, but also profoundly, different resonances from their everyday employments. Earth Heidegger poetically characterizes as "the serving bearer, blossoming and fruiting, spreading out in rock and water, rising up into plant and animal" (351). It is explicitly associated with the ancient Greek notion of *physis,* with the sense of things arising out of themselves, pushing up into and becoming manifest in the phenomenological "world" experienced and appropriated by living beings, like watered seeds. But, Heidegger reminds us, that which emerges to presence is never *fully* disclosed within the phenomenal world. It is only revealed under certain limited modes of appropriation, under certain aspects, in accordance with certain interests, expectations, sensoriums, histories, frames, predetermined interpretations, and so on. In this sense earth *as such,* the mysterious source of material manifestations, always holds something back, it conceals, and ultimately remains concealed, within itself. (To try to lay bare the inner workings of the earth is not to see it whole before us but to force it to appear within one limited and limiting mode of appropriation—for example, that of a technological objectification.) As Heidegger remarks, the world, a self-disclosing openness, "cannot endure anything closed. The earth, however, as sheltering and concealing, tends always to draw the world into itself and keep it there"[9] in life as, ultimately, and quite literally given Thomas's musings, in earth-enclosed death.

What then of the other three aspects of the fourfold? If earth is the creative ground that bears the weight of the world's materializing existence, then sky seems to emphasize the temporal transformations and cycles of life. Sky is "the wandering vault of the sun, the course of the changing moon, the wandering glitter of the stars, the year's seasons and their changes, the light and the dusk of day, the gloom and the glow of night, the clemency and inclemency of the weather, the drifting clouds and the blue depth of the ether" (351). Mortals, as already suggested, are "human beings . . . capable of death *as* death" (352)—that is, having, like

Thomas, intimations and foreknowledge of their often unpredictable but still inevitable demise.[10] Divinities, in their turn, are, somewhat obtusely, described as "beckoning messengers of the [holy sway of the] godhead" (351). To *think* of one of these aspects, Heidegger asserts, is already to think "of the other three along with it but we [generally] give no thought to the simple oneness of the four."[11]

When we stand *exposed* to the rain as it rains down on us, or even if we are physically sheltered from the downfall itself, as in Thomas's (or even Heidegger's) hut, we gather something (glean an inception) of the elemental conditions of our fleeting existence here together, planted on the earth, watered by the vaulting, sometimes inclement, sky. That is to say, we can be momentarily exposed in our very being (verb) to the world's presence, its transience and its transcendence—its becoming and its going on far beyond our sensory capacities and our finite lives. Such exposure to the rain's raining provides for our worldly existence in a protophilosophical as well as a physical sense, not just as a thirst-quenching gift or even through its later suffusion and circulation (its collection and pooling, its flows and forceful springs), but also because it facilitates or calls forth an inception (a transitional phenomenological moment of openness that potentially informs all future worldly appropriations) concerning how we and things are gathered together, opened to the world's worlding. This is why being exposed to rain can be an event that reveals (and perhaps even allows us to revel in) the precarious and precious gift of life.

Perhaps then (whether this veers from Heidegger's own intentions) the term *divinities* might serve to bring to mind each events *bestowal* of such a "revelation," of the inception and understanding given through unpredictable conjunctions beyond any individual's control. Divinities might refer to the transcendent aspects of appropriating events whereby we recall the miraculous gift of life/the world and experience a heartfelt appreciation of such gifts and their occurrence. That is to say, we recall the thanks that should also accompany our giving thought to our being-(t)here. (Such an interpretation might find some support in the ways that Heidegger continually emphasizes the linguistic connection between there is/it gives—*Es gibt*.) Thanking in this sense is not praising, still less worshipping, it does not require bending the knee before any god or gods, whether anthropomorphic or more abstractly conceived, nor is it redolent of any arcane mysticism. Rather, it might be interpreted in a very worldly, down-to-earth fashion, as a mindful and thankful realization of our finite place within the transcendent world's worlding and also as invoking a certain respectfulness toward the things that serve to gather this world. (Not theological divinity but water divining.) *When, as rainfall insists, this respectfulness includes attending to more than just human being(s), this approaches an ecological ethics.*

Veering Away from Heidegger

Despite his explicit critique of humanism, Heidegger's frame of reference is indelibly and purposefully anthropocentric in so very many ways. Only humans are said to see *the open*, to be capable of appropriating the world as world in its "simple oneness." Humans are "world-forming." Animals (and here, as Jacques Derrida[12] points out, Heidegger's generalization includes *all* animals except humans, whether insects, mammals, etc., in one ridiculously all-encompassing category) are "world-poor," merely captivated by those elements of their environments that generate instinctual responses. This anthropocentrism is closely related to other key aspects of Heidegger's position, including his characterizing *Dasein* in terms of mortality—both in terms of self-awareness concerning our own finitude in relation to the world and our being suspended above the abyss of nonexistence, but also in terms of the way such an understanding is a fundamental aspect of the experience of things as things, their gathering together the fourfold and of the event's occurrence.

So, ironically, for Heidegger thinking of the question of being, of existence as such, and giving thought to the ways in which we are gathered together with each other and with things is itself a possibility contingent on positing an unbridgeable gulf, an abyss, *not* between existence and nonexistence, but between different kinds of existence, between those open due to certain capacities and ways of appropriating life and those unknowingly imprisoned in life—that is, between (at least some) humans and *all* other beings. All nonhuman animals, not to mention plants and all the other incredibly diverse lifeforms that compose the world, are, it would seem, unaware of things as things and the world's worlding. From Heidegger's perspective, there is no way in which other beings exposure to the rain's raining can come close to our experience, it is just, so to speak, water off a duck's back.

Despite Heidegger's determination to emphasize Dasein's embodied, emplaced, and worldly existence, this anthropocentric exclusivity risks alienating humans from more-than-just-human nature in very fundamental ways. Denied access to any understanding of world's worlding, the rain's raining, all other beings are placed in the invidious position of merely being, at best, a potential adjunct to an *event's* occurrence—since events are those key *moments* (both in the sense of temporal *instant* and life-altering directional *force*) of individual *human* existence and of humanity in general. If humans are uniquely "world-forming"[13] then without humans there is no sense of the world as world, just a concatenation of vastly different phenomenal registers, each making sense only within their own very limited and biologically specific contexts: a motley "flock" of beings lacking the "shepherd" that draws together their existences as a supplement to her own understanding of being.

Consequently, there is a very real danger here that we also come to contrast our own mortality with a kind of spuriously immortal "nature." Every living creature (not to mention nonliving entities) thereby tends to be viewed not as a fellow transient of worldly/earthly creation but as merely a synecdoche for this reified "Nature"—that is, it is understood as "belonging to" Nature qua an overarching realm from which humans are, in terms of their key differences, excluded. And if human life is defined as a matter of an essential ontological difference between (at least some) humans and all other beings, human death too is regarded as a matter of *exclusivity*—of our being *excluded from* an ontologically (and emotionally, ecologically, etc.) *indifferent* Nature that survives us at our death. Such a view appears to set one apart, to estrange one, both from a purported natural "order" and from other living beings (now reduced to voices of Nature). Thomas seems to suggest as much when he states:

> Once I heard through the rain a bird's questioning watery cry—once only and suddenly. It seemed content, and the solitary note brought up against me the order of nature, all its beauty, exuberance, and everlastingness like an accusation. I am not a part of nature. I am alone. (281)

For Thomas, the order of Nature here subsumes the individual bird and its song. The bird might seem content but only because it cannot apprehend its finitude nor understand that it is imprisoned *within* this order, and because it is effectively awarded a poetic life after death by Thomas as, in its essence, nothing more than just a part of an immortal, "everlasting," Nature. This, of course, devalues the bird as such, restricting the possibility of regarding it as a unique living (and dying) being, a fellow constituent of any ecological community; it exemplifies a sense of human solitude and finitude but only as a form of alienation.

Understanding ourselves as world-forming holds other dangers, especially when applied to the self-understanding of whole segments of humanity that have purportedly achieved or are awaiting such inceptive "enlightenment." Here the "event" can start to take on, as Heidegger's notion often does, a supraindividual mantle of world-historical and world-changing importance, a kind of manifest destiny. Indeed, this was how Heidegger, at least for a while, seems to have envisaged the rise of the Third Reich as a political *event* ushering in a new epoch. Here the event takes on a kind of messianic mantle. The ecological danger thus comes in conflating the necessary critique of a technological ethos of control with Heidegger's accepting a view of history and futurity as a matter of uncontrollable fate and/or destiny—that is, in mistaking life-changing gifts (the incipient presencing beyond any rational/objective control of the transcendent aspects of things and events) with something *given* in the sense of being unalterable and/or predestined (the hurricane, regarded as an act of God or Nature). Here we might think of Heidegger's claim in his (in)famous *Der Spiegel* interview that in a time of "foundering" "only a God can save us now".

Of course, Heidegger's point here concerns the limits on thought's historical effectiveness, especially in the sense that the technological enframing of the world cannot be changed by philosophical or even by intentional human endeavors alone. Its transcendence and overcoming *(aufgehoben)* should be thought, Heidegger argues in this same interview, "in the Hegelian sense, not [as a condition] pushed aside, but transcended, *but not through man alone*."[14] The event initiating a different mode of apprehension, of "releasement," necessarily involves appreciating the limits of human *control* over all events, including the event(s) that would bring about such changes.

This, of course, *is* something political ecology often emphasizes. For example, environmentalists are certainly not immune from the hope that radical alternatives to the current world order might emerge from events outside human *control,* including the increasingly dramatic and frequent storms consequent on human-induced climate change. We might hope that the inundations drowning New Orleans or the rising sea levels inundating Vanuatu could themselves constitute events initiating global changes and a new environmental awareness of the disasters we precipitate. The danger, again reflecting Heidegger's solution, is that such hopes fall prey to a kind of eschatological worldview that emphasizes the apocalyptic, while we see no alternative but to passively await Nature's own quasi-divine interventions to save us all from the end of the world.

Heidegger's post-Hegelian eschatology is thus part of the problem. Not only does it invoke a form of teleological historicism—*destiny* is a term that appears throughout Heidegger's philosophy—but it risks reinforcing a totalizing understanding of the history of human self-understanding in relation to a separate Nature placed over and against "us." It is notable, for example, that Heidegger claims that the "purpose of keeping oneself open for the arrival of or the absence of the god" referred to in this *Der Spiegel* interview should be understood as a "*liberation* of man from what I called "fallenness amidst beings" in Being and Time."[15] But what can this "liberation" consist in if not a doubling of a sense of Dasein's centrality and specialness in standing out from the rest of the world?

Perhaps, though, attending to the event of rainfall, its permeating, saturating, drowning, and/or enlivening might precipitate an ecological mindfulness of our worldly we(a)thering to veer away from the sense of separation, specialness, and predestination that clothes Heidegger's thought.

Rain Falling amid Beings:
Living with (Rather Than Dying of) Exposure

For Heidegger, fallenness is both a case of a falling away from authenticity into a world of facticity *and* an inescapable aspect of Dasein's this-worldly being. That is to say, it refers to the entirely necessary matter of our involvement in mundane (everyday) matters *and* the way that this *ontic* involvement with other beings and things distracts us from thinking about *ontological* questions concerning existence, mortality, and Being as such.[16] Fallenness thus has connotations of both the thoughtless acceptance of the ontic regime that currently prevails—the idle talk that operates to "cover up the entities within-the-world"[17] and suppresses ontological concerns—*and* of a falling away from "authentic" human possibilities. For the later Heidegger at least, "fallenness amidst beings" both submerges Dasein's existential concerns and elides the world's worlding, the rain's raining.

But ecological events open possibilities that Heidegger hardly considered. Indeed, ecology, *understood in an ethico-political rather than just a narrowly scientific sense,* explicitly emphasizes the need to attend most carefully to the ways in which we might find ourselves fallen amid other beings, including nonhuman beings. Its concerns are with these fellow beings and with our shared but precarious existences, balanced above the abyss of extinction.

So understood, ecology is not just a matter of defining beings' and things' ontic forms and functions, it raises questions concerning others ontologies. Far from an uncritical acceptance of the kind of idle talk and everyday activities that "cover up" the existential relations between things, ecology, at least in its more radical forms, *exposes* the thoughtless inanity that currently prevails, the abject failure of the dominant technological (and capitalist) enframing of the world to appreciate things as anything other than objects/resources/commodities. Ecology seeks to *appreciate* the precarious and miraculous gatherings that compose a world that is far from mundane (ordinary), one that has never been just of our creation, or just for our use. It is grounded in an awareness of the evolving temporality and emergent manifestations of the earth, mindful of the intimate codependency of earth and phenomenal world(s), conscious of the ways nothing stays the same forever but that life always gathers anew. Despite, or rather because of, this evolutionarily inceptive (but certainly not eschatological) awareness of our finitude, it seeks to stay with things, to conserve something of the creative compositions of earth and world, things and beings, so long in the making.

Ecology, then, suggests a radical recontextualizing of the event *(Ereignis)* and hence of things and being(s), but not by returning to a scientific form of objectification that might define things as mere inert matter and beings as living entities that utilize these "raw materials"—this straightaway returns us to a technological frame. Rather, ecology's understanding of the (actually far from simple) "onehood" of the world might still conserve key aspects of Heidegger's thought, especially in terms of the advent of concerns with the world's worlding and beings and things as such. After all, an ethical regard for some-thing may also be expressed as concerning oneself with its being in ontological and not just ontic terms. Such ontological concerns (which might be thought necessary if not sufficient to constitute an ethical moment) might also be understood as following upon previous *events* that open an initial apprehension of that thing's or being's uniqueness, its specific form of gathering, staying of, and staying with, earth and sky, an existence "contingent upon a constellation of circumstances that will someday no longer hold together."[18] Ecology is then also a kind of thanking, of being-concerned-for things as things as they transcend any and all human enframings, things that are never just taken for granted as resources for human purposes.

If ecology emphasizes that humans are not the only beings that compose the world, then we need to ask what it means to say that only *Dasein* ever apprehends the world's "simple onehood" and how this is supposed to be intimately connected to a privileged understanding of the world's worlding, the rain's raining. Of course, some have certainly tried to trace the origin of ecological awareness back to a single (and ironically technologically facilitated) event, namely the Apollo photographs revealing the "simple onehood" of the world seen from space—the shifting bands of clouds, the thin atmospheric limits of the vaulting sky, the blue oceans of this "living planet" set against the dark surrounding lifeless void—albeit not the nothing of non-existence.[19] This image certainly elicited considerations of solitude and finitude and generated ethico-political aftershocks. (It also helps picture how humanity might be considered a vexatious "storm sweeping over our planet.")[20]

But *this* is not at all the whole world. The framed simplicity (and the fixity) of the *planet's* photographic image belies the intricacies of both the phenomenal world's relation to the earth and the world's creation as a multiplicity of diverse phenomenal exposures of things and beings to each other. The world in this sense is not the planet per se, it is not some-thing caught in a representation, grasped in its simple unity as one sees the globe from outside, from "empty" space. It is not possible to escape the world by rocket! And perhaps, as Derrida (2011: 264–71) comes close to suggesting, it is even somewhat delusional to think we have really grasped the world's unity at all.[21]

The world *is* what is sensed, what is sensed is the world. In Jean-Luc Nancy's post-Heideggerian terms, the world is the *exposure* of sense.

And following Nancy (2007: 109) the "unity of a world is not one: it is made of diversity, including disparity and opposition. . . . The unity of a world is nothing other than its diversity and its diversity is in turn, a diversity of worlds. . . . The world is a multiplicity of worlds and its unity is the sharing out [partage] and the mutual exposure in this world of all its [diverse phenomenological] worlds. . . . The world is not given. It is itself the gift. The world is its own creation (this is what 'creation' means)."[22] The ecological event then alludes to this compositional diversity in a more-than-just-human sense regarding our sharing the world. It places the anthropically experienced world (sensoria *and* meanings *and* sensibilities) inescapably amid and with those of diverse other beings.[23]

If *Das Ereignis* imbues *Dasein*'s sensorium with a hermeneutic "richness," allowing every-thing in the world to appear existentially, to acquire, or rather to realize, a "deeper" (and we should be wary of this term) ontological meaningfulness, then an *ecological* event opens the world to wonderment concerning thing's and being's precarious existence. It is an apprehension of being's shared world(s): their being-here together, each exposed to others. But we have to think this sharing differently—as not being unified by a sovereign power, a single overarching decision about *what* the world as such is, or even *that* it is. Perhaps the best we can say is that it seems to us "*as if*" the world is a simple onehood.[24] There are myriad animal (and also fungal, botanical, bacteriological, etc.) sensoriums that are, as Heidegger admits, entirely unknown and perhaps unknowable to us. They *may* not be able to grasp the world as world in the same sense that some humans sometimes might, but this does not mean that their phenomenal experiences can be discounted, indeed they too are world creating. The world that is sensed, appropriated, even, perhaps, inceptively apprehended by some of these different forms is a vital aspect of the (natural–historical) composition of the evolutionary and ecological shape of the shared world all beings, including humans, inhabit. The "strife" between earth and phenomenal world, between what is concealed and made manifest, is played out here too in what appears in these sensoriums in often unimaginable and incredibly different ways.

The world then is not just open or closed. It is open in all manner of ways for so long as a diversity of beings exists. The openness of which Heidegger speaks, the openness incited by the event, is "privileged" only in the sense that it is an opening to question the world's creation and existence not in the sense that the question-holder should be viewed as the creator, compositor, or as having an overarching conspectus of the world. It is a po(i)etic insight into the nature of the world, but only a crass form of humanism would think this insight granted us *oversight* of the natural world, still less made us its *overseer*. It is a privilege we should be thankful for being granted but we are not the world's author or even the sole authority on its composition.

Is the Sky Clearing?

Bearing in mind that not all humans are equally responsible for our current situation we can still refer to our contemporary relations to earth and world as vexatious. The storms we now incite, released to sweep over the planet, are out of our control. But perhaps, even so, the rain's raining might expose and cleanse us of some of our own particular anthropocentric illusions as it appropriates and is appropriated by all manner of things and beings in different ways, each according to its own susceptibilities and modes of exposure. Part of rain's revelation of our own existentially exposed situation might be that the things that matter to us may not be things in the same way for other beings but that they matter nonetheless. The rain falls on us all, raining whether *we* are there to experience it. Thankfully it conjoins earth and sky, mortals *and all other beings*. The rain's coming, whether looked or unlooked for, is both a matter of providence (a gift) and ecological provisioning—in watering the earth it provides for and gathers together a living world, whether we are present to experience its sol(ic)itude. That the enervating scent (petrichor) of the earth's steaming after the summer storm is also composed of the exudations of plants and Actinobacteria might remind us of this. Indeed, life itself might appear to be the *eventual* earthly/worldly appreciation of the rain's raining, its falling before, amid, and after all of our lives and deaths.

> And on each twig of every tree in the dell
> Uncountable
> Crystals both dark and bright of the rain
> That begins again.
>
> —EDWARD THOMAS, "After Rain"

Notes

1. Edward Thomas, *The Icknield Way* (London: Constable, 1913). References from this work are given in parentheses.

2. *Poiesis*—defined by Heidegger as a creative mode of bringing-forth the world; a kind of making is also, of course, the root of the word *poetry*.

3. James C. Edwards, "Thinking of the Thing: The Ethic of Conditionality in Heidegger's Later Work," in *A Companion to Heidegger*, ed. Hubert L. Dreyfus and Mark A. Wrathall (Oxford: Blackwell, 2007).

4. Martin Heidegger, "The Question Concerning Technology," in *Basic Writings* (London: Routledge, 1993).

5. Michael E. Zimmerman, *Heidegger's Confrontation with Modernity* (Berkeley: University of California Press, 1990).

6. Richard Polt, "Ereignis," in Dreyfus and Wrathall, *A Companion to Heidegger*, 382.

7. Martin Heidegger, "The Thing," in *Poetry, Language, Thought* (New York: Perennial Classics, 2001), 175.

8. Martin Heidegger, "Building, Dwelling, Thinking," in *Basic Writings*, ed. David Farrell Krell (New York: Harper Collins). References from this work are given in parentheses.

9. Martin Heidegger, "The Origin of the Work of Art," in *Poetry, Language, Thought*, 47.

10. We might certainly argue with the idea that only humans are mortal in the sense that only they (indeed surely only some of them) actually have a fore-understanding of their own inevitable death. However, for Heidegger mortality would also be caught up with an understanding of there being a world *before* my existence, a time before I found myself "thrown" into the world. It may be harder, although perhaps not impossible, to argue that many other species might have this kind of individuated self-understanding.

11. Martin Heidegger, "Building, Dwelling, Thinking," in *Poetry, Language, Thought*, 148.

12. Jacques Derrida, *The Animal That Therefore I Am* (New York: Fordham University Press, 2008).

13. Martin Heidegger, *The Fundamental Concepts of Metaphysics: World, Finitude, Solitude* (Bloomington: Indiana University Press).

14. Martin Heidegger, "'Only a God Can Save Us'. Der Spiegel's Interview with Martin Heidegger on September 23, 1966," *Philosophy Today* 20, no. 4 (1976): 267–84.

15. Ibid., 278, emphasis added.

16. William Blattner, "Temporality," in Dreyfus and Wrathall, *A Companion to Heidegger*, 313.

17. Martin Heidegger, *Being and Time* (New York: Harper and Row, 1962); James J. DiCenso, "Heidegger's Hermeneutic of Fallenness," *Journal of the American Academy of Religion* 56, no. 4 (1988): 667–69.

18. Edwards, "Thinking," 464–65.

19. Denis Cosgrove, "Contested Global Visions: One World, Whole Earth, and the Apollo Space Photographs," *Annals of the Association of American Geographers* 84, no. 2 (1994): 270–94; Timothy Clark, "What on World Is the Earth? The Anthropocene and Fictions of the World," *Oxford Literary Review* 35, no. 1 (2013): 5–24; Kelly Oliver, *Earth and World: Philosophy after the Apollo Missions* (New York: Colombia University Press, 2015).

20. Heidegger, *Fundamental Concepts*, 7.

21. Jacques Derrida, *The Beast and the Sovereign: Volume 2* (Chicago: University of Chicago Press, 2011). 264–71.

22. Jean Luc Nancy, *The Creation of the World; or, Globalization* (Albany: State University of New York Press, 2007).

23. Mick Smith, "Epharmosis: Jean-Luc Nancy and the Political Oecology of Creation," *Environmental Ethics* 32, no. 4 (2010): 385–404.

24. Derrida, *The Beast*, 264–71.

Drown

JEFFREY JEROME COHEN

We are experts at imagining end times.[1]
After four millennia of practice, narratives of worldly obliteration come easily. The *Epic of Gilgamesh* is "a text haunted by rising waters and disaster."[2] The Book of Revelation promises sudden global warming, floods of flame. Millenarianism springs eternal, from the medieval "Fifteen Signs before Doomsday" tradition to the endless Left Behind novels, internet sites, and films.[3] Never out of print since its publication in 1960, Walter M. Miller Jr.'s *A Canticle for Leibowitz* imagines the long aftermath of nuclear winter by arcing time round into a radioactive Middle Ages. A genre dubbed "cli-fi" envisions the drenched and turbulent vagaries of life in the Anthropocene. Venerable in its plotline and conventions, cataclysm is familiar, almost comforting. Every doomsday arrives prenarrated, the latest iteration of a tale from the unsurpassed past.[4] If the world must end in fire or flood, the ecological devastation we foster through every car trip, meal, and vacation ceases to trouble. Maybe we can even stop recycling.

But whereas catastrophe used to arrive in the thunder of heavenly revelation and the unveiling of a divine plan, the ruin of Earth now is born of anthropogenic climate change, ice melt, greenhouse heat, tempest, sea rise. Secular apocalypse is, in the words of Lawrence Buell, "the single most powerful master metaphor that the contemporary environmental imagination has at its disposal."[5] As we brace for denouement in storm and tempest, what might our apocalyptic imagination unveil about the limits of our environmental frames, the limits of the stories that we tell?

Catastrophe seems fitting punishment for our profligacy: heat, drought, hurricane, glacier retreat, ocean acidification, and species loss as nature's remonstrance, the wages for our carbon release. We are sinners in the hands of an angry Gaia, carbon offsets a modern version of indulgences. There is no theology here, we tell ourselves, only the cold science of global warming and the yield unbridled capitalism brings. We long ago smashed the idols, ruptured the bond between human and the divine. Nowadays we can even borrow our apocalypse from nonbiblical sources—maybe place our end times within an imagined version of the Mayan calendar, as in the film *2012*. Publicity for that epic motion picture featured a sudden crack in the Sistine Chapel so that God no longer touches Adam and the toppling into the sea of the statue of Jesus in Rio de Janeiro. Appropriation of a non-Christian ethos, it seems, enables an escape from inherited religious and cultural frames, rendering our doomsday secular.

Yet as anyone who has seen the disaster of a movie *2012* knows, when the world is ending and no god is coming, to survive Earth's obliteration by floods of neutrinos, destabilization of the mantle, and oceanic outpouring we will have to . . . build some arks. As Everest sinks these marvels of technology preserve a small selection of the human population, including anyone who possesses the billion dollars necessary to purchase a ticket. Queen Elizabeth boards her ark with multiple corgis in tow. In other disaster films we launch space *ships* to *sail* to distant stars *(Interstellar)* or get very clever and set Noah's Ark on a train that circles a planet drowned in ice *(Snowpiercer)*. To imagine catastrophe's unfolding we deploy familiar frames, especially those provided by the story of Noah's Flood. This essay contributes to the long history of meditating on what is left behind when we suppose a watery end inevitable and preserve small community. This tradition crosses the centuries and might be called, in homage to Timothy Findley's meditation on the theme, *Not Wanted on the Voyage*.[6] We exclude mightily when we build an ark, or erect a gated community, or construct a wall along a nation's border, as if we could, like Noah, construct a protective chest in which to dwell, some arkitecture of shelter and exclusion to hold against waves of water or of climate refugees, against violence swift or slow.[7] Those barred from ark or enclave are humans whom we refuse to call "fellow," barring them from refuge. Missing from most contemporary accounts of enarkment is consideration of the preservability and potential companionship of most

nonhumans. In the wake of catastrophe suffering is unequally distributed. The failures of our care are vast.

Make Thee an Ark

When we envision ecocatastrophe we quietly return to those biblical
frames we thought we had surpassed. The world is ending through the
melting of polar ice and the rising of the seas, and the Deluge, a disaster
God promised never to send again, crashes anew. We know in advance
the contours of this plot's unfolding and we surrender to its narrative
surge. Yet in that resignation to submergence, to biblical replay in a scientific mode, we lose sight of the actual complexity of the Noah story in
Genesis as well as its vigorous narrative afterlives. Climate change requires
more and better stories than the ones we have been telling. The Genesis
account of Noah and its retellings in the long centuries that followed offer
a diverse and enduring arkive, a source for resilient counternarratives
that do not make of a coming Flood untroubled waters. We typically take
from Genesis the narrative's barest elements (ark, paired animals, scour-
ing flood, dove, rainbow) and its most dangerous affect, a resignation to
sinking things below the waters, an acceptance that threatens to become
an enjoyment. We submit too easily to imagining a world in which global
warming will render the view from St Paul's in London difficult to tell
from the vistas of its former colonies, as Robert Graves and Didier Madoc-
Jones envision in an image from their vivid climate change awareness
project *Postcards from the Future* (www.london-futures.com). In *St Paul's
Monkeys,* simians perch serenely at the top of Christopher Wren's dome,
surveying flooded streets and the remnants of the Tate Modern, as if
England were India or Gibraltar. As the oceans rise, a global connected-
ness that already binds us becomes materially palpable. In other pictures
from the same series, Graves and Madoc-Jones place rice paddies in front
of the houses of Parliament, and shanties around Buckingham Palace
and Trafalgar Square: floods of water, floods of refugees. But linger for a
moment over point of view in these images. Who is the assumed onlooker
of this world in which monkeys, beasts of burden, laborers in rice paddies, shanty and souk dwellers are decorative signifiers of climate indifference, of a world altered environmentally and offered as marvel?
Monkeys, oxen, rice pickers, and the global poor go about doing what
they do, only they are here now, in London, in *our* space, and isn't that a

spectacle? But who is assumed to inhabit that first-person plural? In the wake of catastrophe, suffering is unequally distributed. The flood makes evident a lack of affective connection already present, an everyday inability of sympathy to cross boundaries of nation, race, species, class.

All the fountains of the great deep were broken up, and the flood gates of heaven were open. It's irresistible: projecting ourselves into the future, imagining we can view below us the topography of cities drowned in rising seas. In 2012 the blogger Burrito Justice famously created a series of such maps for San Francisco, detailing the transformation of its hills over the next fifty years into islands, streets into ocean floor.[8] Inspired by this postdeluge cartography, the urban planner Jeffrey Linn fashioned a series of beautiful and clever maps that with seeming accuracy demonstrate the inundation of familiar metropolises in the wake of ice sheet melt.[9] Linn's Drowned Cities Project depicts Manhattan after 100 feet of sea rise. Brooklyn Heights becomes Brooklyn Depths and Midtown is labeled Middrown. Nearby are Central Shark, Hell's Quicksand, and the Upper East Tide. *And the waters increased.* At 240 feet of sea level change, Seattle becomes an archipelago. The outlines of submerged streets are discernable beneath vivid blue ocean, a reminder of what is lost as the Emerald City is rendered a new Atlantis. *And the waters prevailed beyond measure upon the earth: and all the high mountains under the whole heaven were covered.* Portland's urban topography is used to illustrate 250 feet of flood. The city is transformed into a series of artisanally sculpted islands, with the Columbia Gorge an inlet and the Willamette River a new sea. Other maps by Linn depict drowned London, Montreal, Vancouver, Hong Kong. White lines mark where streets once ran and clever new names have been affixed to drowned neighborhoods, but no other human trace marks these aerial views. *And he destroyed all the substance that was upon the earth, from man to beast, and the creeping things and fowls of the air, and they were destroyed from the earth, and Noah only remained, and they that were with him in the ark.*

Arkaism

Maps and altered images of drowned cities enact what Donna Haraway has called "the god trick."[10] They assume a perspective serenely floating about observed facts. At such critical distance, truth (disembodied, viewable only from an outside) appears. Catastrophe becomes conceptual and foregone, something we witness as we gaze down from the clouds. Deluge conveys delusion. But what about on-the-ground, entangled knowledge? Distant perspective abstracts us from forging (in Stacy Alaimo's words) "more complex epistemological, ontological, ethical and political perspectives in which the human can no longer retreat into separation and denial or proceed as if it were possible to secure an inert, discrete, externalized this or that."[11] In the midst of things, it is muddy, messy, and uncomfortable. You'll get soaked. You might get stuck. You may even drown. But environmentality is a mode of material and ethical saturation, promising no dry heaven from which to view in safety what unfolds during any cataclysm (from the Greek *kataklusmos*, "deluge," a word invented to describe the Genesis Flood). When we imagine that we can behold Earth from a distance we render ourselves divine.[12] Thomas Burnet created for his *Sacred Theory of the Earth* (1690) an illustration of Noah's Ark afloat on a wholly inundated globe, the tiny vessel guarded by two small angels but barely visible on a wave-covered Earth. Perspective can recede so far from anything palpable, from anything sensible, that submerged expanses cease to trouble. It's kind of peaceful looking down, imagining ourselves the spirit of God moving over the waters. It's also beautiful. Yet such perspective disembodies, singularizes, and deprives.[13]

To create the images of London in their *Postcards from the Future* project, Graves and Madoc-Jones digitally manipulated pictures of the city to portray the varied effects of climate change: drought, tropical incursion, severe cold (and thereby the return of Frost Fairs on a frozen Thames). Most of their pictures portray a sinking metropolis. In the breathtaking *London as Venice,* the Thames barrier has failed and most of the city is submerged. The Houses of Parliament and Westminster Abbey are radiant in the sunset, encircled by a river of shimmering blue. The London Eye spins peacefully above flood. The image gives the effect of looking down on this watery expanse from a great distance, perhaps an airplane. The city seems serene.

Graves and Madoc-Jones participate in a long history of sinking London beneath the sea. In what might be the first example of CliFi, J. G. Ballard's *Drowned World* (1962) imagines that after the ice of Greenland melts from solar activity, England's capital has been rendered a tropical city in which only skyscrapers remain above the waves (fortunately the air conditioning and cocktail bar at the Ritz Carlton still function). The

Figure 1. *London as Venice.* Image copyright Robert Graves and
Didier Madoc-Jones. Background photography copyright Jason Hawkes.
www.postcardsfromthefuture.com.

London Magazine in 1899 printed an altered photograph of the city in
which the streets were transformed into canals, gondolas gliding what
had been busy lanes. Entitled *If London Were Like Venice: Oh! That It
Were,* the picture was created back in the days when drowning a city
could still seem lighthearted rather than calamitous. Or maybe it still
does. The following text appears on the *Postcards from the Future* website, describing the creation of the *London as Venice* image:

> Like a modern day Canaletto, this disturbing yet strangely peaceful aerial
> view of a flooded Thames was inspired by shots of New Orleans sub-
> merged under the floodwaters of Hurricane Katrina. Curious to know
> how London would appear under similar conditions, Graves and Madoc-
> Jones transposed projection of a 7.2 metre flooded river on to their digital
> 3D model of London and aligned with a photograph of the Thames shot by
> Jason Hawkes. 7.2 metres is the level at which flood waters would breach

the Thames Barrier. The low light of the photograph creates an evocative sense of dimension to the view, forming the impression that we are looking at a partially submerged stage-set.

Graves and Madoc-Jones sink London to render the city "disturbing yet strangely beautiful," "evocative" like a "partially submerged stage-set" or a painting (Canaletto). Their inspiration: New Orleans in the wake of Hurricane Katrina.

More than ten years after the hurricane, pictures of a flooded New Orleans have not lost their ability to haunt. They require no alteration to make them evocative, and they should never suggest a stage set or Canaletto, a painter renowned for combining the real and the imaginary in radiant water scenes. Beneath the overspreading waters of the Mississippi are people who lost their lives when the levees broke, people left to drown. In the wake of catastrophe, suffering is unequally distributed. Do you remember how tourists boarded buses to view in air conditioned comfort the devastation of that hurricane? Do you remember that Katrina revealed the swift violence of ecological catastrophe as well as the slow violence of persistent, racialized inequality? Some locals used their ruined houses as message boards, spray-painting rebukes like "USA Talk About Race and Poverty!" or this poignant reproach:

TOURISTS Shame On You
Driving By Without Stopping
PAYING TO SEE MY PAIN
1,600+ DIED HERE

Katrina's aftermath might spur us to ask what watery London would look like if beheld not through the god trick of celestial and disembodied view, not through the windows of a tall bus or some other ark that floats over suffering, but from the midst of the sea swell, through the eyes of those in peril in the waters, those left to perish in the flood.

Noah's Arkive

Most of us know the story of Noah and the Flood not from Genesis, where the narrative is complicated, but from a children's bible, puppet show, toy ark, or other reductive restaging. A righteous Noah who sort

Figure 2. New Orleans after Katrina. AP Photo/U.S. Coast Guard, Petty Officer 2nd Class Kyle Niemi.

of looks like Charlton Heston builds a large boat during a sinful age. Animals enter two by two, lions mingling with zebras and peacocks. Rain falls while Noah's family is snug against the storm. The story ends with a raven, a dove, an olive branch, and a rainbow, celebration of a cleansed world. Absent from this version of the Flood is the strange reference in Genesis to the giants who dwelled on the earth in those days (the mysterious Nephilim) and the oblique suggestion of a primal miscegenation behind their arrival. Genesis 6 contains a divine command that two of every beast enter the ark, while the following chapter states only unclean beasts be taken in pairs; the clean come "seven and seven, the male and the female." The waters prevail for 150 days and then only gradually recede, ensuring that Noah is arkbound for more than a year. When the family emerges after their long sojourn, they sacrifice some of the animals and eat others. Vegetarian before the Flood, Noah and his kin become the first carnivores, devouring their ark-mates. Animals and humans are henceforth to struggle against each other. The rainbow in the sky as sign against future cataclysm is an actual bow (Latin *arcus*, related to *archery*), a weapon that shoots a lethal arrow, a suspended promise-threat. Shortly after he reclaims the world, Noah becomes so drunk on the wine he makes that he passes out. When Cham laughs at his father's nakedness, Noah curses his son's descendants to eternal slavery—a biblical scene that will eventually be used to justify the reduction of humans into merchandise that engendered the Middle Passage and fueled the American plantation system.

An ark is not a ship but a chest, deriving from the same Greek word that gives us *archive* (a place for keeping records and stories safe and a source of authority). The great etymologist Isidore of Seville connects the ark to memory, mystery, preservation and exclusion:

> A strongbox (*arca* [that is, an ark]) is so called because it prevents (*arcere*) and prohibits seeing inside. From this term also derive "archives" (*arcivum*, i.e., *archivum*), and "mystery" (*arcanum*), that is, a secret, from which other people are "fended off" (*arcere*).[14]

Not all of the stories collected in Noah's arkive cohere: sons of God and daughters of men, giants, inebriated nudity, a threat within a promise, a movement from cross-species companionship to animal sacrifice and consumption. Nor is Noah's vessel necessarily the gated community it becomes through translation into the Latin *arca*. The Hebrew word *tebah* seems to mean a box, boat, or basket. The word is used only one other time in the Torah, to describe the floating reeds on which Moses as a baby is conveyed from death. Noah's arkive is a whirlpool of heterogeneous narratives, filled with dissonance and counterstories from the start, a word or chest or basket preserving all kinds of forgotten tales and alternative plots. If we realized better the complexity of the Noah narrative and its long history of augmentation and reinvention, we might not be so resigned to climate change, to allowing the world to drown: an ark not as container but generative spur, arkiving as story forging and future making.

Patriarky

Noah was obedient to God. Commanded to build an ark, he constructed the vessel to precise specifications. Christian tradition for the most part praises his perfect obedience. Islamic and Jewish interpreters could be more ambivalent. Rashi, for example, held that "in relation to his generation [Noah] was righteous, but had he been in Abraham's generation, he would not have been regarded as anything," and the Zohar suggests that Noah is culpable for the flood "because he did not appeal for mercy on the world's behalf."[15] When God declares the impending destruction of Sodom and Gomorrah in a flood of fire, Abraham demands: "Will you sweep away the righteous with the wicked?" Recalling the promises made to Abraham, Isaac, and Jacob, an impassioned Moses on Sinai refuses to allow an angry deity to destroy the wayward Israelites and start again. Yet Noah is told to build an ark against flood and complies, leaving the earth to drown.

Medieval depictions of Noah's Ark surface some of the intricacies of the Genesis narrative. In Hebrew and Christian manuscripts as well as sculpture, the ark may be represented as a castle, church, longboat, house, rectangular box, or floating orb. Animals and people share the same tranquil expressions, and fish sometimes swim in oceanic peace below the vessel. Survival is for the determined, and those on board are in it together, a community of men, women, horses, owls, deer, and the occasional unicorn. The ark itself is often lively, with a zoomorphic prow or rudder. Such lush depiction gets at the vibrancy of objects in medieval art.[16] Noah's Ark was often read as an allegory, prefiguring Christ's resurrection and the founding of a new order: the Flood as universal baptism.[17] A thirteenth-century English manuscript of Peter of Poitiers's *Compendium Historiae in Genealogia Christi* features an ark resembling a gothic cathedral of the seas. But even as allegory burgeons, the natural world continues to exert its material presence. The green waters beneath the church-ark are nearly opaque, but fish swim under the boat, a dynamic world rather than a sea of death. The illustration stresses the intimacy of humans and animals, their shared equanimity as ark-mates.

Arkitectures

Stories remain alive by mutating into new forms, drawing to themselves roiling subplots and strange characters, taking unexpected detours. An illustration of Noah's Ark from the fourteenth-century Queen Mary Psalter (British Library Royal MS 2 B.vii) features an ark that floats on transparent waters, revealing the devil making a secret escape through its bottom. Satan pulls the tail of the snake behind him to close the hole he bored through the planks. Meanwhile in the floodwaters humans and animals swim, founder, die. Intent on his business with the dove, Noah does not look at the storied sea, just as in the upper-left corner a raven is intent on its business with the flesh of a dead horse. Noah thinks that through obedience to God he has cleansed the world, but the devil's underwater flight suggests how evil survives the deluge.[18] A possibility this illustration raises is that Noah might have come to fuller and more sober knowledge of the postdiluvian world were he only to look down, were he to behold the devil he has himself sheltered in the boat that he built against a fallen era, were he to witness the men, women, and animals excluded from the ark perishing in the sea.

Likewise composed in accessible Norman French, the contemporary Holkham Bible is a richly illustrated manual for instructing priests on how to teach biblical stories. The manuscript depicts Noah in his ark releasing a raven and dove. Below him swirl aquamarine waves, beautifully transparent. The corpses of a man, woman, and ox are suspended in the waters, while a dead horse rests on a protruding rock—food for the raven. The human and animal bodies drifting through the ocean are in positions never possible on land, a weightless underwater dance. When through the sea drift sensually entwined corpses, elegant in aqueous suspension, while Noah looks resolutely forward, enraptured in avian business, what exactly does the image teach priests to teach their parishioners? In the

Figure 3. Queen Mary Psalter. British Library Royal MS 2 B VII. f. 7r. Courtesy of the British Library and reproduced as part of the public domain.

Figure 4. Holkham Bible, British Library Add. MS 47682, f. 8r. Courtesy of the British Library and copyright the British Library Board.

wake of catastrophe suffering is unequally distributed. Noah is serene as he tends the birds and assesses the livability of the flooded world for those he has preserved. But we are forced to look below the waters, to linger with the submerged. Think for a moment of the words of Abraham to God at the promised destruction of Sodom and Gomorrah, "Will you sweep away the righteous with the wicked?" Is everyone at the bottom of the sea in the Holkham portrait wicked? The ox? The horse? Are we allowed to tarry over such questions? We could bear in mind that in both Jewish tradition and the Chester Play of Noah's Flood, Noah takes one hundred and twenty years to complete the ark, hoping that if he stretches the labor over so long a period some of the doomed might repent.[19] The Chester Play's Noah reveals a sympathy not often seen in the figure, a man usually content to dream of olive branches while the vault of heaven and the abyss pour forth their waters.

In a fifteenth-century manuscript from Rouen of St. Augustine's *De civitate dei* (Bibliothèque nationale de France, Français 28. f. 66v), the ark offers room enough for Noah's peaceful family, some green devils, and a unicorn. The boat is at the front bottom of the picture. The oceanic background is vast and vibrantly blue. Behind the ark a water wheel spins uselessly. Swimmers seek security in home, church, and castle, sinking structures built against the elements that now overwhelm them, architectures meant to preserve. The water is full of detritus. One floating corpse is fresh, another has gone grey. One uprooted tree possesses leaves, another is a kind of arboreal cadaver. An ox swims; a dog drowns. Yet within this mesh of shared immiseration crows, ducks, and swans swim as they always have. Blunt rocks indifferently protrude. The deep blues of the scene are stunning. It is hard to say if we are supposed to feel the peace within the ark, the frustration of the swimmers who seek a place of rest, the inevitability of bodies and trees becoming flotsam. Something changes, perhaps, when we notice that just above Noah's vessel and to the left is a cradle that floats like a miniature ark, empty of its occupant. Is it possible to see that cradle and *not* fill it with an overwhelming story of loss?

My punishment is greater than I can bear. These are the words of Cain, the first in a long line of complainers against God's justice. Cain's declaration might also translate as "My sin is greater than I can bear"—and maybe he means both, that killing his highly favored brother and being exiled from community are unbearable. Either way he protests his state to God and receives in return a mark that will preserve him from harm. Cain is the first builder of cities, of those homes and churches and castles that in this illustration are overwhelmed by the waters on which Noah, his kin, and the animals float in peace. *Drown.* That would seem the command hurled against those not wanted on the ark, an imperative that Noah does not protest when directed at those who are not his family, an injunction we repeat ritualistically as we envision climate change. A lively world is stilled into death, corpses below a churning sea, while an ark of the saved floats in safety.

In a harrowing thirteenth-century illustration of the deluge by William de Brailes, no ark appears and thereby little hope (Walters Art Museum W.106 3r.). Scalding waters pour from the heavens. Layers of the dead accumulate like sediments: the land animals, the beasts of the air, men and women. The exterminated demand examination: piles of faces, human and animal, layered but not separate. No god trick here. This illustration makes insistently visible what happens when we surrender the world to submergence. It refuses to hide what unfolds beneath that blue-green sea. No escape to a transcendent point of view, just immersion in waters that do not cease to flood. William provides no peace, no refuge, no floating vessel. *Anarky.* He paints what's missing from those pictures of a submerged London, Seattle, New York as seen from the sky. In the wake of catastrophe suffering is unequally distributed. This suffering binds humans to hares, falcons, pigs, ravens, dogs. The figures on top reach for those below, their bodies aligned in a downward vector, a post-mortem embrace that is strangely touching, difficult to receive as mere allegory, difficult not to feel. Something here arcs the ages.

Truth can be a little colder when viewed from above—or even from within. Accepting the literal factuality of the Bible, John Wilkins in 1668 attempted to map how so much animal diversity could have been preserved inside a single boat. Wilkins converted every animal collected by Noah into an equivalent number of cows (any beast feeding on hay), sheep (beasts feeding on fruit, roots, insects) or wolves (carnivores). Each creature becomes a storable life-unit with determinate survival needs. Humans are not part of his tally, but it is interesting to note that if they were they would enter the ark as sheep and emerge as wolves. To accomplish his animal calculus, Wilkins thought through the effects of both fair accommodation (in which each animal is lodged comfortably) versus immiserating confinement (granting animals only the minimal space

Figure 5. William de Brailes, *The Flood of Noah (Genesis 7:11–24)*. Ink and pigment on parchment. Walters Art Museum, by bequest. W.106 3r.

required to sustain life during the voyage). He created a massive floor plan for the ark to demonstrate scientifically how Noah preserved the animals during the Flood. Wilkins does not wonder about those left outside. He could not have known that his ark also offered, as Laurie Shannon has powerfully shown, the blueprint for ships in which humans are reduced to livestock, enabling a trans-Atlantic slave trade in which some of those on board hurled themselves into the sea rather than remain in confinement.[20] Another future for this arkitecture is a landlocked one, the factory farm. The ark that saves is also carceral.

We like to think that people in the Middle Ages or before the Enlightenment or a few decades ago were nothing like us. As an inheritance of the flood story we want strong moments of demarcation, secure punctuation of change. Entanglement is difficult. Despite an abiding love in literary and cultural history for sharp periodizations and catastrophism, an affective relationship of viewer to the drowned has always been possible. In the Vienna Genesis, a sixth-century illustrated codex of the Septuagint from Syria and the first illustrated Bible we possess, those who have not been admitted to a pyramid-shaped ark struggle against the rising waters and cling to what stone has not yet been swallowed. We cannot always be sure if we are supposed to feel a sympathetic inclination toward those who fight against the rising waters or take pleasure in divine justice enacted. Maybe both. But what matters about such immersive illustrations is that a potential for compassion, of suffering-with, exists—even if as affective misreading. Sympathy's arc offers a connection that overleaps resignation to loss, affirming other futures to forge. A bulwark against fatalism, sympathy renders grim and reflexive bracing for catastrophe difficult to take seriously in its endless iterations. Apocalypse begins to operate (as Greg Garrard has shown) in a comic mode, in a mode that exults in the fact that even cataclysm fails to offer an obliterating totality, an imperative without exception, a story not to be modified.[21] Complacency and resignation are discarded for endurance, struggle, strange community, the surfacing of hope.

Anarky

Noah was obedient to God. He built the ark and never questioned that the waters must arrive, that all outside must drown. He believed that the world unfolds in a downward turning, a drownward turning, better things arriving only after a foundational apocalypse wipes away what has been. Catastrophe is, quite literally, that downward turn (*kata-* "down" + *strophē* "turning," from *strephein* "to turn"). But can we turn catastrophe down? Or can we at least not be resigned to stories about small communities safe inside their arks? We must embrace the fact that we have become postsustainable. No doubt we must desist in attempting to abstract ourselves or float above the drowning world—we must learn immersion, must learn (as Steve Mentz has argued) how to swim.[22] That's life in the waterlogged Anthropocene. But swimming can seem too heroic, masculine (it's how Beowulf proves himself in youth worthy of great destiny)—as well as too solitary an endeavor, every man for himself, an embrace of a waterworld in which it is impossible to keep anyone but yourself afloat. It's possible even to love that inundation as a kind of rebirth, the Anthropocene as return to swampy, amniotic prehistory. J. G. Ballard's *Drowned World,* the first work of science fiction to imagine climate catastrophe driven by ice melt, delights in obliterative individualism. But Beowulf was heroic not for swimming best, nor for being at sea alone, but because he refused to abandon his competitor to drown during a storm. Only rough waters could part them.

What about those who cannot swim? What about those barred from the ark? What about a community of the unrelated, or at least affinities that exceed near family? In the Chester Play of Noah's Flood, a late medieval drama that reenacts the Genesis story for a city audience, Noah's wife refuses to board the boat and imagines an affective gathering of those about to drown. She knows that what is demanded of women in the ark is not necessarily a way of life to be preserved.[23] She remains with her drinking buddies, the good gossips, as the waters rise. The song of these women as the waves engulf them resounds as powerfully as the holy hymn sung later on the ark as it lifts above their drowned bodies. *Sinken or swimmen.* We might take some solace from the fact that *swim* in Middle English means to float and to glide the waves. *Swim* describes what boats, humans, dolphins, and ducks do in the water, together. If they all swim in unexpected, perhaps even unwanted, togetherness, what communities might arise? Who might join them? What modes of life may be imagined without or outside arks?

I don't know what the future holds, but I suspect the frameworks we have internalized from our meager version of the biblical Flood are not

Figure 6. *Og, Riding Gaily on the Unicorn behind the Ark, Was Quite Happy.*
From Gertrude Landa, *Jewish Fairy Tales and Legends* (Bloch Publishing, 1919),
26. Reproduced courtesy of Project Gutenberg. This image is for the use of
anyone anywhere at no cost and with almost no restrictions whatsoever. You
may copy it, give it away, or re-use it under the terms of the Project Gutenberg
License included online at www.gutenberg.org.

serving us well in imagining the contours of life—*all* life—in the Anthropocene. Let's open the arkive. Let's cease to be resigned to allowing people or animals or even olive trees and rocks to drown. Let's keep in mind that a future of submerged cities is a future of unequally distributed suffering, of environmental injustice. Katrina and New Orleans taught us that. So does the Noah story in its fullness. By not embracing resignation we *can* turn down catastrophe—even if we cannot escape watery perturbations. An ark's value may not reside in its walls so much as in their breaching, in their ability even as flotsam to enable as wide a collective as possible not to drown. So, build an ark if you must, but keep in mind that its fellowship will gather a community of humans and nonhumans alike, an arkive of the diverse that offers little stability. Let your ark have many windows. Let its occupants go for the occasional swim, mingle with the sea. And if (as a Jewish folktale of the nineteenth century insisted) the giant Og of Bashan should ride a unicorn through rising waters and decide to join you—well, that's OK too. The world is always wider than we expect.

Notes

This essay would not have been possible without the collaboration, encouragement, and provocation of Lowell Duckert. I am also deeply grateful to Julian Yates, who read the piece and recognized how deeply in dialogue with his own work it is. I thank audiences at Washington and Jefferson College, Emory University, the University of Auckland, Southern Methodist University, Rice University, the V-A-C Foundation in Moscow, and the University of Basel, who listened to versions of the project and provided invaluable feedback.

1. On the long history of dreaming the apocalypse in the West, see Greg Garrard, *Ecocriticism,* 2nd ed. (London: Routledge, 2012), 93–116. Garrard writes perceptively of what he calls the "secular apocalypse."

2. Dan Brayton, "Writ in Water: *Far Tortuga* and the Crisis of the Marine Environment," *PMLA* 127, no. 3 (2012): 565–71 (570). For a consideration of the long history of imagining the world ending in flood, see Norman Cohn, *Noah's Flood: The Genesis Story in Western Thought* (New Haven: Yale University Press, 1996), esp. 1–21.

3. On the enduring tradition of the signs that will betray the world's end (most of which are environmental changes such as earthquakes, fire, and flood), see William W. Heist's classic study *The Fifteen Signs before Doomsday* (East Lansing: Michigan State College Press, 1952). A portal to the Left Behind media industry may be found at http://www.leftbehind.com/.

4. Srinivas Aravamudan labels this time-swirl effect around disaster *cata-chronism* (catastrophe + anachronism). Catastrophe becomes "oddly comfort-able" because "the sped-up time of lurching toward a cataclysmic event allows for many grand clichés around life and death and the intoxicating spectatorial sense produced by an aesthetic return to the grand canvas of epic." See "The Catachronism of Climate Change," *Diacritics* 41, no. 3 (2013): 7–30.

5. Lawrence Buell, *The Environmental Imagination: Thoreau, Nature Writing, and the Formation of American Culture* (Cambridge, Mass.: Harvard University Press, 1995), 285.

6. A weaving together of medieval and contemporary materials, *Not Wanted on the Voyage* (New York: Viking, 1984) has been a spur to this project. I thank Glenn Burger for giving me a copy of the novel in 1992.

7. In *Slow Violence and the Environmentalism of the Poor* (Cambridge, Mass.: Harvard University Press, 2011), Rob Nixon writes that "Neoliberalism's prolifer-ating walls concretize a short-term psychology of denial: the delusion that we can survive long term in a world whose resources are increasingly unshared. The wall, read in terms of neoliberalism and environmental slow justice, materializes temporal as well as spatial denial through a literal concretizing of out of sight out of mind" (20; cf. 265, on walled communities).

8. A variety of the blog posts at the Burrito Justice website detail the proj-ect; for a starting point, see http://burritojustice.com/2012/03/20/san-francisco-archipelago/.

9. All of the maps I discuss may be viewed at the website Spatialities, www.spatialities.com.

10. Donna Haraway, "Situated Knowledges: The Science Question in Femi-nism and the Privilege of Partial Perspective," *Feminist Studies* 14, no. 3 (1988): 575–99. "The god trick" of pretending an infinite, detached view is possible is defined at p. 582.

11. Stacy Alaimo, "Sustainable This, Sustainable That: New Materialisms, Posthumanism, and Unknown Futures," *PMLA* 127, no. 3 (2012): 558–64 (563). See also Alaimo's work on transcorporeality in *Bodily Natures: Science, Environ-ment, and the Material Self* (Bloomington: Indiana University Press, 2010) and her work on perspective and environmental justice in *Exposed: Environmental Politics and Pleasures in Posthuman Times* (Minneapolis: University of Minne-sota Press, 2016). In this collection, Teresa Shewry describes attunement to such entanglement as *hope*.

12. And dry. See Steve Mentz on the saturated perspectives demanded in a world of constant sea surge and wreck: *Shipwreck Modernity: Ecologies of Global-ization, 1550–1719* (Minneapolis: University of Minnesota Press, 2015). I am not certain it took modernity's shipwrecks to drench us; perhaps with these sea catastrophes comes the knowledge that we have always been diluvian.

13. Lindy Elkins-Tanton and I describe the long history of this desire to view the globe from above throughout *Earth* (London: Bloomsbury, 2017).

14. *The Etymologies of Isidore of Seville,* trans. Stephen A. Barney et al. (Cambridge, Mass.: Cambridge University Press, 2006), 20.9.2. For a compelling reading of Noah's Ark as real archive, see Sarah Elliott Novacich, "*Uxor Noe* and the Animal Inventory," *New Medieval Literatures* 12 (2010): 169–77.

15. For a convenient collation of sources see http://www.chabad.org/parshah/ in-depth/plainBody_cdo/AID/2599 and http://www.hebrew4christians.com/Scrip ture/Parashah/Summaries/Noach/Noah_and_Tradition/noah_and_tradition .html.

16. On the lush materiality of medieval objects, see Anne Harris and Karen Overbey, "Field Change/Discipline Change," in *Burn after Reading, Volume 2: The Future We Want,* ed. Jeffrey Jerome Cohen (Washington, D.C.: Oliphaunt/ punctum books, 2014), 127–43.

17. For a thorough allegorization of the ark see especially Hugh of St Victor's *De Arca Noe Morali* and *De Vanitate Mundi,* in *Hugh of St-Victor: Selected Spiritual Writings,* trans. Anonymous (New York: Harper and Row, 1962), 45–182.

18. And he does so, not surprisingly, with the assistance of Noah's wife. For a consideration of the story see V. A. Kolve, *Chaucer and the Imagery of Narrative: The First Five Canterbury Tales* (Stanford: Stanford University Press, 1984), 201–3.

19. "Noyes Flood," in *The Chester Mystery Cycle,* ed. R. M. Lumiansky and David Mills (Oxford: Oxford University Press, 1974). That Noah warned of the Flood for 120 years, hoping some would repent, is a story also told in the Midrash. See Norman Cohn, *Noah's Flood: The Genesis Story in Western Thought* (New Haven: Yale University Press 1996), 33.

20. See Laurie Shannon, *The Accommodated Animal: Cosmopoly in Shakespearean Locales* (Chicago: University of Chicago Press, 2013).

21. Greg Garrard, *Ecocriticism,* 2nd. ed. (New York: Routledge, 2012), 93–116, especially 113ff.

22. See Steve Mentz, "'Making the Green One Red': Dynamic Ecologies in *Macbeth,* Edward Barlow's Journal, and *Robinson Crusoe,*" *Journal for Early Modern Cultural Studies* 13, no. 3 (2013): 67–84, esp. 81–82 ("Sailors must become swimmers").

23. Sarah Elliott Novacich compellingly argues that Noah's wife is positioned in the Wakefield and Chester flood plays as "co-archivist, as subversive archivist, as archived human animal" whose stories travel to the "emergent earth" (*Uxor Noe,* 171). She describes the Chester "good gossips" as "unsinkable glitches in Noah's archival project and in the performance of salvation: they are characters who are un-saved, un-archived, and yet un-forgotten" (172).

Haunt

COLL THRUSH

Haunted places are the only ones people can live in.
> —MICHEL DE CERTEAU, *The Practice of Everyday Life*

I will tell you something. It is to the thought of the river's banks that I
most frequently return.
> —BARRY LOPEZ, *River Notes: The Dance of Herons*

About an hour south-southeast of Seattle, thirty miles as the heron
flies. One of the outermost suburbs of the city, and one of its poor-
est, Auburn was founded in the mid-nineteenth century and was an
important farming and railroad center by the turn of the twentieth century. Thirty years ago, when I was growing up there, it was a small town
surrounded by dwindling farmland; now it is firmly held within the
concreted fabric of suburbia, the nighttime darkness between towns that
could once be seen from the hill where I lived now transformed into a
seamless constellation of lights.[1]

I am here a quarter century after moving away, in search of a lost river.
There were once three that flowed here. The first, the Green, runs brown
green and lazy out of the mountains to the east, turning north toward
Seattle. The second was the White, a cloudy blue-grey torrent emerging
from the glaciers of Mount Rainier and merging with the Green at what

---

All hauntings have an inception; this one begins with dynamite. This ghost is born
in a blast and a collapsing hillside, in hope and regret. This revenant is a river.

268

became Auburn. The third, the Stuck, was once only an intermittent river, branching off of the White and heading south toward Tacoma. Their ancient names in Whulshootseed, the first language of the local Coast Salish peoples, describe the nature of each: Skwúp (The Rising and Falling Thing) for the Green, Sbálqwu (Turbulent Water) for the White, and Stúkh (Plowed Through) for the Stuck.[2] The merging and unraveling of these rivers is the focus of this story, of this reckoning with place and the past (and the onrushing future). It is necessarily a ghost story.

I am here with three friends: a geographer, a librarian, and a documentarian.[3] We are in search of the ruins of a moment that irrevocably changed the landscape of Auburn and the surrounding valley. This is the short version of the story: in 1906, local residents, tired of disastrous flooding, dynamited a hillside and diverted the White south into the channel of the Stuck, meaning that the White and Green would never again meet. This might seem like a small thing; after all, the stretch of the White that was lost was only some five miles in length. But such minor reengineerings are constitutive of our current ecological predicament and highlight the interpolation of past, present, and place. They initiate and intensify a kind of haunting, a legacy of uncertainty and anxiety that confounds temporality and teleology.

As the Australian anthropologist Peter Read has written, "Those untroubled, those unhaunted, by the ghosts of the past have missed something profound."[4] And so we are here, following Ann Laura Stoler's admonishment to explore "the corroded hollows of landscapes . . . the gutted infrastructures of segregated cityscapes and . . . the microecologies of matter and mind."[5] It may not amount to much, but such small places—even the ones no one seems to care much about—offer insight to the spectral nature of our historical moment. The history of the White River and its ghostly channel through the heart of one small town reveals much about the nature of how we dwell in a capitalist, settler-colonial state, and gesture toward the larger problem of our ecological present. Despite the everyday-ness of the White River and its banal vernacularity, this place matters, and it does so because it haunts. As Indigenous scholars Angie Morrill, Eve Tuck, and their colleagues argue, to haunt is "a materializing. Haunting is a mattering."[6]

~

In 1867 American surveyors penned a detailed description of the valley where the White ignored the Stuck and flowed on to braid itself with the Green: stands of fir and cedar, groves of alder and cottonwood, underbrush of vine maple and wild rose, interspersed with root-rich prairies cultivated by fire. This was land that seemed designed for resettlement by Americans. As one surveyor noted, the area contained "land of first rate quality" and would "admit a large number of settlers."[7] Within a decade, the valley was already in the midst of a wholesale transformation. One 1877 account described a bucolic and hopeful scene:

> The forests are slowly giving way to farms.... Houses, barns, and fences
> show a good degree of comfort and thrift on part of their owners, yet the
> depth and richness of the soil is the most marked feature. What has been
> cleared and cultivated in patches reveals the wealth[,] ... a continuous bed
> of choicest soil for the grasses, the cereals, the vegetables and the fruits,
> which grow large and abundant wherever tested.[8]

By the beginning of the twentieth century, Auburn had a population of several hundred and evinced a "spirit of progressiveness and public spirit among its citizens.... There is no reason in the world why the city of Auburn cannot grow and flourish and become a great and influential trade and industrial center."[9]

At the same time, Auburn was in constant peril. Fall and winter floods often devastated local towns, inundating crops, washing away homes and businesses, killing livestock and the occasional human, and even unearthing the dead in the small cemeteries of the valley. "It made us all miserable," recalled one early settler. "I, for one, detested it and always shall." Another resident described the White cutting a new channel across his property: "Every few minutes we could hear a big piece of the bank go 'Chug,' and by the time supper was over the river was nearly to the house. We put in the night watching and the next day it was worse.... A little snow had fallen and it was a desolate looking place."[10]

The floods of November and December 1906 would be the final straw for local residents. Seattle jurist Clarence Bagley recounted the destruction of this particular torrent: "Highways and railroad bridges were destroyed or badly damaged; grades were washed out; many buildings were destroyed; hundreds of acres of land were cut away by the currents

and many thousands of acres covered with deposits of sand and silt. . . .
An area of not less than fifty square miles between Seattle and Tacoma
was inundated." Two men drowned.[11] One engineer's report described the
White's channel as "completely demoralized," suggesting that there was a
more than purely material element to the disaster.[12] Unruly and destructive, the White River asserted its turbulent self, its identity as Sbálqwu,
on the settlers in its valley. It haunted their dreams of progress, so they
took dynamite and blasting caps and forever changed it.

~

My three companions and I are carrying copies of the maps the surveyors made in 1867, ones made by engineers in 1907 just after the diversion, and aerial photos from the 1930s, and we are following the route of
the White through the neighborhoods on the south side of town. The
south side has always been a bit mysterious to me. Certainly, I never had
a clear sense of the river's path through this postwar street grid and its
landscape of low-slung ramblers. But with the maps and photos, we find
the riverbed. It's there behind a house, a low spot with a little decorative
bridge across it; its bank remains in a high spot next to the main road. I
imagine it haunting these homes in rising damp in crawlspaces, in sodden patches of lawn, maybe even in dreams of rushing water.

~

In 1907 one magazine's portrayal of Auburn noted that the town "offers
much and has a future large and bright."[13] Two years later Auburn boasted
a population of 1,500 with land selling for up to $1,200 per acre. At the
same time, Sbálqwu continued to haunt the city. At the very heart of
town—indeed, on Main Street—a bridge still stood across the empty riverbed. As one Coast Salish elder recalled, "Yeah, my first trip up through
there, you know, there was a bridge there, sit on that old river bed there,
when the White River used to run and get hit the Green River. . . . I says,
'How come they have bridge here and no water and no river?'"[14]

    Meanwhile, flooding continued to affect the valley, and the White River
was subject to numerous engineering projects: riprap along riverbanks,
traps to catch drift logs, a huge earthen dam that blocked salmon runs,
and other interventions meant that by the mid-twentieth century the
White River had become what environmental historian Richard White

has called an "organic machine."[15] In this it represented what had happened to many American rivers with the rise of technorationalist regimes of watershed management. But if the history of environmental engineering teaches us anything, it is that such attempts to control nature rarely work; indeed, their shortsightednesses produce a kind of ecological haunting, in which rivers still struggle to stay within their banks and a town fusses over its own future.

~

"Experience, like memory itself, is written on water," writes essayist John Jerome. "Happily," he continues, "water has a way of giving memory back. For some of us water is memory, and memory water."[16] However, when the water goes, amnesia can ensue. Now we are well along the old channel of the White, continuing north toward the heart of Auburn, and there's virtually nothing to see. A sewer construction project includes piles of river cobbles waiting to be buried with the pipes; we imagine for a moment that these are White River stones, but perhaps they aren't. We find stumps of old cottonwoods in the grounds of a middle school; at first they seem likely relics of the old riverside groves, but they are probably too young. But near another school, one that two of us attended as young boys, we find a depression amid the cul-de-sacs, with a children's playground nestled in it. Here again is the river. And our minds turn to the schools and the amnesia they inculcated: never once had we heard that a river had flowed under these classrooms and recess yards, even as we built our own memories in its path.

~

Decades after the diversion of the White, local residents argued about what had actually happened, and it is here that I see another haunting. Town historian Roberta C. Morley pointed out in the 1960s that the topic remained a "touchy subject."[17] Daughter of early settlers Ettie Ward Roberts Calkins, meanwhile, had her own telling: "What changed it? Well, the men did, that's what. Whey did they do it? They wanted to change the course of the river, that's why. How did they do it? They dynamited it, that's how. Who did it? Everybody did it. . . . Let's not have any more lies about it." Another settler daughter, Meta Trager Jensen, "laugh[ed] about it mysteriously," and Morley admitted that "there is much conflicting evidence, but little argument since people just don't want to talk, out loud that is." Chester Crisp, also from an early family, noted that the diversion "frightened [him] because it seemed secret."[18] One gets the sense that, for all its benefits to the town, Auburn residents struggled to come to terms with what had happened to the White River. "Belonging and possessing," historian Judith Richardson has noted, are "issues that perennially trouble American minds,"[19] and we can see that troubling at work in these ongoing debates and evasions, even six decades after the dynamiting. Coast Salish elder Ed Davis described the moment as, "That's where the dirty work was coming in,"[20] and the dirtiness of the decision continued to stain historical consciousness in Auburn, a kind of revenance that left open the consequences of environmental change. "Ghosts pass so quickly," writes Nicholas Royle; what is remarkable about the White River is that its demise was as sudden as its story has been persistent, at that same time that it was always already half-forgotten.[21]

～

Walter Benjamin once wrote of "crossroads where ghostly signals flash from the traffic, and inconceivable analogies and connections between events are the order of the day."[22] Now we are on Main Street, and the traffic signals here seem pointless, given that there are virtually no cars: on a Sunday afternoon in the Christmas season, downtown Auburn is, pun intended, a ghost town. Where one early settler remembered a landscape "marshy and swampy, fallen trees, a few yellow apple trees, buttercups and beautiful even to my memory,"[23] a town with its brick facades and strolling sidewalks had risen up, but now all that seems dead: the SuperMall on the edge of town has sapped the vitality from Auburn's old shopping district. It seems half the storefronts are empty; department stores that once anchored the local economy are shuttered. It is remarkably silent.

The SuperMall, though, is only an extension of the environmental changes in this place. As Auburn grew over the twentieth century, spurred by the destruction of the White, the rich wetlands of its valley were drained or diked or filled for houses, businesses, and, in the later parts of the century, the installation of vast numbers of warehouses built atop some of the region's richest, blackest soils. The arrival of this suburban and globalized landscape spelled the end of the clouds of waterfowl that once called the valley home. I remember seeing as a child five thousand widgeons in a single field on the edge of town; now that place is a shipping terminal. Where dozens of herons once searched for voles and frogs, there is a Wal-Mart parking lot. And again, the quiet. "The silence and stillness of birds killed by pesticides or motor vehicles, or habitats depleted by logging or drainage," writes Andrew Whitehouse, "promote an anxious semiotics of death and loss typical of the Anthropocene."[24] More on the Anthropocene later; for now, suffice it to say that birds once co-created this human place, but now what Thom van Dooren calls the "dull edge of extinction," the quotidian loss of our nonhuman kin, has transformed the valley into a place with little space for marsh wrens or migrating tundra swans.[25] All of this is linked to the diversion of the White: the same settler ideologies that changed the river changed the whole valley. They haunt this place still, and in the search for the river's course through an emptied downtown, I find myself wondering how best to grieve.

~

Then we are at the place where the White and Green once merged. This place was known as Ilálqwu, a term that has been glossed as Striped Water but more often simply as Confluence. Here, there was a large Coast Salish settlement, whose residents controlled much of the valley that would become known as the White. It was from here, for example, that warriors went out to confront—and in a few cases, kill—settlers who had encroached on their river-bottom prairies.[26] Eventually, the people of Ilálqwu would be relocated to the Muckleshoot Reservation, and the site today is entirely devoid of any acknowledgment of its Indigenous past.

But this part of the valley is a place rife with stories, many of them traumatic. The years in the 1940s when the valley was faced with a sudden absence, as hundreds of Japanese and Japanese American residents were sent away to internment camps.[27] Gary Ridgway, the Green River Killer, whose first four victims were found in 1982 just downstream, where my mother fished for steelhead.[28] And the time in the late 1980s when my cousin's friend, drunk behind the wheel, ran down and killed three small children playing at the old confluence. These stories demand a question: Do places hold trauma? Are certain places fated to carry pain and loss? And what are my own implications and investments in these hauntings?

⌒

The traumatized silence of Ilálqwu suggests an Indigenous haunting. But the local Coast Salish peoples of the White, Green, and Stuck rivers—the Sbálqwuabsh, the Skwúpabsh, and the Stúkhabsh[29]—remain. The descendants of Ilálqwu are still present, most living on the Muckleshoot Indian Reservation east of town. They and their government have the ear of the federal government, unlike the Auburn city council; as participants in an 1855 treaty, they are comanagers of what is left of the salmon; they are, I am told, the largest employer in town after the Boeing Company. Indigenous haunting requires Indigenous absence, which is simply not the story here.

This notion of an Indigenous unhaunting runs counter to most writing on Indigenous peoples in the American imaginary. "Motifs of dispossession recur again and again in early American descriptions of Native Americans," claims literary scholar Renée Bergland. "Indians are figures of melancholy and loss, homelessness and death."[30] Eve Tuck and her collaborator C. Ree, meanwhile, take a more confrontational approach, noting that "the United States is permanently haunted by the slavery, genocide, and violence entwined in its first, present, and future days."[31] The invisibility of Ilálqwu would suggest that this place is haunted by an unfinished process of evasion and erasure.

To be sure, local Coast Salish people have their own sense of ancestral presences in the landscape, including stories about Ilálqwu. But my project here is not one of ethnographic revelation; indeed, it is more one of ethnographic refusal, leaving Coast Salish beliefs and ways of knowing aside as they are not my stories to tell.[32] That said, elders did tell stories in the late twentieth century of the diversion of the White River and its lost salmon runs. Ed Davis told one of his relatives that "they dynamited the river . . . they dried it up."[33] This memoried landscape remains as a kind of haunting, if not one involving ghosts themselves. As Judith Richardson has argued, "A sense of hauntedness is not necessarily reliant on actual apparitions."[34] Indeed, in a place where the Muckleshoot Tribe, made up of the descendants of the peoples of the three rivers, is so strong and so visible, to speak of an Indigenous haunting is to totally miss the point of survivance. This is not to say, however, that this place does not haunt.[35]

~

We have completed our transect across town and now we are back where we started, at the place where they blew up the hillside. From here on up, the White is largely inaccessible, so I have to imagine a journey toward its source: along the uninhabited valley below the reservation, past the dam, through the clear-cuts of the foothills, all the way to its source on Mount Rainier. It is here that two other hauntings rise out of the living earth: one from the distant past and another from the emerging future.

First, there is an ancient apocalypse. Around 5,600 years ago, Mount Rainier lost about 1,500 feet of its summit in an earthquake or eruption. That massive lahar of rock, ice, and earth tore down the channel of Sbálqwu, eventually making its way to the valley where Auburn stands today, which at that time was still an arm of Puget Sound. It turned saltwater into solid ground, through which the Green, White, and Stuck rivers would eventually cut their way.[36] Coast Salish stories of whales that once inhabited the valley speak to the memory of this ancient history, even as they evade the destruction that must have destroyed settlements and hundreds of lives.[37] And today, volcano evacuation route signs signal the specter of this ancient, but potentially future, reality.

Now it is a second apocalypse that haunts the mountain and the valley of the White: the onrushing trauma of climate change. On Mount Rainier, almost all of the glaciers are receding. The Emmons Glacier, primary source of the White, bucks this trend, its lower edge continuing to move down the slopes into the river valley. But this is not the growth of a glacier; this is instead a wave, the entire icy structure slowly sliding downhill as it degrades. This apocalypse is a ghost from the future, a developing relic of the Anthropocene, evinced on the mountain not just in shrinking glaciers but in storm-wrecked roads, dying forests, and disappearing alpine habitats.[38] Timothy Morton has famously framed climate change as a "hyperobject," something that we struggle to observe directly and can only understand through computer models and other abstractions.[39] I'm not so sure; in my own family photos, taken repeatedly over decades at an overlook above the Emmons and the White, I can see the glacier changing and moving. Either way, though, this is a ghost story, and we are moved to a kind of preemptive grieving for, in the thinking of Ashley Cunsolo Willox, "losses expected to come, but not yet arrived. . . . This anticipatory memory of loss is a mourning."[40] This is the basic truth of haunting, that it can come from both past and present, and the story of the White River conflates the two. The dynamiting of the river lies at the confluence of an apocalypse that came before and an apocalypse yet to come. As David Abrams notes, "The past and the future are absences that by their very absence concern us, and so make themselves felt within the present."[41] That is the kind of story I want to tell, the kind of haunting I desire.

~

"Why must America write itself as haunted?" writes Renée Bergland, and I have to ask the question of myself as well.[42] Certainly, I am haunted by my own history here; my mother's ashes lie on a high ridge that looks down on Emmons Glacier. I might follow Judith Butler's lead: "Something takes hold of you: Where does it come from? What sense does it make? What claims us at such moments, such that we are not the masters of ourselves? To what are we tied? And by what are we seized?"[43] I am seized for reasons I cannot yet articulate; this essay is my first attempt to reckon with what happened in the place I am from, to wrestle with the inchoate sense that there is a Big Story here.

Yet that story cannot simply be about melancholia and nostalgia. Indeed, as Ann Stoler has argued, "Melancholy . . . nourish[es] imperial sensibilities of destruction and the redemptive satisfaction of chronicling loss. . . . Nor is it the wistful gaze of imperial nostalgia to which we turn."[44] Instead, like others who write about our ecological moment, I propose mourning. Ashlee Cunsolo Willox argues that the work of mourning can challenge and disrupt the desensitization to our intractable and uncentered environmental predicament, itself built out of small moments such as the dynamiting of the White River.[45] And it is tied to action. "Relearning the world," Thomas Attig writes, "requires that we make changes."[46] I don't yet know what change I want to make; as a scholar, writing is my first impulse. After all, as Paul Ricoeur argues, "the work of narrative constitutes an essential element of the work of mourning."[47] But then what? Rituals of some sort make sense; what exactly those might be remains an open question beyond the scope of this first foray.

Perhaps the answer is in fact to move beyond haunting. "There is more to social life than haunting," Avery Gordon warns. "It would be foolhardy and self-serving to suggest that haunting is the only measure of the world." At the same time, she writes, *haunt* as a verb opens us to a critical interdisciplinary engagement with ourselves and our place in the world, a process that, while unnerving, leads to new ways of knowing and being.[48] Central to that project is the cultivation of care, which, Donna Haraway argues, "means becoming subject to the unsettling obligation of curiosity, which requires knowing more at the end of the day than at the beginning."[49] In that curiosity, we find more than we bargained for: survivance, persistence, even celebration. "We can acknowledge the extraordinariness of colonialism, in its violence, abjection, and duration," notes historian Christopher Clements, "but we can also reckon with the fact that it need not always haunt our stories."[50] As we walked the old course of Sbálqwu, these were the things that preoccupied me, and if this essay means anything, it is that rivers are good to think with, that herons and downtowns may be mourned, that Indigenous peoples survive, and that "to be haunted" means to be actively entangled in the past, present, and future, seeking traces among the living of what might at first be understood as dead.

Notes

1. For general histories of Auburn, see Josie Emmons Turner, *Auburn: A Look Down Main Street* (Auburn, Wa.: City of Auburn, 1990); Hilary Pittenger, *Images of America: Auburn* (Charleston, S.C.: Arcadia, 2014).

2. Dawn Bates, Thom Hess, and Vi Hilbert, *Lushootseed Dictionary* (Seattle: University of Washington Press, 1994), 33, 132; Arthur C. Ballard, *Mythology of Southern Puget Sound* (Seattle: University of Washington Press, 1929), 88; *City of Auburn, 1891–1976*, ed. Roberta C. Morley (n.p.: Washington State Historical Society, n.d.).

3. I wish to thank Wes Pope, Amir Sheikh, and Justin Wadland for their companionship and insights on this transect across time, place, and memory.

4. Peter Read, *Haunted Earth* (Sydney: University of New South Wales Press, 2003), 59.

5. Ann Laura Stoler, "Imperial Debris: Reflections on Ruins and Ruination," *Cultural Anthropology* 23, no. 2 (2008): 194.

6. Angie Morrill, Eve Tuck, and the Super Futures Haunt Collective, "Before Dispossession, or Surviving It," *Liminalities: A Journal of Performance Studies* 12, no. 1 (2016): 3.

7. Cadastral survey of Township 21 North Range 5 East, 1867, General Land Office. Available at www.blm.gov/or/landrecords/survey/ySrvy1.php.

8. G. H. Atkinson, "White River, King County, Washington Territory," *The West Shore* 3, no. 1 (October 1877): 1. Until 1893 Auburn was known as Slaughter, after an American soldier killed during the Treaty War of 1855–1856.

9. "Northwest Cities and Towns," *Coast* 10, no. 1 (July 1905): 32–33.

10. *City of Auburn*, n.p.

11. Clarence B. Bagley, *History of King County Washington* (Chicago: S. J. Clarke, 1929), 708–11.

12. M. Banks, *Report of an Investigation by a Board of Engineers of the Means of Controlling Floods in the Duwamish-Puyallup Valleys and Their Tributaries in the State of Washington* (Seattle: Lowman and Hanford S. and P., 1907), 11.

13. "Cities and Towns of the Northwest," *Coast* 13, no. 4 (April 1907): 268–69.

14. *Elders Dialog: Ed Davis and Vi Hilbert Discuss Native Puget Sound Language, Culture, and Heritage*, ed. Jay Miller (n.p.: Jay Miller, 2014), 122, University of Washington Manuscripts, Archives, and Special Collections.

15. Richard White, *The Organic Machine: The Remaking of the Columbia River* (New York: Hill and Wang, 1996).

16. John Jerome, *Blue Rooms: Ripples, Rivers, Pools, and Other Waters* (New York: Henry Holt, 1997), 8.

17. *City of Auburn*, n.p.

18. Roberta C. Morley Papers, Notes and Stories file, n.p., White River Valley Museum.

19. Judith Richardson, *Possessions: The History and Uses of Haunting in the Hudson Valley* (Cambridge, Mass.: Harvard University Press, 2003), 8.

20. *Elders Dialog*, 122.

21. Nicholas Royle, *Veering: A Theory of Literature* (Edinburgh: Edinburgh University Press, 2011), 120.

22. Walter Benjamin, *Reflections: Essays, Aphorisms, Autobiographical Writings*, ed. Peter Demetz, trans. Edmund Jephcott (New York: Harcourt Brace Jovanovich, 1978), 183.

23. Morley Papers, notes and stories file.

24. Andrew Whitehouse, "Listening to Birds in the Anthropocene: The Anxious Semiotics of Sound in a Human-Dominated World," *Environmental Humanities* 6 (2015): 53–71.

25. Thom van Dooren, *Flight Ways: Life and Loss at the Edge of Extinction* (New York: Columbia University Press, 2014), 12–13, 63–85.

26. For one account of the war (based almost entirely on settler archives), see J. A. Eckrom, *Remembered Drums: A History of the Puget Sound War* (Walla Walla, Wa.: Pioneer Press, 1988).

27. For accounts of the Japanese community in the valley, see Stan Flewelling, *Shirakawa: Stories from a Pacific Northwest Japanese Community* (Auburn, Wa.:

White River Valley Museum, 2002) and *A Pictorial Album of the History of the Japanese of the White River Valley,* ed. Koji Norikane (Auburn, Wa.: White River Chapter of the Japanese American Citizens League, 1986).

28. For one of the most comprehensive, victim-centered, and least sensational accounts of the Green River Killer, see Ann Rule, *Green River Running Red* (New York: Pocket Star, 2005).

29. The Whulshootseed–absh suffix here means, literally, *people of*; when people said *who* they were, they were also saying *where* they were.

30. Renée Bergland, *The National Uncanny: Indian Ghosts and American Subjects* (Hanover, N.H.: University Press of New England, 2000), 3.

31. Eve Tuck and C. Ree, "A Glossary of Haunting," in *Handbook of Autoethnography,* ed. Stacey Holman Jones, Tony E. Adams, and Carolyn Ellis (Walnut Creek, Calif.: Left Coast Press, 2013), 642.

32. For one justification of this refusal, see Audra Simpson, "On Ethnographic Refusal: Indigeneity, 'Voice,' and Colonial Citizenship," *Junctures* 9 (2007): 67–80.

33. *Elders Dialog,* 121–22.

34. Richardson, *Possessions,* 5.

35. For this term, see Gerald Vizenor, *Survivance: Narratives of Native Presence* (Lincoln: University of Nebraska Press, 2008).

36. For one analysis of this lahar, see James W. Vallance and Kevin M. Scott, "The Osceola Mudflow from Mount Rainier: Sedimentology and Hazard Implications of a Huge Clay-Rich Debris Flow," *Geological Society of America Bulletin* 109, no. 2 (February 1997): 143–63.

37. Ballard, *Mythology,* 87–89.

38. For an in-depth journalistic account of climate-related changes at Mount Rainier, see the *Tacoma News-Tribune*'s special report at media.thenewstribune.com/static/pages/rainier/.

39. Timothy Morton, *Hyperobjects: Philosophy and Ecology after the End of the World* (Minneapolis: University of Minnesota Press, 2013).

40. Ashlee Cunsolo Willox, "Climate Change as the Work of Mourning," *Ethics and the Environment* 17, no. 2 (Fall 2012), 140.

41. David Abram, *The Spell of the Sensuous: Perception and Language in a More-Than-Human World* (New York: Vintage Books, 1996), 212.

42. Bergland, *The National Uncanny,* 4.

43. Judith Butler, *Precarious Life: The Powers of Mourning and Violence* (London: Verso, 2004), 21.

44. Stoler, "Imperial Debris," 194, 197–98.

45. Willox, "Climate Change," 145.

46. Thomas Attig, *How We Grieve: Relearning the World* (New York: Oxford University Press, 1996), 107–8.

47. Paul Ricoeur, "On Stories and Mourning," in *Traversing the Imaginary: Richard Kearney and the Postmodern Challenge,* ed. Peter Gratton and John Panteleimon Manoussakis (Evanston: Northwestern University Press, 2007), 8.

48. Avery F. Gordon, *Ghostly Matters: Haunting and the Sociological Imagination* (Minneapolis: University of Minnesota Press, 1997), 206, 27.

49. Donna J. Haraway, *When Species Meet* (Minneapolis: University of Minnesota Press, 2008), 36.

50. Christopher Clements, "Between Affect and History: Sovereignty and Ordinary Life at Akwesasne, 1929–1942," *History and Theory* 54 (2015): 123.

Seep

————▶ ◀————

STEVE MENTZ

It seeps out from the edges of something, and when it's done there are no clear edges. All the lines between the things have blurred. Liquid exchange replaces definition. The result is fuzzy, but not so indistinct that you can't still make out separate shapes. When things seep together, they infiltrate each other but don't merge. The process is a slow, inexorable, grinding-into. It's hard to see it happening. By the time you notice any movement, it has seeped its way all through one thing to another, inch by inch, slowly. What was once a stain or the slightest tangent of contact seeps into mutual contamination. There's no way to keep these things out once they're in, and no way to stop them getting in. It's not so much process as force, or maybe law of force. There's no inside or outside: just slow, inexorable seeping.

Seeping names an ecological truth that all borders must be crossed and all boundaries spanned. The pressure of this process makes visible the corrosive force that lurks inside ecologist and activist Barry Commoner's celebrated "first law of ecology," in which "everything is connected to everything else." Seeping overflows Commoner's happy dictum.[1] Everything seeps out of itself and stains everything else. For Commoner and his fellow activists and early theorists of sustainability, ecological connection presented a way to imagine alternatives to corporate agriculture and runaway growth. Ecological thinking still does that—but its liquid forces flood green politics. Taking inspiration from Stacy Alaimo's "transcorporality" and the tradition of materialist ecofeminism from which Alaimo writes, this essay seeps into melancholy and viscous connections.[2]

Stains from seeping permeate today's postgreen and postsustainable world, now that we are almost a half-century past Commoner's epoch-defining laws.[3] Seeping shows that ecological connections may not provide happy green answers to the problems of the Anthropocene. Mutual stainings and toxicities mark the harsher palate of a postgreen world.

This essay seeps across and between two saturated spaces, one of literary matter and the other of ecotheoretical practice. Trying to reveal the process through which literary and theoretical discourses flow together, my thinking seeps out from two literary texts: Edmund Spenser's Elizabethan verse epic *The Faerie Queene* (1594) and Thomas Pynchon's postmodern novel *Inherent Vice* (2009). These unlike texts share environmental structures built out of allegory. In both cases allegorical narratives provide structural models for environmental interseeping that resonate with our era of entangled catastrophe and crisis. My hope in lingering over the fluid processes of contact and exchange is to erase boundaries that were once deemed solid while emphasizing that, given the slow diffusion of seeping's action, *rough differences and separations remain*. The resulting model troubles boundaries but does not dispense with them entirely. Seeping describes an accumulated exchange of fluids. This model emphasizes that all substances, from water to rock to flame, flow and seep as liquids under the proper conditions and pressures. The leakage of seeping is not a failure or departure from an ideal solid materiality but an essential fluid property of matter itself. Everything seeps. Seeping stains. We can learn to live with it but not until we face its corrosive movement.

Seep names movement and countermovement, a both-ways-turning process of mutual exchange between unlike bodies. In an ecotheoretical context, *seep*'s porosity straddles the divide between the radical separation of OOO (Object-oriented ontology) and the dynamism of Deleuzian "becoming."[4] The arc of this interchanging ecology seeps through the three volumes of Jeffrey Cohen and Lowell Duckert's ecomaterialist trilogy, from *Prismatic Ecology* (2013) to *Elemental Ecocriticism* (2015) and now *Veer Ecology* (2017).[5] *Seeping* describes patterns of exchange between material and fictional bodies. Responding to Alaimo's "Elemental Love in the Anthropocene," her essay in *Elemental Ecocriticism*, I suggest that *seep ecology* names a continual counterflow that abrades and distorts familiar ecological habits of thought. My object, following Alaimo, is to imagine an ecotheory "not corralled by dualisms between matter and metaphor, world and text, scientific capture and humanistic interrogation."[6] I agree with Alaimo that such methods remain "in their infancy" (305), but I also hazard that a return to premodern literature, in particular allegorical literature, can help rethink the porous confines of dualism. The project of allegory is not to overturn or discard the divide between symbol and meaning but rather to populate meaningful richness as widely as possible. An allegorical landscape invests its materiality with maximum metaphorical meaning. Seeping ecological force thus models both a postsustainable reimagining of mutability at the heart of the Anthropocene and also the allegorical abundance long present in literary responses to nonhuman nature. I seep toward reconciliation, not displacement, looking for a seep-stained ecology spanning metaphor and materiality and a dynamic vision of "storied matter" and material stories.[7]

The slow seeping pressure of ecological exchange finds an unlikely parallel in the fraught duality that structures the ancient tradition of literary allegory. Seep ecology proposes that allegory represents a crucial, if currently understudied, ecocritical genre. Ecological criticism often operates on two levels: it engages with the material interchange of things, and it usually offers implicit ideological or human meanings. Allegory provides a rich literary tradition that operates on multiple levels in an analogous way. The (at least) doubled nature of signification through which allegory functions presupposes continuous seeping between layers of meaning. The letter always seeps into the spirit, and vice versa—this contaminating truth sits near the heart of all allegorical traditions. In this sense allegory, though deeply invested in dualism, provides a model to think through and past dualistic limitations, in the manner that Alaimo has suggested. By redoubling allegory's simultaneous connection and separation between symbol and material agent, this literary mode can generate a robust new language for ecomaterialist theory.

This short essay lacks space to consider the full panoply of theories of allegory from Plato and Dante to Angus Fletcher and Paul de Man and instead will demonstrate the value of allegorical literary texts for ecocritics through three episodes from two ornate and seep-full allegorical fictions. Two bleeding episodes from *The Faerie Queene* demonstrate that this epic can meaningfully contribute to the Shakespeare-dominated field of early modern ecocriticism. To wander the meaning-filled valleys and streams of Faerie Land with Spenser's allegorical figures means seeping from narrative action across multiple levels of symbolic meaning, with theological, ethical, legal, and philosophical conceptions bleeding together. I shall consider two episodes in Spenser's epic, the bleeding bush / Fradubio episode in book 1 (1.2.28–45) and the more theologically inflected story of the bloody babe, Ruddymane, in book 2 (2.1.25–62). Sandwiched between these moments in Elizabethan verse epic, I proffer the contemporary novelist Thomas Pynchon's surfing allegory in his hippie private-eye novel, *Inherent Vice* (2009). Thinking about the novel's engagement with fluid structures, especially surfing and fog, suggests that Pynchon, like Spenser, recognizes what Shakespeare's sonnet 64 calls an "interchange of state" as an essential feature of both physical matter and meaning-bearing narratives.[8] For Spenser and Pynchon, liquids allegorize material exchanges both inside and outside human bodies. These two literary–ecological allegories represent the mutual interpenetration of human meanings and nonhuman matter. Material actions and emotional forces seep together. Spenser's verse allegory and Pynchon's postmodern novel elaborate intricate literary systems that model ecotheoretical engagements. Intertwining fictional narratives with philosophical systems generates multiple seepings and multiple stains.

My project interrogates seeping as process that moves in two directions at once. Seep ecology lays outlines, draws boundaries, and imagines a somewhat legible exchange from one thing to another, while at the same time denying or blurring such lines. It's precisely my own habitual system making that seep ecology muddies, though perhaps it's the mud itself that makes meaning, as Sharon O'Dair suggests in her "Muddy Thinking" contribution to *Elemental Ecocriticism*.[9] Might Spenser's allegorical blood, in either its classical or Christian iterations, seep into Pynchon's hippie-surf? Might materialist ecotheory and the "storied matter" theorized by Serenella Iovino and Serpil Opperman reveal itself to be vulnerable to a seepage that contaminates our dearest wishes? One can hope. None of these texts or modes or objects remains isolated from the others. The period at the end of this sentence lies if it purports to mark a final end. There remains a liquid presence inside of and seeping out from the edges of all matter and all symbols. Seeping evades closure, always moves around or across barriers. We can't see or feel it in progress, but we know that it's happening. Seeping stains.

Seeping into Ecotheory

The great shared insight of the theoretical turn in twenty-first-century ecostudies has been revealing the vast implications that flow from the radical decision to no longer separate the "human" from "nature." From different notions of "natureculture" in Donna Haraway and Bruno Latour to Timothy Morton's "ecology without nature" to Graham Harman's dictum that "nature is not natural and can never be naturalized," the shared task of theoretical ecologists has been rethinking the separation of the human from the nonhuman that Latour influentially names the defining act of modernity.[10] The OOO-strain of this philosophical tradition returns human bodies and selves to being simply another set of objects among objects. Seep ecology accepts the unity of natureculture and the inextricable enmeshment of humans in the nonhuman environment. The slight stain that seeping makes on the tapestry of theoretical ecology may resemble a human body, but that's only because that's the kind of object reading and writing these words.

To describe the process as clearly as possible: seep ecology presumes two separate bodies, a source and a destination, but the process of seeping ensures that the two bodies end up less separate than they began. Like allegory, metaphor, and (one hopes) education, seep ecology presumes a system that progressively reduces separations, while never quite eliminating them. The process has two stages:

1. Seeping changes the body from which fluid leaks.

Large things get smaller through this not-only-physical process. Blood seeps out of bodies, and meanings seep out of once-substantial conceptual bodies, such as the anthropos in the Anthropocene. It takes a long time to deflate these conceptions entirely, but diminished versions of once-dominant concepts may become receptive to progressive challenges. In this sense seep ecology is explicitly posthuman: it takes an idealized Man and lets his precious fluids out.

2. Seeping *colors* the body into which fluids flow.

The outflow of fluids and authority stains and inflates formerly marginal bodies into which they flow. The hero's blood colors the world, so that in place of the besieged isolation of the *White Male Hero,* our eyes seep into myriad colors and sexes and narratives. Seep ecology suggests

that the true heroes of Spenser's romance epic might be plants, and the
central activity of Pynchon's private-eye mystery might be surfing rather
than sleuthing. A world marked by fluid motion becomes less enduring
but more resonant. Endings remain problems for seep narratives; it's not
coincidental that neither Spenser's nor Pynchon's epic adventures lead
to anything as substantial as Virgil's Rome. But ecotheorists such as
Haraway and Latour remind us that the epic drive for monumental form
and imperial arrival may represent the "modern" detour that helped birth
the Anthropocene.

As these principles suggest, the paradigmatic seeping fluid for humans
is blood, which marks everything it touches and represents a liquid
female intrusion on male fantasies of bodily integrity. (The corresponding fluid for the earth may be petroleum—but that's for a different
essay.)[11] Menstruation may be the prototypical activity of an ecology
of seep, alongside which the myriad cultural forms built around the masculine erection-tumescence narrative model seem boyish fantasies.[12]
The two depictions of allegorical blood I offer from Spenser show how
blood typifies seep ecology. All bodies hold themselves together and
crave dissolution, and they express their desire to become liquid through
slow seeping pressure. Pynchon's Pacific seeps from human borderline
to global historical pattern maker. Seeping always continues, and no exchange is the last one.

Bleeding Bushes:
Fradubio and the Knight of Holiness (1.2.28–45)

In nominating Spenser as a key Elizabethan writer for ecocriticism, I
recognize that I am swimming against a powerful tide of Shakespeare
studies, not to mention a smaller but distinct current that highlights
both John Milton and Andrew Marvell in ecocritical terms.[13] Treating
The Faerie Queene as an ecological text has an early if relatively under-
theorized precursor in Sean Kane's 1983 essay "Spenserian Ecology,"
which smartly observes that the dense "world of interrelationship" that
characterizes the middle books of Spenser's epic parallels the natural
models that Ernst Haeckel was exploring when he coined the word *ecology*.[14] More recently critics have begun to deploy twenty-first-century
ecocriticism to produce sophisticated readings of Spenser. Alf Siewers

has observed that ecocritical methods can update Northrop Frye's sixty-year-old reading of Spenser's "green world" so that *The Faerie Queene* becomes a text that challenges "both medieval Scholasticism and . . . secular science as a metaphysical system."[15] Gail Kern Paster has deployed ecological models to rethink Spenserian bodies in a humoral context.[16] A 2015 special issue of *Spenser Studies* focused on "Spenser and the Human," coedited by Melissa Sanchez and Ayesha Ramachandran, suggests that this poet's "vital, yet oddly neglected archive" may come to revitalize early modern ecostudies as that discourse matures past its mostly Shakespearean initial phase.[17]

Searching the living landscape of Spenser's Faerie Land for traces of seep ecology leads, first of all, to the bleeding bough that speaks with the voice of Fradubio, the Redcrosse Knight's brother in doubt and fear (1.2.28–45). Spenser's version of the man-into-plant trope, which he adapts from Virgil, Ovid, Dante, Ariosto, and Tasso, presents Fradubio and the Knight who encounters him as symbols of masculine credulity and victims of Duessa's wiles.[18] Traveling with the disguised Duessa, the Knight of Holinesse picks a bough of "two goodly trees" (1.2.28.3) and is rewarded by seeping: "Small drops of gory blood, that trickled down the same" (1.2.30.9).[19] Fradubio's blood, equally tree and human, demonstrates how the allegorical magic of Faerie Land operates: meaning and blood seep out of the physical landscape to assume symbolic visibility. Fradubio, the doubting brother, parallels the Redcrosse Knight—both follow Duessa to their peril—and the tree-man informs the hero and reader of the dangers of false belief. The bough warns the man who plucked it: "O spare with guilty hands to teare / My tender sides in this round rynd embard, / But fly, ah fly far hence away, for feare / Least to you hap, that happened to me heare" (1.2.31.2–5). Fradubio speaks for doubt and fear; in a triplicate nature peculiar to Spenser's allegorical landscape, he represents moral failure, threatens Redcrosse with that failure, and himself has failed in precisely that way. Bleeding boughs seep false faith, and the resulting flows stain hero, episode, and allegorical structures.

The liquid turn signaled by Fradubio's blood opens up fluid connections between this figure and the poem's hero. While the capacious symbolic force of liquidity in the *Faerie Queene* remains outside my purview in this essay, Fradubio as seeping symbol resonantly combines partial liquidity with transformative movement.[20] In retelling the story of his transformation from man to tree, Fradubio emphasizes the not-entirely-solid means through which Duessa transformed him: "With wicked herbes and oyntments [she] did besmeare / My body all" (1.2.42.3–4). Smeared with ointments as well as more solid herbs, the plant-man moves to "this desert waste" (1.2.42.6), where his tree-body assumes its place alongside the woman he had abandoned to follow Duessa. His new form—substantial and solid but filled with blood-sap eager to seep forth—can only be countermanded by "a living well" (1.2.43.4). The well of holiness trumps Duessa's thick ointments with clear flow. A series of puns on the word *well*, which means living well (1.2.43.7) and also "knew well" (1.2.44.3) in the following stanzas, suggests that these multiple wells will not easily be reconciled into the clear flow of holinesse into which Redcrosse seeks to transform himself.

The episode closes with a symbolic quenching of seeping flow. The Redcrosse Knight takes "the bleeding bough [and] did thrust [it] into the ground" (1.2.44.6), thus hiding his own continuing performance of Fradubio's signature doubt. Like the tree-man, Redcrosse continues to be deceived by Duessa, with whom he "forward forth did beare" (1.2.45.9). A puzzling line suggests, however, that Redcrosse recognizes that Fradubio's blood contaminates; he buries the bough "that from the blood he might be innocent" (1.2.44.6). The term *innocent* here may refer to moral knowledge—the knight wishes to forget Fradubio's error—but it also gestures forward to the more obviously theological symbolism of Ruddymane's bloody hands in book 2. Duessa's brief fainting fit (1.2.44–45) parallels Redcrosse's efforts to hide his continuing guilt. Like Anthropocene Man, the knight refuses to face that the pollution marring his landscape derives from himself.

Surfing Something: *Inherent Vice*

Centuries and oceans away from Faerie Land, Pynchon's portrait of surf culture in faux-seventies southern Cal allegorizes the seeping pressure of fluids on almost-solid structures. Near the core of a typically busy panorama of Pynchonian crazies sits a figure who would not be out of place in Spenser's poem: Saint Flip of Lawndale, "for whom Jesus Christ was not only personal savior but surfing consultant as well" (99).[21] While the Saint appears only tangentially in the main plot and the story's main hero, Doc, is himself a private investigator, not a surfer, Flip's slightly off-center splash through *Inherent Vice* carries an intriguing Spenserian flavor: "What was 'walking on water,' if it wasn't Bible talk for surfing?" (99). In Australia, the wave-wanderer even purchased "a fragment of the True Board" (99), which allegorical token would not be out of place in Faerie Land. In his search for the "gnarliest break in the world" (99), St. Flip eschews the flatland entanglements of money, sex, drugs, and politics that drive the paranoid plots of Pynchon's novel. Instead he finds a hidden environmental enmeshment: "surfing where no surf should've been, a figure in white baggy trunks, whiter than the prevailing light could really account for" (99). Flip is not the only mystic in the novel, though he may be the least false of California's many prophets.

The otherworldly focus that marks Flip as what Doc calls "one of those advanced spirits" (100) points toward Devil's Doorsill, a mysterious and massive offshore break that only the Saint can find. Before he skips to Maui, Flip surfs the hidden break daily, but the novel does not directly encounter this oceanic magic until its climactic episode. Doc and his lawyer buddy Sauncho Smilax are at sea, seeking the good ship Golden Fang, which promises to explain so many mysteries:

> They rounded Palos Verdes Point, and there in the distance, out from San Pedro with all her staysails and jibs set, blooming like a cubist rose, came the schooner. The look on Sauncho's face was of pure unrequited love. (355–56)

Like Spenser's allegorical figures and, truthfully, like every plot-mystery in Pynchon since *The Crying of Lot 49*, the ship hovers ambiguously between double meanings; it is both the Golden Fang of an international heroin cartel and the lost schooner of pre-70s American idealism. Fantasy is always pursued by History in Pynchon—or is it the other way around?—so it's not surprising that a Coast Guard cutter impounds the ship before Doc and Sauncho get there. As usual, the mystery vanishes into fog and mist.

The key encounter, however, isn't with the Man but the Wave. Seeking the Golden Fang's historical conspiracy, Doc and Sauncho instead encounter St. Flip's secret offshore break:

> Doc put the sets rolling in at them from the northwest at thirty and maybe even thirty-five feet from crest to trough—curling massively, flaring in the

sun, breaking in repeated explosions.... It was St. Flip of Lawndale's mythical break, also known to old-timers as Death's Doorsill. (357–58)

This water doesn't seep but curls, and its impact soaks the novel's final pages. As with all of Pynchon's novels, the plot mysteries never quite untangle, but the enigmatic final pages, which lose Doc in highway fog somewhere east of the beach, obliquely recall St. Flip's waves. The key echoing term in both passages is *something*: Doc sits immobilized in the final scene, waiting "for the fog to burn away, and for something else this time, somewhere, to be there instead" (369). When Doc and Sauncho are about to glimpse Death's Doorsill, Pynchon sneaks the word *something* in his prose three times, and supplements this repetition by also repeating the closing lines' other opaque term, *somewhere* (357, 369). At sea, Doc, knight of the Beach, asks his maritime squire, "Saunch, do you hear something?" "Something," the lawyer replies (357). The narrator triples the word: "Something was also happening to the light, as if the air ahead of them were thickening with unknown weather" (357). Fog and wave, ocean and beach come together to create an allegory of seep, in which the paranoid histories of modern California and ancient Lemuria meet heroin cartels and dezombified sax players. The moist environment of the beach becomes a site of temporary transcendence and a glimpse past the veil, as Pynchon's postmodern California briefly becomes as allegorical as Faerie Land. Seeping ecological force unites these disparate visions.

Bleeding Birth: Ruddymane and the Knight of Temperance (2.1.25–62)

Neither Pynchon's surf-fog nor Spenser's bleeding bush quite captures the full contaminating force of seep ecology. Both Doc and the Redcrosse Knight, each of whom in different ways embarks on a voyage of education across an ecofantastic landscape, treat their encounters with magical liquids as opportunities for insight into human and symbolic environments. Spenser's next hero, Guyon, the Knight of Temperance, faces a more complex fluid exchange. He seeks balance, not transcendence. In the early stages of his adventure, Guyon finds the bloody counterfamily of the dying knight Mortdant, Amavia, and their child Ruddymane (2.1.35–2.2.10). Their gruesome tableau recalls and exceeds Fradubio: "Pitifull spectacle of a deadly smart, / Beside a bubbling fountaine low she lay, / Which she increased with her bleeding hart, / And the cleane waues with purple gore did ray" (2.1.40.1–4). More shocking than Amavia's bleeding, however, is the babe who plunges himself into the seeping fluid: "For in [Amavia's] streaming blood he did embay / His little hands" (2.1.40.7–8). The child Ruddymane, whose hands can never be washed clean, transforms the failure to perceive falsehood that was Fradubio's crime into an allegory for original sin. Guyon sees the child as the "ymage of mortalitie" (2.1.57.2), but the Palmer insists that the entire family represents passionate excess: "But temperance (said he) with a golden squire / Betwixt them both can measure out a meane" (2.2.58.1–2). As countermodel for the Knight of Temperance, the babe's parents represent excessive physicality and the wages of lust. In the material allegory of Faerie Land, these passions spill into liquidity. Their seeping stains even apparently innocent bodies, such as Ruddymane's.

In addition to extending blood's corrupting power from moral to theological failures, the dilemma of Ruddymane shows Spenser's allegorical landscape operating through what we might call *seep logic*. The babe is innocent but stained by his parents' sin, and this marked condition, according to theological structures Spenser inherited from Paul's letter to the Romans and commentaries on it from St. Augustine, Calvin, Luther, and others, typifies postlapserian humanity.[22] Sin seeps and contaminates whatever it touches. No mortal bodies remain unsaturated. The poetic innovation of Spenser's verse epic uses partial repetitions in its stanzaic structure to half-respond to seeping pressure. In two of the ten stanzas of the second canto of book 2 in which the poem features Ruddymane, Spenser twirls his poetry around the ambiguous conjunction *or*, a word Milton would later turn into a structural feature of his post-Spenserian epic.[23] In Spenser's earlier stanza, *or* mediates between two possible meanings of the babe's smile: "As carelesse of his woe, or innocent / Of that was doen" (2.2.1.6–7). From the imagined and inaccessible point of view of Ruddymane, the smile reflects lack of knowledge of his physical plight ("his woe") or its spiritual implications ("Of that was doen"). The blood's stickiness implies that this child of passion cannot be truly innocent, as does the narrator's theological pronouncement: "Such is the state of man" (2.2.2.8). The poet's *or* adds another level of seeping, in which alternative meanings of the bloody hands interpenetrate each other. Having set up the *or* machine, Spenser expands it several stanzas later. After Guyon fails to clean the babe's hands, stanza 4 provides a triple alternative: the stained hands represent either "blott of fowle offense" (2.2.4.1) or "bloodguiltinesse" (2.2.4.5) or "the charme and venome" of the water. Even though this episode, like that of Fradubio, turns on the figure of a well, in this case the magic waters are not clear life but a monument to complex entanglements, especially Amavia's chastity and her husband Mortdant's concupiscence. "Let them still be bloody" (2.2.10.4) pronounces the Palmer about the babe's hands, "and be for all chaste Dames an endlesse monument" (2.2.10.9). Like Fradubio's blood and all of Faerie Land's ecological symbols, this monument faces both ways as warning and lure. Guyon, embarking on the journey that will lead him to vanquish the seductress Acrasia, who originally corrupted Mortdant, bears the memory of Ruddymane's hands with him as allegory of sin, crime to be revenged, and evidence of an ecology of seeping exchange.

Conclusion: Seep or Swim?

What can we do in a world of seep? Some might say we can only endure, and perhaps cease to wish that seeping would stop happening. But in an effort to get somewhat ahead of such dismal prognostications, I'll offer a perhaps counterintuitive suggestion: learn to swim. It's the only way to engage an increasingly fluid world.

A seeping ecology conjures up potent fears of groundlessness, and I think it's important to recognize the disorientation of that feeling. Abandoning dreams of stability can be frightening. But it's also—and this strikes me as a crucial point—familiar. Every child who's learned to tread water has learned to exchange fear of groundedlessness for a temporary dynamic system of activity that enables one to stay afloat. Our legs and arms churn, and we manage to keep our faces out of the water so we can breathe. It's not a promise of endless safety—we'll need to come onto land eventually—but the precarity of the swimmer models a way to endure in a liquefying, seeping world. Swimming is not sustainable but temporary; not solid but fluid; not green but blue. As oceans rise, this practice asks us to remodel our relationship with our environment. Active acceptance and skilled labor can generate a productive attitude for a seeping world.

It's a long way from the slow stains caused by seepage to a drowned world in which we only survive by swimming. In choosing swimming as a response to seeping, I might be overreacting, or at least getting ahead of myself. It's also hard to swim in toxic waters, as I remember each time coastal storms flush wastewater into my favorite Long Island Sound swimming hole. But the shared physical intimacy of seeping and swimming provides glimmers of opportunity. The world isn't drowning yet, *except* in some human-dense areas near its warmer coasts. It's seeping: places we thought were dry are becoming moist and unreliable. Seep ecology may feel like a transition, a halfway house in between solid and liquid. But ecological history suggests that we never get all the way to one side or the other; there is no firm respite, either in green land or blue water. There is only seeping and exchange, dry things becoming wet, and wet things moving their moisture outside themselves. Seeping names an endless ecological process that violates boundaries without dissolving them entirely. In that moist exchange we live, if we would only admit it.

Notes

1. As Timothy Morton riffs off Commoner's law: "Everything is connected. And it sucks." *The Ecological Thought* (Cambridge, Mass.: Harvard University Press, 2010), 33.

2. Stacy Alaimo, *Bodily Nature: Science, Environment, and the Material Self* (Bloomington: Indiana University Press, 2010), 2–4 passim. See also Serenella Iovino and Serpil Oppermann, eds., *Material Ecocriticism* (Bloomington: Indiana University Press, 2014).

3. Commoner first introduced the "four laws of ecology" in *The Closing Circle: Nature, Man, and Technology* (New York: Random House, 1971).

4. On OOO as a corrective to Deleuzian becoming, see Ian Bogost, *Alien Phenomenology; or, What Is It Like to Be a Thing?* (Minneapolis: University of Minnesota Press, 2012), 6–9.

5. "Brown," my essay in *Prismatic Ecology* (2013), suggests that the color of sand, swamp, and shit defines ecosystems of decay that trouble the boundaries of object thinking. Jeffrey Jerome Cohen, ed., *Prismatic Ecology: Ecotheory beyond Green* (Minneapolis: University of Minnesota Press, 2013), 193–212. Exploring the unreal element "Phlogiston" in *Elemental Ecocriticism* (2015) shifted my analysis toward fictionality: "Phlogisitication mixes matter and the imagination." Jeffrey Jerome Cohen and Lowell Duckert, eds., *Elemental Ecocriticism* (Minneapolis: University of Minnesota Press, 2015), 55–76, 56.

6. Stacy Alaimo, "Elemental Love in the Anthropocene," in Cohen and Duckert, *Elemental Ecocriticism*, 298–309, 305.

7. On "storied matter," see Serenella Iovino, "Bodies of Naples: Stories, Matter, and the Landscapes of Porosity," in Iovino and Opperman, *Material Ecocriticism*, 97–113.

8. William Shakespeare, sonnet 64, in *The Riverside Shakespeare*, 2nd ed., ed. G. Blakemore Evans (Boston: Houghton Mifflin, 1997), 1854–55.

9. Sharon O'Dair, "Muddy Thinking," in Cohen and Duckert, *Elemental Ecocriticism*, 134–57.

10. See Donna Haraway, *When Species Meet* (Minneapolis: University of Minnesota Press, 2007); Timothy Morton, *Ecology without Nature: Rethinking Environmental Aesthetics* (Cambridge, Mass.: Harvard University Press, 2009); Graham Harman, *Guerilla Metaphysics: Phenomenology and the Carpentry of Things* (Chicago: Open Court, 2005); Bruno Latour, *We Have Never Been Modern*, trans. Catherine Porter (Cambridge, Mass.: Harvard University Press, 1993).

11. On petroculture, see Stephanie LeManager, *Living Oil: Petroleum Culture in the American Century* (Oxford: Oxford University Press, 2016); Mathew Schneider-Mayerson, *Peak Oil: Environmentalism and Libertarian Political Culture* (Chicago: University of Chicago Press, 2015).

12. For a fuller feminist treatment of abjection and liquidity, see Julia Kristeva, *Powers of Horror: An Essay on Abjection*, trans. Leon S. Roudiez (New York: Columbia University Press, 1982).

13. As in so much of early modern studies, Shakespeare dominates. See Gabriel Egan, *Green Shakespeare: From Ecopolitics to Ecocriticism* (London: Routledge, 2006); Lynne Bruckner and Daniel Brayton, eds., *Ecocritical Shakespeare* (Aldershot, UK: Ashgate, 2011); Simon Estok, *Ecocriticism and Shakespeare: Reading Ecophobia* (London: Palgrave, 2011); and the forum "Shakespeare and Ecology," edited by Julian Yates and Garrett Sullivan, *Shakespeare Studies* 39, no. 3 (2011): 23–116. On Milton see Ken Hiltner, *Milton and Ecology* (Cambridge: Cambridge University Press, 2003). On Marvell see Dianne McColley, *Poetry and Ecology in the Age of Milton and Marvell* (Aldershot, UK: Ashgate, 2007).

14. Sean Kane, "Spenserian Ecology," *ELH* 50, no. 3 (1983): 461–83, 461.

15. Alf Siewers, "Spenser's Green World," *Early English Studies* 3 (2010): 1–43, 2.

16. Gail Kern Paster, "Becoming the Landscape: The Ecology of the Passions in the Legend of Temperance," in *Environment and Embodiment in Early Modern England*, ed. Mary Floyd-Wilson and Garrett Sullivan (London: Palgrave, 2007), 137–52.

17. Ayesha Ramachandran and Melissa Sanchez, "Spenser and 'the Human': An Introduction," *Spenser Studies* 30 (2015). My thanks to Melissa Sanchez for sharing a preprint version of this essay with me.

18. On the five poetic precedents, see William Kennedy, "Fradubio," in *The Spenser Encyclopedia*, ed. A. C. Hamilton (Toronto: University of Toronto Press, 1990), 318.

19. All quotations from *The Faerie Queene,* 2nd ed., ed. A. C. Hamilton (London: Pearson Longman, 2001) are cited by book, canto, stanza, and line numbers in the text.

20. On Spenser's coastal poetics, see Elizabeth Jane Bellamy, *Dire Straits: The Perils of Writing the Early Modern English Coastline from Leland to Milton* (Toronto: University of Toronto Press, 2013), esp. 46–87.

21. Thomas Pynchon, *Inherent Vice* (New York: Penguin, 2009). All citations in the text are by page number.

22. For a brief summary of how Spenser engages with this theological tradition, see Carol Kaske, "Amavia, Mortdant, Ruddymane," in Hamilton, *Spenser Encyclopedia,* 25–27.

23. See, for example, Peter C. Herman, "*Paradise Lost,* the Miltonic 'Or,' and the Poetics of Uncertitude," *SEL* 43, no. 1 (Winter 2003): 181–211.

Saturate

LAURA OGDEN

The islands of the Fuegian archipelago are like fragments of land, broken off from South America's continental tip. They remind me of the way the jutting bits of family heirlooms are always vulnerable to neglectful care: teacup handles, the outstretched arm of a porcelain ballerina. But when you are on the islands, it is clear that there will be no putting things back together. Instead, the islands of Tierra del Fuego and Cape Horn seem to be barely holding their ground against the rough marriage of the Pacific and Atlantic Oceans. Generations of slow-moving glaciers created the archipelago's topographic features, which one early explorer described as "an inconceivable labyrinth of tortuous, storm-swept waterways."[1] Here, particularly around Cape Horn, enormous tanker ships are dwarfed and battered by tremendous walls of water. These are seas that make worlds and take them too.

Within certain traditions of nature, wonder is tinged with uneasy precarity. In the writing of early naturalists, for example, marvel over other lifeworlds is often simultaneous with feelings of insignificance in the face of the vast unknown that is nature. The Fuegian archipelago is one of several generative sites for this line of thinking. Charles Darwin, to illustrate, was so enthralled by Tierra del Fuego that he wrote, "No one can stand in these solitudes unmoved, and not feel that there is more in man than the mere breath of his body."[2] For the young naturalist, and countless others after him, the Fuegian archipelago offered a mode of nature resonant with transcendent possibility, while at the same time serving as a laboratory for thinking about nature's laws and change in

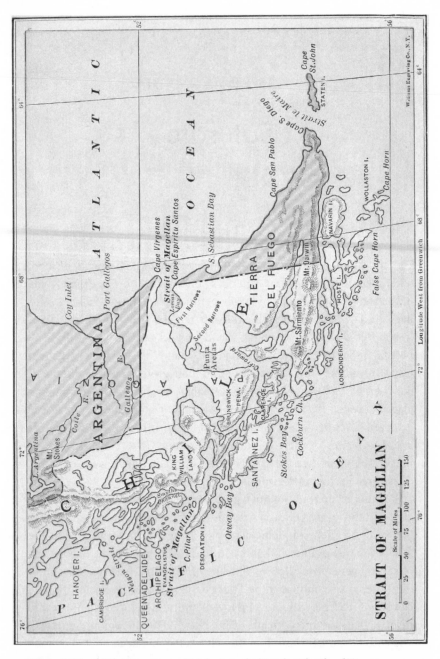

Figure 1. Map of the Fuegian archipelago, including the islands of Tierra del Fuego and Cape Horn. By James Bryce, First Viscount Brice, 1838–1922.

species. These are familiar tropes in the contradictory logics of modernity. Instead, let us pause on the image of an awe-inspired Darwin standing alone on the beach of this windswept archipelago—a reminder of the ways in which metaphysical uncertainties permeate claims to the universal ("the origin of species," for example).

Today, North American beavers are remaking Tierra de Fuego's sub-Antarctic forests at a rate that would have taken Darwin's breath away. Where water once flowed through the forests of Cordillera Darwin, beavers have now dammed the rivers, forming vast ponds that seep into the surrounding forest. These are sodden worlds of smothered roots and submerged life. Toppled trees form jumbles of silvery-grey timber, resembling a discarded game of pick-up sticks (Figure 2).

I traveled to the Fuegian archipelago to understand the politics of environmental change there, particularly the world-making power of beavers. To help with this, I spent months learning from ranchers and shepherds on the great sheep farms called *estancias* that span the interior steppe of Isla Grande, in Tierra del Fuego, as well as talking with scientists, foresters, and others who are deeply concerned about the ways beavers are transforming the landscape. More broadly, my collaborators on this project, Christy Gast, a U.S.-based artist, and Camila Marambio, a Chilean curator, and I have been inspired by Isabelle Stengers's speculative

Figure 2. Film still from *The Dreamworlds of Beaver* (2014), by Christy Gast, Laura Ogden, and Camila Marambio.

approach to environmental ethics.[3] For Stengers, the speculative entails "the power to make practioners think, feel, and hesitate."[4] Her approach encompasses similar ethical obligations as Donna Haraway's call to "response-ability," a commitment that has been equally significant to this project.[5] The speculative, as I explore in this essay, can destabilize naturalized categories of social–ecological difference—in this case, ideas about "native" and "invasive" species, though this is not my primary intention here. Instead, I am interested in using the speculative to evoke a more historically situated, as opposed to universal, ethics and politics of living and dying in a time of ecological crises.

Multiple modes of saturation are implicated in the making of the Anthropocene, as we call this time of ecological crisis: rising tides seep across shorelines, carbon emissions soak the atmosphere, invasive species infiltrate new lands and seas. This essay maps several modes of saturation specific to the Fuegian archipelago, such as the global trade in animal skin, colonial settlement in the region, and scientific ideas about nature. While we often think of "saturation" as an additive and fluid process, it is a process of loss as well (Figure 3).

Skin

European colonial settlement in North America was sustained by a ready supply of animal skins, particularly the North American beaver, or that is a possible reading of Eric Wolf's majestic *Europe and the People without History*.[6] Like a war machine fueled by beaver pelts, competing French, English, and Dutch trading companies staked claims to beaver territories along interior waterways, such as the Hudson and St. Lawrence Rivers. Imagine a network of rapidly expanding trap lines, spurred westward by the serial collapse of beaver populations. As Wolf shows, this territorial assemblage, centered on an economy of skin, led to the dramatic reorganization of the social, spiritual, and political life of native communities in the Americas. Yet his work also reminds us of the role of animal life in the production of empire. In this case, the beaver fur trade became the logic and means for world making at the frontier.

Skin is the membrane where worlds meet.

In 1946, decades after the near extinction of beaver communities in North America, the Argentine government imported twenty Canadian beaver pairs to Tierra del Fuego in the hopes of establishing a fur trade.

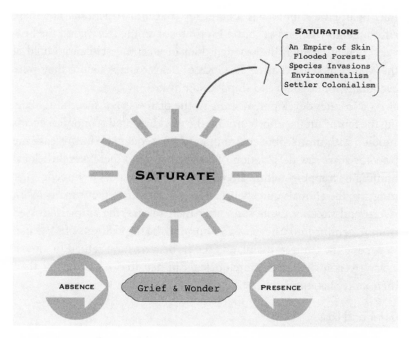

Figure 3. Saturation is a process of becoming through absence and presence. This essay maps the tensions of absence and presence (grief and wonder) in the Fuegian Archipelago.

Today, stocking Tierra del Fuego with Canadian beavers seems like an unlikely economic development proposition. But at the time, Tierra del Fuego's limited infrastructure and connection to Buenos Aires posed significant challenges to the Perón administration's dreams of a united, economically independent nation. From the standpoint of Buenos Aires, Tierra del Fuego was worlds away, a periphery particularly suited to the logic of frontier world making. This was also a time when U.S. and Canadian wildlife agencies were restocking lakes with beavers throughout North America, suggesting there was an apparatus in place for handling the logistics of capturing, transporting, and reintroducing beavers into new territories.

We know something of the beavers' long journey to the southern hemisphere. First, Thomas Lamb, a Canadian bush pilot, helped trap the beavers from lakes in northern Manitoba, Canada. Next, Lamb accompanied the animals to New York City by rail, then by air to Miami, where they boarded a clipper ship bound for southern Argentina. For these efforts Lamb was paid $13,000.[7] En route, or so the story goes, the beavers chewed through the wooden door of a refrigerated cargo hold at the Miami airport, causing all kinds of beaver chaos before they were eventually rounded up and shipped southward.

As a frontier development strategy, the plan was not successful, sharing the fate of similar efforts around the world: limited economic success coupled with unwelcome environmental consequences. In this case the postwar years saw devaluation of beaver pelts and there was little local hunting or trapping culture anyway. With no predators to impede their progress, the animals quickly began to colonize the region's many lakes, rivers, and streams. Quite soon after their release, the animals crossed from Argentina into Chile, even swimming the frigid Straits of Magellan to occupy the Chilean mainland. Today, beavers have settled in almost every river in the Fuegian archipelago, in densities much higher than their native habitats.[8]

Grief and Loss

My field notes are filled with observations about birds, a habit I picked up from my ornithologist father. When I travel to places that feel like nature, as surely Tierra del Fuego does, there are always moments when I realize that my father's bird wonder is edging out my duties as an anthropologist. For example, during the second week of November 2011, I was so thrilled by the Lesser rhea *(Rhea pennata),* a small ostrich-type bird, that I recorded little else about my visit to the Patagonian mainland. Instead of ethnographic observations, my field notes carefully catalog the ways the rheas dart and flutter across the steppe, with "crazy dodgy, jerky, leapy movements." These are flightless birds, yet when spooked they spread their wings wide and speed across the grasslands "like an airplane too heavy to take off," or so my notes describe. Male rheas spend forty days incubating the eggs of multiple females, a reversal of things that I found particularly compelling. Bird folks know that the lesser rhea enchanted Darwin as well. *Darwin's rhea,* as they are also called, provoked his curiosity about the origin and difference of species.

But grief and loss can extinguish wonder about other worlds, as I discovered in Tierra del Fuego. Within two months of my return to the United States from my first research trip to the archipelago, my father became very ill and died. I have few words to describe his abrupt loss except to say that when I returned to the islands a few months later, Tierra del Fuego had changed in my absence. It was as if an ambivalent haze had cloaked the landscape. I could not bear to even think about birds. I could not scribble lists in my bird guides. Andean condors filled me with rage.

Species loss and species invasions are contingent processes of world making in the Anthropocene, with grief over loss the affective register of our time. In many conservation communities, grief over loss (habitat, species, wilderness), or at least the fear of loss, dominates the ways in which landscapes and related practices of care are produced and contested. Losses related to species "invasions" seem to be particularly unsettling, contaminating the distinctions between wild nature and social nature. Certainly, the eradication of beings considered nonnative continues to be a top priority for land managers around the world, including in the Fuegian archipelago. There, grief and loss seem to overwhelm any ethical or political nuance, limiting the boundaries of concern about the beaver "problem" to the practicalities and probabilities of eradication: different kinds of traps, possible hunting incentives, various biological controls, and the like.

Forest

In the forest, beavers are strangely invisible and simultaneously ever-present. They are fleeting. Beavers spend their days hidden below tannin-dark waters. I have rarely seen a beaver, even after waiting hours in the cold beside their ponds. Instead, only trace impressions of being: rows of chew marks in the bark of a fallen tree, castoreum's oily residue at the water's edge, draglines in the mud (Figure 4).

While one rarely sees a beaver, radiating networks of dead and dying trees signal their advance across the landscape. Ecologists have compared the beavers' impacts in Tierra del Fuego to the largest landscape-scale change "since the retreat of the last ice age."[9] This is an apt comparison, since the Fuegian forests were formed shortly after the retreat of the ice sheets 10,000 years ago.[10]

While I was visiting the Argentine Centro Austral de Investigaciones Científicas (CADIC) in Ushuaia, I spoke with Guillermo Martínez Pastur, a scientist interested in practices of sustainable forestry. Martínez

Figure 4. Rows of chew marks on a fallen tree. Photograph by Laura Ogden at
Vicuña, Karukina Nature Park, Tierra del Fuego, Chile.

Pastur estimates that beavers have caused the loss of about 700,000 hectares of Tierra del Fuego's forests, with much of this loss occurring along
the region's rivers. Unlike other forms of selective forestry, Martínez Pastur expects these losses to be long term; forest regeneration, if it occurs,
will take about a hundred years in the flooded areas. The refrain of uncertainty reverberates in discussions with environmental scientists, as
forest loss caused by widespread flooding is without analogue.

For environmentalists and forest scientists, the Fuegian forests are
rare and understudied. Ecologists are still trying to understand the ecosystem dynamics of these forests, as the region's thin soils and harsh conditions challenge accepted ideas about forest nutrient cycling and other
processes. At first glance, the forests seem simple. There are only a few
species of trees here, with the lenga *(Nothofagus pumilio)*, a member of
the southern beech family, serving as the forest's backbone. To survive
the strong sub-Antarctic winds, which can blow at around ninety miles
per hour, lenga trees grow very close together, their trunks and branches
forming an entangled mass. Within these entanglements, lush miniature
forests emerge composed of liverwort, moss, and lichen. Some lichens

seem to live forever, quite possibly the very lichen in the forests today were there when Darwin arrived over a century ago. Today, they too are being swept away as beavers reengineer the landscape.

For Chilean forest lovers, the steady incursion of beavers into Chilean Patagonia resonates with a longer history of territorial disputes with Argentina. Conflicts over Patagonia's geopolitical boundaries arose soon after independence from Spain in 1810, and tensions over these boundaries continue to this day. In an interview a rancher I spoke with recalled the beavers' arrival in Chile in 1962. His father, who was living at their coastal estancia at the time, heard word that "little men" were cutting the forest trees, as if a diminutive army had crossed the border in an effort to despoil the Chilean landscape. This rancher, who has a real affection for the forest, used the term *plague* to describe the beavers' unchecked expansion.

An Empire of Skin

Beavers are not the only recent arrivals colonizing the Fuegian archipelago. Missionaries brought European rabbits over from the Falklands, which then spread dramatically, creating an "outbreak" in Tierra del Fuego comparable to the Australian experience.[11] Combatting these rabbits by introducing potential predators, including grey foxes, produced a familiar cascade of unanticipated problems. Now grey foxes can be found in greater density in Tierra del Fuego than in the rest of Chile. Minks, muskrats, and wild pigs roam Tierra del Fuego as well, though they are also relatively recent residents.

Yet, pound per pound, sheep are the most dominant form of introduced animal life in the region. This is an empire of skin. For the few tourists that drive the interior roads of Isla Grande on Tierra del Fuego, scattered historical markers interpret the grand era of the estancias. At the turn of the last century, two large companies managed the majority of lands within Chilean Isla Grande for sheep ranching.[12] Over the past century, changes in market prices for wool and the impacts of agrarian reform have shaped estancia life. Still, for most of this century, two to three million sheep have grazed the pastures of the Chilean estancias (Figure 5).

In Tierra del Fuego, work is done with horseback and dogs, and labor is hardly touched by electrification. Sheep are entangled in complicated webs of animals and humans, where the global circulation of sheep mirrors the circuits of empire. Farmers import sheep stock from New Zealand, the Falklands, and South Africa. After being tended to by Chilean migrant laborers (mainly from the island of Chiloe), their wool is shipped to New Zealand where it is graded, then sold through multinational brokers. Some of this wool ends up in Merino base layers sold by Patagonia, the American outdoor company. While sheep are ever-present, they are largely invisible in the debates about invasive species, indicative of the ways we value animal life.

Forest Grief and Loss

The introduction of sheep ranching to southern Patagonia in 1877 disrupted the fairly amicable relations that seemed to have existed between early settlers in Punta Arenas, the penal colony capital of southern Patagonia, and the local Indigenous communities.[13] Within the grasslands of northern Tierra del Fuego, the Selk'nam (also called Ona) hunted guanaco, foxes, rodents, and birds and also collected berries, birds' eggs, and mushrooms.[14] The Yaghan and Kawésqar (sometimes referred to as the "canoe peoples") lived along the coastal inlets and channels of Tierra del

Figure 5. A shepherd at Estancia Serena, Tierra del Fuego, Chile. Photograph by Laura Ogden.

Fuego and were dependent on marine mammals and shellfish for their livelihoods. Of course, much of the ethnographic information we have about these people coincides with the grief and loss associated with rapid colonization.

Indeed, European settlement brought rapid usurpation of Indigenous land for ranching, vicious cycles of escalating violence, introduced diseases, and missionization, eventually leading to the near extermination of all of the Indigenous communities in the Fuegian archipelago, except for a small Yaghan community who still live in Chilean Puerto Williams and a community of Kawesqar families living in Puerto Eden, Wellington Island.[15] As Martinic Beros chillingly describes, settlers on Isla Grande enacted clandestine raids of such "homicidal violence" that the Selk'nam were annihilated within two decades.[16]

In Tierra del Fuego, one is always grappling with the dialects of absence and presence. Sepia-toned images of exterminated Selk'nam obligingly decorate the walls of almost every café, as well as mugs, t-shirts, and book covers (Figure 6). During a community meeting about the ethics of using these images, an older Yaghan woman told us, with reserved

Figure 6. Mural depicting the Selk'nam Hain ritual, at the Hain Hotel in Punta Arenas, Chile. Photograph by Christy Gast.

sadness, about a jigsaw puzzle of her face sold at the Buenos Aires air-port. These images, often of painted semiclothed bodies, have become the icon of Tierra del Fuego's "wild" nature. While these forms of touris-tic commodification are not particularly complicated, I continue to be surprised by how important the Selk'nam are to the thinking and feel-ing of Chilean forest lovers. In casual conversations, for instance, the Selk'nam come up all the time, as if their presence continued to enliven the forest landscape.

Many environmental activists in Chile first became radicalized as human rights and peace activists in Santiago during the Pinochet dicta-torship. In other words, they became radicalized around commitments for life. When things got "too hot" in Santiago, as one activist explained, she and others fled southward, seeking refuge in the southernmost periphery. Years later, Tierra del Fuego's forests became an object of con-cern after U.S. forest activists alerted the local community to an Ameri-can company's deforestation plans. From my interviews, it is clear how indebted Chilean activists were to their counterparts in the United States. American forest activists, all veterans of the antilogging wars of the Pacific Northwest, taught Chile's emergent environmental activist com-munity the strategies of territorial inscription—or marking the presence of life in ways to make political claims. These inscription practices in-cluded protests and making signs as well as particular ways of framing debates, such as using the courts.

Yet Chilean environmentalists also brought their own commitments to the fight to protect Tierra del Fuego's forests. For these activists, the Fuegian forests are significant as ecosystems, but they are also landscapes marked by multiple cycles of violence (including the violence of invasive species). For environmentalists in Tierra del Fuego, the Selk'nam have become another mode of inscription. Instead of the Selk'nam represent-ing the purity of pure nature, for Chilean environmentalists the Selk'nam signal and mark the fragility of life at world's end.

Speculative Wonder

Grief can calcify our curiosity about other worlds. This I learned in the forests of the Fuegian archipelago, a landscape resonant with multiple histories of loss, all more significant than my own. The philosophers Gilles Deleuze and Félix Guattari use the term *arborescence* to describe a world-making logic that is static and hierarchical and ignores the messiness of entangled lives.[17] If wilderness is an abstract ideal of nature cleansed of the messiness of politics and history,[18] forests have become the epicenter of the purification campaign. But really, "tree logic" speaks less about the moss-rich assemblages of real trees, such as those of the Fuegian forests, and more about the process of petrification. Let us refuse the ways loss and grief can turn us to stone and instead pause for a moment of speculative wonder.

A beaver's life acquires form in the water. Over the millennia, beavers have evolved with land *and* water. At some point, they learned forestry. Later, they learned to build underwater dens. We all make spaces that comfort us. Beaver social relations, as the anthropologist Lewis H. Morgan described, are manifest in the "dams, lodges, burrows, and canals for objects which are common to them as a family."[19] Piles of rough timber and tree limbs: *really,* this is an architecture of kinship. Kinship resists the logic of trees. It is a much messier entanglement of relations. Like other forms of domestic life, in beaver worlds, home comes into being through routinized labor. Chewing and dragging, chewing and dragging. Such repetition.

In the woods beavers are graceless and lumbering. Yet their ponds offer a lightness of being, a safe tranquility. Here, light captures and enlivens

Figure 7. A film still from *The Dreamworlds of Beaver* (2014), by Christy Gast, Laura Ogden, and Camila Marambio.

microscopic particles. With the calm of a snow globe, these sediments drift along the water column. Silt, decades old, hovers and skips along every surface, like the landscape of the moon. For astronauts, water signals life's potential, though this seems a low bar for encountering the magic of other worlds. Let us resist the indications of industry. "Busy as a beaver." Instead, let us contemplate the intimacy of water.

Notes

1. Charles Wellington Furlong, "The Alaculoofs and Yahgans, the World's Southernmost Inhabitants," in *Proceedings of the Nineteenth International Congress of Americanists*, ed. F. W. Hodge (Washington, D.C.: ICA, 1917), 224–34.

2. Charles Darwin, *Journal of Researches into the Natural History and Geology of the Countries Visited during the Voyage of* H. M. S. *Beagle round the World* (1839), 455.

3. Isabelle Stengers, *Cosmopolitics I,* trans. Robert Bonnono (Minneapolis: University of Minnesota Press, 2010).

4. Ibid., 15.

5. Donna Haraway, *When Species Meet* (Minneapolis: University of Minnesota Press, 2008).

6. Eric R. Wolf, *Europe and the People without History* (Berkeley: University of California Press, 1982).

7. Alejandro G. Pietrek and Laura Fasola, "Origin and History of the Beaver Introduction in South America," *Mastozoología Neotropical* 21 (2014): 355–59.

8. C. A. Silva and B. Saavedra, "Knowing for Controlling: Ecological Effects of Invasive Vertebrates in Tierra del Fuego," *Revista Chilena de Historia Natural* 81 (2008): 127.

9. C. B. Anderson et al., "Do Introduced North American Beavers *Castor canadensis* Engineer Differently in Southern South America? An Overview with Implications for Restoration," *Mammal Society* 39 (2008): 33–52.

10. J. Rabassa et al., "Quaternary of Tierra del Fuego, Southernmost South America: An Updated Review," *Nat Rutter Honorarium* (2008): 217–40.

11. F. M. Jaksic and J. R. Yanez, "Rabbit and Fox Introductions in Tierra del Fuego: History and Assessment of the Attempts at Biological Control of the Rabbit Infestation," *Biological Conservation* 26 (1986): 368.

12. G. Butland, *The Human Geography of Southern Chile* (London: Institute of British Geographers, 1957); M. Martinic Beros, "The Meeting of Two Cultures: Indians and Colonists in the Magellan Region," in *Patagonia: Natural History, Prehistory, and Ethnography at the Uttermost End of the Earth*, ed. L. A. Borrero, C. McEwan, and A. Prieto (Princeton, N.J.: Princeton University Press, 1997).

13. Beros, "The Meeting," 215.

14. Anne Chapman, *European Encounters with the Yamana People of Cape Horn before and after Darwin* (Cambridge: Cambridge University Press, 2010).

15. Ibid.

16. Beros, "The Meeting".

17. Gilles Deleuze and Félix Guattari, *A Thousand Plateaus: Capitalism and Schizophrenia*, trans. B. Masumi (Minneapolis: University of Minnesota Press, 1987).

18. William Cronon, "The Trouble with Wilderness; or, Getting Back to the Wrong Nature," *Environmental History* 1 (1996): 7–28.

19. Lewis H. Morgan, *The American Beaver and His Works* (Philadelphia: J. B. Lippincott, 1868), 134.

Behold

SERENELLA IOVINO

Behold moves us out from where we stand. Closed in the solitude of self-contemplation, the I is thumped by a call—*behold*—that smashes its self-sufficient silence, a silence that concentrates all presence in the close proximity of the present or in the apparent distance of things. *Behold* forces us to look elsewhere—or just to swerve our mind beyond what seemed familiar, and conceals instead landscapes unseen. Because *behold* is a call that draws the eye/I to something that has always already been there—unheeded, undetected, or unrecognized—or to something that there will be, in a time that will come, and, once it will be here, it will be impossible for us to disregard. Many things and beings inhabit these landscapes: nonhuman natures, marginal persons, gulfs of injustice, impure inhuman lives, beauty, the future, the earth, darkness, the countless hyperobjects that mark, as some say, the end of the world. Or, what might be the greatest hyperobject of all—God.

As I think of *behold,* my veering verb in this collective passage across the natures of nature that this book attempts, echoes of the ways we humans have been summoned by the call of the divine come to my literary memory. Because—even if God, to my earthly ears, more often speaks with the endless speeches of this planet where we live and have our being—*behold* traditionally carries in itself that very voice. It is the voice of something both present and unattainable, the voice of all the invisible mundane enmeshments and—at the same time—of the world's distance from itself. And so, two routes open in front of me. One is the route of immanence—the route of a here-and-now full with hidden but luminous further dimensions. The other is the route of transcendence—the route to an elsewhere that might enlighten or blind you, depending on which side you stand on while you gaze at it. Both routes carry you away from where you are, and both carry you back to where you were. But, in doing so, they both trigger a swirl: because, if you really *behold* what these routes hint at, *you* will be the one who change. *You* will be veered. Because *behold* calls you to veer from the used paths. Because *behold* is a veering call.

First Route: Beholding Immanence

Rio de Janeiro, 1963

There is a cockroach crawling out the closet. You slam the door. She is smashed. She is dying. Behold. Her eyes are open. These eyes are watching you. They want you to behold this crushed body. You see, you witness. This crushed body is there, her eyes are asking you to taste her moistness and fullness and otherness. This crushed body—her eyes wet and open—is a face. The face of the radically other. The face of the one being that is all beings. You see, you witness, you behold. This crushed body it is you, too. This seeing body you now behold is the face of god.

The Passion according to G.H., by Clarice Lispector, is a book that, once it enters your life, there it will stay. Buried as it might seem, it periodically reemerges, carrying with itself the "difficult pleasure" that, as its author understood, was yet a pleasure. Clarice was well aware of this complexity as, addressing the novel's "Potential Readers," she pointed out that she "would be happy if it were read only by people whose outlook is fully formed. People who know that an approach—to anything whatsoever—must . . . traverse even the very opposite of what is being approached" (v).[1] People, in other words, able to see—and veer. Forming one's *outlook*—like forming one's soul—is indeed challenging.[2] It means to be able to look outside yourself and give a frame to what you see. It means to be virtually ready to organize your experience in relation to your (inner) I/eye, while at the same time being ready to forget all this and "traverse" from the self to its very opposite. And so, you really need to be fully prepared when you face a story where the inside and the outside of your I/eye *impurely* merge into one another—a story that says: "What I have seen is unorganizable" (60).

In this decisive novel, written "in a quick outburst" at the end of 1963, Clarice captured "the full violence, the physical disgust of her encounter with God."[3] Through the character of G.H.—whose initials might be an abbreviation of the Portuguese *gênero humano,* "humankind"—she completely shuttered the standpoints of anthropocentric discourse, inviting readers to *behold* what is beyond and beneath the disenchantment of the nonhuman and the humanization of God. She knew what this was all about. Born in Ukraine in 1920, her family emigrated to Brazil when she was two months old. As her biographer reminds us, she "emerged from the world of the Eastern European Jews," a vanishing society whose stubborn mysticism and religiousness she brought "into a new world, a world in which God was dead."[4] In this world, loaded with the inquietudes of two continents and her personal history of loss, violence, and staggering talent, the beautiful Clarice was to become not only a national monument but also the catalyzer of a worldwide attention for her capacity to reshape the language, categories, and style of South American literature. Topics such as maternity (an ancestral, prelinguistic, and *material* motherhood), the ontological revelation of sameness-in-difference, the unremitting observation of the personal and apersonal embodiments of life, have contributed to fasten her bond with feminist thinkers and with philosophy in general.[5] Underlying all these themes is Clarice's tendency to swerve the gaze from an alleged human superiority, and to restore the connection—almost an eye contact—with a matter/God that permeates all forms of being.

Incubating all this, *The Passion* is a long monologue in which G.H., a bourgeois and wealthy professional from Rio de Janeiro, tells the story of how—faced with a dying cockroach—she "converted" to the understanding "that the world is not human, and that we are not human" (61). As in all mystical tales, the plot, evoked in my "lyrical" synthesis, is minimal—and yet abyssal. Alone in her penthouse, G.H. is tidying up a room previously occupied by her servant. In the closet of this unadorned and almost empty space, she discovers a cockroach and, after observing it for a while, crushes it. By beholding the insect's smashed body, she realizes that the cockroach itself (or better, *herself*: G.H. views it as a female and an Ur-mother) is life in its primary, "impure" form. And this impure matter—a "prehuman divine life" (93–94) that *looks back* through this broken body—is God. In the final scene, G.H.'s pantheist *passion* is completed as she eats the matter coming out from the roach's body, here again

in a ritual of inherent mutuality: "Then, through the door of condemna-
tion, I ate life and was eaten by life. I understood that my kingdom is of
this world" (112). The traditional order of transcendence is thus reversed
in a revelation of pure immanence, and the dualism of matter and spirit is
rejected for a mystical Spinozism: an ecological vision of the divine based
on the intimate, bodily, and prelogical reciprocity of being in which
"everything looks at everything, everything experiences the other" (58).

In this unexpected ontological trial, seeing and beholding play the
major role. Their meaning, however, is not limited to the visual dimension: overcoming the realm of reason, they monistically condense all the
experiences occurring to G.H. as the narrating fragment of a vast body–
mind continuum. It is by beholding this "crude, raw glory of nature" (58)
that G.H. progressively overturns the organized discipline of the ego,
opening the cataracts for a mystical seeing in which everything is "tra-
versed" and becomes its other. G.H in fact "becomes (with) what she
sees": "The desire for intimacy is the desire to surrender to immanence,
to experience continuity with the world, the desire to become-animal,
to become-flesh. It is through vision that G.H. surrenders . . . and emp-
ties herself from humanity."[6] In an impure narrative–ontological swirl,
the Jew Clarice Lispector revives the passage of the Exodus (a word that
means "a way out") implying that the face of the divine cannot be beheld
and survived (Exodus 33:20). By beholding the face of God—a God that
is here and now—G.H. takes her "way out," veering away from her human
self and relinquishing to the world. And so, emptied of her humanity,
she can eventually find herself "face-to-face with the dusty being that
was looking back" (49) at her. In its immediate, intimate manifestation,
the roach's body indeed *is* a face, a "shapeless face" (47) that touches and
beckons from the immemorial neutrality of being. Emmanuel Levinas
has famously suggested that the other's face—*le visage d'autrui*—carries
an endless appellation to recognize the Other in its naked, "absolute exposure."[7] In *The Passion* the *visage* addressing us—a neutral, inexpressive,
but at the same time naked and absolutely exposed face—is plunged even
deeper in the ontological abyss of copresence. The kind of reciprocity
that emerges in this revelation of mutual belonging and unfathomabil-
ity, in fact, is not accidental but primordial; it is rooted in the womb of
matter, *mater materia*. In this ur-material dimension, existence is the
transitive proximity of all beings that *are* as they *see* each other:

> The cockroach . . . was looking at me. I don't know if it saw me. . . . But if
> its eyes didn't see me, its existence existed me: in the primary world that I
> had entered, beings exist other beings as a way of seeing one another. . . .
> The cockroach saw me not with its eyes but with its body. (68)

This bodily distributed seeing is a "chiasm," an "intertwining," as Maurice
Merleau-Ponty would say. If we abandon the human-centered categories
structured around the divide of a seeing I/eye and a visible other, we reach
a dimension where "there is a reversibility of the seeing and the visible,"[8]
a deep and carnal intermingling that marks each being as "enmeshed
within the visible present and [as] both seeing and seen, touching and
touched by the world and the things around us."[9] This dimension is inhabited by things that see, by worlds that behold each other as their way
of existing. This side—the Greeks imagined—is the realm where nature/
physis shines in the plenitude of light/*phōs,* and all creation is nothing
else but a seeing, a seeing that pierces the darkness and emerges in all
appearing things. Here *phainesthai* is the law of being, and coming into
light—alighting in this world—does not simply happen *in front of* the
subject but *along with* it.[10] In this dimension, lateral to the order of reason,
G.H. sees, beholds, and is one with everything: empires' cycles, pyramids
surging and decaying in the appearing desert, turns of elements and extinct species, glaciers arising and melting, proteins and protozoa, voices
of hieroglyphs and magmatic darkness—all this she witnesses and beholds
through the roach's body, feeling the horror of knowing that her organized self depends on the very disorder of being, spiraled with time. Stability in fact is only an illusion, whereas reality is everything which veers
and returns our gaze—it is the roach's cilia, it is moistness and nausea, an
inferno of matter, madness—it is raw life. But at the same time, G.H.'s
experience is also the poetic translation of another vision: that all life is
a dynamic unfurling of forms, a blind, unrelenting movement that, over
eons, continues to evoke kinships. Because, in Lynn Margulis's words,
"all beings alive today are equally evolved. All have survived over three
thousand million years of evolution from common bacterial ancestors.
There are no 'higher' beings, no 'lower animals,' no angels and no gods."[11]

If one can behold all this, then one is ready to understand that the
"deepest life identity" of the human is not simply intertwined with the

nonhuman but is one with the *in-human*, this "estranged interiority" that forces the I to change sides and accept the "discomfort of unfamiliar intimacy" with the Other.[12] This "unfamiliar intimacy" is—biologically as well as ontologically—the cypher of the human itself:

> I had looked upon the live cockroach and had discovered in it my deepest life identity. . . . Listen, in the presence of the living cockroach, the worst discovery was that the world is not human, and that we are not human. . . . The inhuman is our better part, is the thing, the thing part of people. (49–50, 61)

The "thing part" inhabiting the human—be it the blindness of evolution, the magmatic Ur-mother of forms and bodies, or "God-matter" (61)—at once levels and disorganizes everything, undermining all articulation, including understanding and language: "The world interdepended with me [and] never again shall I understand what I say. . . . Life is itself for me, and I don't understand what I am saying. And, therefore, I adore." (173). "I adore": which means, standing in awe, I behold. Because beholding is an experience that leaves you at a loss for words—it leaves language itself at a loss for words. Here all the intellectual counterforts built to ensure our essential "purity" by seceding the human from all the rest are nothing else but a way to contain our contact with matter—a matter stigmatized as "impure" only because it takes us back to our original chaos. In G.H.'s story this very chaos is redeemed by redeeming matter itself and by beholding in the roach's body the divine who dwells in the copresence of things: a "dynamically unfolding process of open-ended interactivity"[13] that is the *creation* of God-matter. This explains why such words as *hope* or *forgiving* sound meaningless now: if matter is *already* a manifestation of God, it is not necessary to imagine an elsewhere in which the present will be redeemed: "The present is God's today face. The horror is that we know that it is right in life that we see God. It is with our eyes truly open that we see God" (141).

   If salvation has to be sought, it is not "from matter but within it, within our delicate, difficult interactivities."[14] In the immanent body of being, God is there, in the open, looking back to us with all H* ever-evolving creaturely eyes.

Second Route: Transcend, Behold
Auschwitz, 1943–44

> From the point of view . . . of substances that you could steal with profit,
> that laboratory was virgin territory. . . . There was gasoline and alcohol,
> banal and inconvenient loot: . . . the offer was high and also the risk, since
> liquids require receptacles. This is the great problem of packaging, which
> every . . . chemist knows: and it was well known to God Almighty, who
> solved it brilliantly . . . with cellular membranes, eggshells, the multiple peel
> of oranges, and our own skins, because after all we too are liquids. Now,
> at that time, there did not exist polyethylene, which would have suited
> me perfectly since it is flexible, light, and splendidly impermeable: but it
> is also a bit too incorruptible, and not by chance God Almighty himself,
> although he is a master of polymerization, abstained from patenting it:
> He does not like incorruptible things.[15]

All those who know Primo Levi know that the author of *If This Is a Man*
was a chemist. Familiar with the elements' "lyrical drift and continuous
conjoining," in a beautiful book titled *The Periodic Table* (1975) he followed them in their "stormy cultural and material intermixes,"[16] letting
them speak and act along with the events and using their laws and swirls
to shed some light onto one of the darkest times of European history. In
the tale "Cerium," Levi contrasted the way humans manipulate chemicals and the way God—traditionally, more farsighted—does. The image
of "God Almighty" as a "master of polymerization" facing the unwanted
consequences of polyethylene resonates deeply with our discourse on
behold. We can picture this God as a chemist who observes all substances
through a microscope, while at the same time beholding their effects
in the vast, macroscopic horizon of creation. Provident by definition,
this Ultimate Senior Chemist—Levi pinpoints—does not simply abstain
from *creating* polyethylene, but from *patenting* it: S/He leaves to humans
the *choice* (and *responsibility*) to do so, just as S/He leaves to them the
choice (and responsibility) to do all the rest.

As this scene of a divine scientist in a lab suggests, the movement
inbuilt in the gaze of Levi's God is *transcendent*: by considering new elements, God beholds beyond singularities and individuals, surmounting
(*id est*, trans-scending) the here-and-now from a higher point that "bear[s]

vision of things to come over the terrestrial yonder."[17] But the gaze of ecology is also transcendent: it is so certainly not because it calls us to behold an alleged metaphysical truth beyond the deceptiveness of phenomena but because it regards the more-than-human world by *hovering over* its systemic fabric of causes and effects. Surpassing the horizon of the present and of unconnected presences, the "transcendent" dimension of ecology is indeed entrenched in the time-space-matter field in which we are all entangled, whether we grasp it or not. This entanglement transcends singularities, it transcends the punctual being of things, and creates flows of excess—it overflows, transcending the landscapes that might be contained by our gaze. This ecological transcendence is to us another veering call; it too says: *behold*. And maybe also Levi's Almighty God, master of polymerization, wants to tells us this: *behold, you*—you all, and you humans with your patented creations—are in and of this excess. Behold this unfurling view that "grounds human beings within the continuum of life, and . . . situates the history of their embodied skills within the unfolding of that continuum."[18] But in order to understand the becoming of this excess within the continuum of life, we need another way of seeing things, we need a *prospect,* a dynamic "consideration of something in the future to be gained from viewing it, and the action of facing it."[19] This *prospect* is the standpoint of a moral imagination that would allow us to envisage the scope of our actions, as they are combined with the agency of beings, things, and elements. The new materialisms and the ethical vistas that this movement has opened have said much about these themes. But the path we are moving along now takes another tour, leading us back to the roots of environmental ethics.

Here we encounter a philosopher named Hans Jonas. A Jew like Levi and Lispector, in the 1970s Jonas helped veer our ethical gaze away from the here-and-now and from the enclosures of human-centeredness. Like the title of his famous book, his "imperative of responsibility" is a call to *behold* both the future of our actions and the life of the ontologically Other, thus transcending the categories of traditional ethical discourse.[20] The difference here is simple: in their classical formulations, ethics of virtue or of duty, for example, invite us to mold our acts upon the insight of a superior principle, not to *behold* the world in which these acts are to be performed. And these acts are by definition ethically synchronic and species-specific: accomplished by *human* subjects, they *immediately* reflect on the subject itself. Jonas's responsibility ethics, instead, asks us to act by observing the upshots of our acts from a vaster—more-than-human and transtemporal—perspective. This ethical shift is understandable if one considers that its historical context is the one in which the whole world is irrevocably called to behold the systemic facets—at once social, political, and bio-geo-chemical—of the ecological crisis. To a philosophical debate struggling to envision new moral horizons, Jonas bequeathed the idea of a "commanding solidarity with the rest of the animate world" (138–39) and of a "*real* future as an open-ended dimension of our responsibility" (12). To his eyes, the triumph of *technē* and of human destructive power over the biosphere "reveal" (138), like in a mystical disclosure, the link between the "critical vulnerability" (6) of natural systems and the necessity of new extended duties:

> That which had always been the most elementary of the givens . . . —that there are [humans], that there is life, that there is a world for both—this suddenly stands forth, as if lit up by lightning, in its stark peril through human deed. In this very light this responsibility appears. (138–39)

Unlike those who maintained the impossibility—both ethical and juridical—of a moral imperative to safeguard natural balances in the face of future generations, for Jonas such an imperative is instead a real obligation and does not require reciprocity. If *behold* is a veering call, to behold the future of our interconnected being-there is a blind *behold*—a *behold* based not on an act of faith in the future but on acting so as to let the future be. In the perspective of responsibility, the future of humankind and that of nonhuman nature converge in one single gaze.

Although it would be improper to describe Jonas as an advocate of radical anti-anthropocentrism, reading *The Imperative of Responsibility* we see that the compass of his ethics becomes gradually broader. As if *in crescendo* his vision is progressively veered, *anthropodiscentered*. Moving beyond the instrumentalism embedded in traditional humanism, he explicitly speaks of "nature's own dignity" (137). The human/nonhuman juxtaposition is therefore transcended: seen in this light, the nonhuman is a mirror through which the human might discover and behold its most authentic face, Jonas maintains. The instrumental destruction of "the rest of nature," in fact, can only produce "the dehumanization of [the human], the atrophy of [its] essence even in the lucky case of biological survival" (136): recognizing the dignity of nonhuman natures is not only the condition of our own existence but a path toward a richer humanization of the human itself. This idea of ontological porosity and ethical mirroring reappears in the way Jonas considers our exploitation of the earth and the living. With his gaze turned toward monocultures and industrial farming, almost echoing Levi, Jonas asks if this allegedly "humanized nature" is still nature. And his response is unambiguous:

> The ultimate degradation of feeling organisms, eager to live and endowed with sensibility and capability to move, deprived of their habitats, imprisoned for all of their life, . . . transformed into egg- and meat-producing machines, does not have anything in common with nature.[21]

As this quote suggests, Jonas's philosophy, including his environmental ethics, is a way to *see through holocausts*. It is hence unsurprising that as his gaze, a few years after *The Imperative of Responsibility*, veers from ethics to theology, *the* Holocaust is the setting of this veering. As Levi also knew, in fact, beholding the inhumane necessarily implies a reflection on God—not only on God's existence but also on God's nature. This question is the subject of *The Concept of God after Auschwitz*, a lecture held by Jonas in 1984, whose argument is at once simple and vertiginous.[22] After Auschwitz, he maintains, everything has to be reviewed because Auschwitz calls us to *behold*: to behold the human world as well as the divine. In particular, Auschwitz has shown that the three attributes that Jewish theology assigned to God—"absolute goodness, absolute power, and intelligibility" (9)—are no longer compossible, they do not hold together anymore. It is then necessary to sacrifice one of them, and Jonas's choice falls on omnipotence:

> The *Deus absconditus*, the hidden God . . . is a profoundly un-Jewish conception. Our teaching, the Torah, rests on the premise and insists that we can understand God, not completely, to be sure, but something of him—of his will, intentions, and even nature—because he has told us. . . . But he would have to be precisely [unintelligible] if together with being good he were conceived as all powerful. After Auschwitz, we can assert with greater force than ever before that an omnipotent deity would have to be either

not good or . . . totally unintelligible. But if God is to be intelligible in some manner and to some extent (and to this we must hold), then his goodness must be compatible with the existence of evil, and this it is only if he is not *all* powerful. Only then can we uphold that he is intelligible and good, and there is yet evil in the world. (9–10)

Auschwitz—and all the other Auschwitzes of history (for both humans and nonhumans)—expose the existence of a "divine fate bound up with the coming-to-be of a world" (12). Following Gerschom Scholem, Jonas goes back to a very influential Kabbalistic doctrine, formulated in the sixteenth century by the mystic Isaac Luria. According to Luria's theory of *Tzimtzum* (self-limitation, contraction) in the moment of creation God withdraws from the world:

To make room for the world, the *En-Sof* . . . had to contract himself so that, vacated by him, empty space could expand outside of him. . . . Without this retreat into himself, there could be no "other" outside God, and only his continued holding-himself-in preserves the finite things from losing their separate being again into the divine "all in all." (12)

Being at once infinitely absent and yet categorically existent, God compresses H*self in order to allow a magnified expression of the world. From the outside of this absolute transcendence, God beholds the world, letting the world free to behold God. The freedom of this mutual beholding comes at the price of God's omnipotence over the creation: "Creation was that act of absolute sovereignty with which it consented, for the sake of self-determined finitude, to be absolute no more—an act, therefore, of divine self-restriction" (11). This has an important consequence: "Having given himself whole to the becoming world, God has no more to give: it is [hu]man's now to give to him" (12). From ethics to theology and then back to ethics: an ethics of responsibility based on the absolute transcendence of a God that beholds us from the external edge of creation, leaving us free—and bonded—to move within the finitude of these limits. This is God, the "master of polymerization" who does not patent polyethylene. Indeed, only a horizon defined by an act of "divine self-restriction" can enable an ethics of human responsibility—one calling us to *behold* creation and *hold to* its fragility.

At the end of his lecture, Jonas declares that we have to contend with "Auschwitz rather than the earthquake of Lisbon"—namely, with "deliberate evil rather than the inflictions of blind, natural causality" (11). But here is the point: if the earthquake that smashed Western Europe in 1755 was the outcome of "*blind*, natural causality," then the causality that triggers catastrophes that interlock human fate with the fate of nonhuman natures is not *blind*—or ought not to be.[23] Still, in an epoch that we call "Anthropocene" the challenge is just this: how to see the innumerable entanglements of human causality and natural agencies, how to hold and *behold* them in space, time, and matter. This *visualizing* challenge is one with the call to embrace the future in our ethical prospect, also because we need a farseeing moral imagination to face the almost undetectable forms of "slow violence" coming from pollutions and contaminations, the hidden aggression that "occurs gradually and *out of sight*," whose repercussions are played out "across a range of temporal scales" and through geographies that gradually become interconnected.[24] This *transcending* violence requires a different ontology, an ontology that outdoes individuals and enables us to *behold* the world from another perspective: not that of its future, but that of its end. Indeed, as Tim Morton notably insinuates, the world's end—the end of the way we are used to beholding our world—is already here, already now. And this apocalypse—this *revelation*—is the Anthropocene. In our postgeological age, objects finally reveal themselves for what they are: not things individualized in the here-and-now but processes and substances spread in quagmires of space-time-matter that merge with our existences: *hyperobjects*. How otherwise could one picture "things that are massively distributed in time and space relative to humans" such as the biosphere, global warming, "the sum total of all the nuclear materials on Earth," or simply a polystyrene cup?[25] Hyperobjects reside in this permanent interference across individuals; they are not punctual but radial and viscous. Not only, then, do they falsify "the idea that time and space are empty containers that entities sit in"[26] but, as *blind* outcomes of human causality, they are directly responsible for the end of the world—of the world in which we used to see space and time as linearly interrelated. They—really—are "unorganizable."

In this world of hyperobjects, in which our moral imagination is challenged by the impotence of God and the vulnerability of creation, we have to start seeing our life as a hyperobject itself. *We* are ourselves a wider being—a "we" that is immanent in the human and yet transcends it: *we* are "life looking back," Clarice would say. It is across and within the body of this huge hyperobject that the slow violence of irresponsible practices takes place. Because, in the *Tzimtzum*-world, "both creation and destruction are always on the horizon."[27]

So, at the end of all these stories—of roaches and chemistry, of human freedom and divine restraint—what is the lesson? Maybe this one: that we need to swerve our gaze and respond to the world's calling faces, even if this means the risk to lose (or loosen) our human uniqueness. And we need a mystical stance to let things and beings reveal themselves to us. Away from the solitude of the I/eye and from the abstraction of isolated singularities, *behold* is an invitation to reenchant the world by holding together the entangled, luminous fragility of all beings. Hyperobjects, slow violence, and all the invisible bonds that make us one with everything in space and in time, coiling together immanence and transcendence as in a Möbius strip, forcing us to veer from our usual paths to embrace the swirling pace of our worldly sacredness: this is what we must behold. So, *behold* this veering world, now. Yes, *behold*: this—is—a—veering—call.

Notes

1. Clarice Lispector, *The Passion according to G.H.*, trans. Ronald W. Sousa (Minneapolis: University of Minnesota Press, 1988). Further citations of this work appear parenthetically in the text.

2. The original Portuguese text reads "pessoas de alma já formada." See Clarice Lispector, *A Paixão Segundo G.H.* (Rio de Janeiro: Rocco Digital, 2009), Kindle edition. I do, however, endorse Ronald W. Sousa's translation of *alma* with "outlook," a choice which mirrors the idea—clearly expressed in this novel—that to possess a "fully formed soul" means to be able to see (and frame things) in unpredictable ways.

3. Benjamin Moser, *Why This World: A Biography of Clarice Lispector* (New York: Oxford University Press, 2009), 261–62.

4. Ibid., 5.

5. Particularly important here are Hélène Cixous's contributions, especially *Reading with Clarice Lispector* (Minneapolis: University of Minnesota Press, 1990) and *L'heure de Clarice Lispector* (Paris: Editions des Femmes, 1989).

6. Chrysanti Nigianni, "The Taste of the Living," in *The Animal Catalyst: Towards Ahuman Theory*, ed. Patricia MacCormac (London: Bloomsbury, 2014), 123.

7. Emmanuel Levinas, *Otherwise Than Being or Beyond Essence*, trans. Alphonso Lingis (Dordrecht, Netherlands: Kluwer, 1994), 90.

8. Maurice Merleau-Ponty, "The Intertwining—The Chiasm," in *The Merleau-Ponty Reader*, ed. Ted Toadvine and Leonard Lawlor (Evanston, Ill.: Northwestern University Press, 2007), 406, 412.

9. Louise Westling, *The Logos of the Living World: Merleau-Ponty, Animals, and Language* (New York: Fordham University Press, 2014), 33.

10. There is an etymological connection between the root of the words *physis* (nature, generation), *phos* (light), and *phainesthai, phainomena* (appear, appearances). This connection continues with the terms related to the semantic sphere of knowing/thinking. *Idea*, for example, comes from the root *(v)id-*, which is common to the verb *orao/eidon* and to the Latin *video*, "see." Like all nature is light, all knowing is seeing.

11. Lynn Margulis, *Symbiotic Planet: A New Look at Evolution* (New York: Basic Books, 1998), 3.

12. Jeffrey Jerome Cohen, *Stone: An Ecology of the Inhuman* (Minneapolis: University of Minnesota Press, 2014), 10, 24.

13. Kate Rigby, "Spirits That Matter: Pathways Toward a Rematerialization of Religion and Spirituality," in *Material Ecocriticism,* ed. Serenella Iovino and Serpil Oppermann (Bloomington: Indiana University Press, 2014), 288.

14. Catherine Keller, *On the Mystery: Discerning Divinity in Process* (Minneapolis: Fortress Press, 2008), 173.

15. Primo Levi, *The Periodic Table*, trans. Raymond Rosenthal (New York: Schocken Books, 1984), 143–44.

16. Jeffrey Jerome Cohen and Lowell Duckert, "Introduction: Eleven Principles of the Elements," in *Elemental Ecocriticism*, ed. Jeffrey Jerome Cohen and Lowell Duckert (Minneapolis: University of Minnesota Press, 2015), 11, 10.

17. Lowell Duckert, "Earth's Prospects," in Cohen and Duckert, *Elemental Ecocriticism,* 240.

18. Tim Ingold, *Being Alive: Essays on Movement, Knowledge, and Description* (New York: Routledge, 2011), 49–50.

19. Duckert, "Earth's Prospects," 241.

20. Hans Jonas, *The Imperative of Responsibility: In Search of an Ethics for the Technological Age* (Chicago: University of Chicago Press, 1985). Further citations of this work appear parenthetically in the text.

21. This crucial passage belongs to a paragraph titled *Die humanisierte Natur* (Humanized nature), which was omitted from the abridged English translation. My translation is based on the original German edition. Hans Jonas, *Das Prinzip Verantwortung: Versuch einer Ethik für die technologische Zivilisation* (Frankfurt am Main: Suhrkamp, 1984), 372.

22. Hans Jonas, "The Concept of God after Auschwitz: A Jewish Voice," *Journal of Religion*, 67, no. 1 (1987): 1–13. Further citations of this work appear parenthetically in the text.

23. Kate Rigby's *Dancing with Disaster: Environmental Histories, Narratives, and Ethics for Perilous Times* (Charlottesville: University of Virginia Press, 2015) is here a mandatory reference.

24. Rob Nixon, *Slow Violence and the Environmentalism of the Poor* (Cambridge, Mass.: Harvard University Press, 2011), 1, emphasis added.

25. Timothy Morton, *Hyperobjects: Philosophy and Ecology after the End of the World* (Minneapolis: University of Minnesota Press, 2013), 1.

26. Ibid., 65.

27. Duckert, "Earth's Prospects," 260.

Wait

CHRISTOPHER SCHABERG

What will compensate me for this wait, which is so much longer than the wait expected?

—JOANNA WALSH, *Vertigo*

10. When asked to choose a title verb for my chapter of this volume, I immediately thought of *wait*. Which is odd, in a way—I didn't have to wait for it, at least not for long. This choice was partly due to my penchant for thinking about airports as environmentally rich sites. Places where, inevitably, one must *wait*. The surroundings press in on one, or float nebulously around, detached. Either way, the simple space of the airport is always also a tacky time, ready to latch on to the traveler at any moment. I like to contemplate airport temporalities: killing time, dead time, in-between time. For me, these strange experiences of time are evidence of a conceptual hang-up: how it becomes difficult to entertain ecological thoughts when one is actively attempting to annihilate space via time, utilizing the various technologies and architecture of air travel. When one is forced to wait, to endure the time that stretches out like a piece of taffy in the mind, pulled by repetitive overhead announcements and the indifferent bustle of other flights bound elsewhere, the zone of the airport can quickly become abject, uncomfortable, numbing. And yet, this too is a place on the earth. To wait at the airport may strangely open up to a profound glimpse of ecology. Or this is my wager: I want to get from airport waiting to something bigger, something ecological—

something that veers away from the airport, that veers into the experience itself—the expandable, inescapable injunction *to wait*.

9. I realized quickly that I didn't really want to write only about airport waiting. For one thing, my book *The Textual Life of Airports* includes a chapter called "Ecology in Waiting," which takes up this topic at some length.[1] For another, *waiting* is one thing—an act, if a seemingly passive one—but *to wait* is quite another thing. *Wait* implies a kind of ongoing presence and an indefinite postponement—and also the idea that one might dive into this experience, headlong. It might also be a command, an omniscient call to be still. A command to just exist, however awkwardly or in suspense. Still, more often than not, the word gets stretched into its gerund form. Consider, for example, this sign in the Traverse City, Michigan, airport, a small terminal I pass through a couple times a year, en route to visit my family and my first (now second) home:

Figure 1. Sign in TVC Airport.

Here the operative word is *waiting,* spatialized by the adjoined if redundant term *lobby.* As if a consolation prize to this abject phenomenological comportment, a specific *space* is dedicated to this time. Even a specific comportment is given by the sign, a certain angle of repose for the waiting body resting on the minimally signified seat. Then, too, the clock orb floating above the person's lap says that it is some minutes in relation to some hour . . . time is passing. But this sign bespeaks *waiting,* the thing I said I wasn't going to talk about. So what if this sign simply said *wait?* That is, after all, the subtext. What if the sign just looked like this:

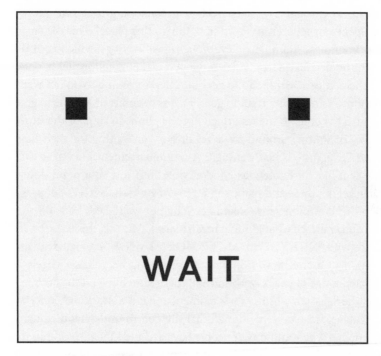

Figure 2. Latent content.

This command, this verbal reality, might not appease as well in the constraining space/time of the airport. But it is, perhaps, more honest to the experience and necessity, within and beyond the terminal doors. We have to get from the one to other, like how Joanna Walsh arrives there in one sentence of her book *Vertigo:* "In a few hours I will travel back to the airport to take another plane. Sitting here I am already waiting to wait."² This line is from a story about attuning to the minutiae of everyday life, such that the surroundings of a vacation begin to decompose into ordinariness. The destination deteriorates into the space of the same, the recognizable, the legible. To wait at the restaurant (where the narrator is when she thinks these thoughts) blurs into the anticipated wait at the airport—utterly annihilating the necessary distance between the as-if bracketed time of vacation and the mundane time of the ordinary. And this is the ecological paradox of *to wait*. It is *now*, always now. Look around—this is it.

8. Midway through Dr. Seuss's classic graduation gift *Oh, the Places You'll Go!* lies an apparently vexed site: "the waiting place." This is a dark, weirdly boundless room where everyone is just *waiting*—which according to the book is an unavoidable yet dangerously entangling realm: once you're there, it can be difficult to get out. There's nothing to do but wait. People are "Waiting for a train to go / or a bus to come, or a plane to go / or the mail to come, or the rain to go / or the phone to ring, or the snow to snow / or waiting around for a Yes or No / or waiting for their hair to grow. / Everyone is just waiting."³ As commencement guidance, this might seem like fairly obvious good advice: don't just sit around doing nothing; get out there and make something of yourself. But let's put pressure on this common sense mandate. Why not wait? Let's back off the gerund and embrace what it might mean *to wait*. After all, this is the book that ends with "KID, YOU'LL MOVE MOUNTAINS!"—replete with an illustration of a contraption literally transporting on rickety wheels a decapitated range of peaks, with soil and rocks crumbling off the bottom. Bracket the Seussian whimsy for a moment and it is a shockingly unecological image. It is suddenly not all that difficult to understand climate change denial when millions of people have accepted the idea of mountains as mere things to be moved—and this being a *desirable* thing. Instead of the forward-barreling mentality championed in Dr. Seuss's book, we should reverse this, for the sake of being more ecological. Ponder a revision of this book called, *Oh, the Time You Will Wait*. What you do with this time—or *not* do—becomes precisely what matters. I am reminded of a couple lines from Timothy Morton's *The Ecological Thought*: "Don't just do something—sit there. But in the meantime, sitting there will upgrade your version of doing and of sitting."⁴ How might waiting around

"upgrade" our ecological consciousness? Could Morton's book have been alternatively titled, simply, *Wait*? What would it mean to *wait* for ecology to arrive? Or is *to wait* itself somehow ecological?

7. I want to attempt to merge this line of thought with something more recognizably in the realm of nature: fishing. I love to fish. It latched on to me when I was around eleven years old. I can't remember the exact date. But I remember the day. We were headed home from church on a Sunday, and somehow I convinced my parents to stop at Meijer and buy me a Zebco rod and reel, a simple single-level tackle box, a package of six hooks, a bag of lead sinkers, and a couple plastic bobbers—red and white spheres with spring-loaded line clasps. Upon arriving home, my father went out to the vegetable garden in the backyard and dug up some worms for me. Then I crossed the street and headed to the pond in the middle of the neighborhood. I had seen people fishing from the bank. I don't know what happened next, or how. Who tied on my hook? Did my father accompany me and do it from muscle memory, or did I somehow figure out how to do it myself? Who threaded the worm onto the hook, a horrific impaling in miniature? The next thing I remember is catching fish: several bluegill sunfish, broad sides pulling my rod in a dancing arc, and I was hooked. Later, I would learn to coax giant smallmouth bass out of this same pond—some of the most thrilling fishing I've ever done. I became entranced with the intricacies of bait presentation and structure mapping, reel mechanics and rod action. The church I was raised going to was for Christian Scientists. Looking back, I think that fishing created the first significant crack in the foundation of my belief in that religion, which eschews the material world for a Neoplatonic realm of spiritual perfection. When I began fishing, I gave myself over to the great outdoors—to a world of matter that included fish, fingernails, hooks, humans, monofilament, titanium, snapping turtles, clouds, gasoline, water, wind, worms. These were the things that I would learn to wait for, and live with, through my angling forays in my teenage years and beyond.

6. Fishing left my life for several years—during the end of high school and throughout college, turbulent times of formation and angst—but it found me again in Bozeman, Montana, during my early graduate studies in English and environmental studies. There, curiously enough, in retrospect, I also was working at the nearby airport. I would go fishing often with my professor Greg Keeler, who knew every bend and hole in the

Gallatin, Madison, and Jefferson Rivers. Sometimes we would go to the headwaters of the Missouri River and use worms laden with large sinkers bumping along the bottom of the river to catch suckers—which we would then cut up into cubes to use as bait for rainbow trout, which we would go up the Gallatin River to fish for. This kind of fishing demanded us to wait, to be sensitized to both the line in the water and the nearly imperceptible bounce of the rod tip when a fish would strike. There was one tight bend on the Gallatin that was a few feet from a train track, on a steep boulder embankment, and as we would be balanced perilously on the rocks freight trains would hurtle by—often hauling the green, wingless fuselages of Boeing 737s en route to Everett, Washington, for final assembly. Other times we would fish in the more recognizably romantic Western way, with fly rods and tiny surface flies. These were hot summer days when the trout would be rising at caddis flies and grasshoppers, and the icy water felt good to stand in up to the waist. *To wade* is *to wait*, to feel time in the form of water curl around feet and legs.

5. I am conflicted about fishing these days. I cannot recommend it as a simple act of nature loving or as an innocent back-to-the-land endeavor. It is inherently violent—even if the subsuming act of food preparation and consumption somewhat justifies this violence. (And since I often catch and release, I cannot deploy this justification, anyway.) Still, fishing makes me care deeply about my place—my place as it exists apart from me, and my place within this place, this ecosystem. Myself as part of *place,* however artificial and blurry this designation is. Up in Michigan I awake in the earliest hours before dawn: before the sun even begins to marble pale the black horizon. I slip outside and into my car, which I have aimed away from the house, and coast down the gravel drive toward M-22, the main road that wraps around the peninsula. Passing the old cherry orchard, I flick on the headlights, and I start my drive in earnest to one of my favorite lakes. To drive is to wait. This is all part of it, this strange activity that is couched under the label *fishing* but is really something much bigger, much more encompassing. The fishing is only part of it. In *The Arcades Project,* Walter Benjamin writes briefly about "he who waits. He takes in the time and renders it up in altered form—that of expectation."[5] Wresting Benjamin's formulation out of context, this might be me as I am moving toward fishing—as I wait to fish, full of expectation but knowing that the very fun of the experience is to wait

once I am *there*, once I am hip deep in the water, scoping the surface around me, looking for the slightest ripple or V suggesting a submarine fin veering in pursuit of prey. I leave my car at the end of a blocked overgrown two-track in the national lakeshore, whereupon I have to hike back to the remote lake. The hike too serves as an openness to wait. I pass a plump toad on the ground, perched near narrow puddles in the mud made by the hooves of deer who passed by earlier in the morning. At the edge of the lake I hear two sandhill cranes crooning their haunting calls into the dawn. A beaver paddles across a small bay, toward an aspen grove on the far side of the lake. I kneel at the water's edge and assemble my rod and reel, thread the line through the guides, and tie on a frog pattern popper. I hear fish rising at the surface, after dragonflies that dart and dip recklessly close to the water. The scene surrounds me and includes me, and as I wait to fish I feel deliciously attuned to this place.

4. Landscape ecology involves the practice of determining a given scale so as to map relations, cycles, and effects over time. It is a way of marking place, a way of knowing what exactly to wait for, what to measure. It is a method for gathering information about an ecosystem, while knowing that the "system" in question really has porous borders and can be zoomed in or out on, to focus on different changes, tensions, or stresses. It is "landscape" because it fixes a scale, if then to suggest far vaster reaches of ecological impact as well as subtler dynamics within. I have been learning about landscape ecology from a colleague of mine in the biological sciences who has been studying aquatic vegetation patterns in the Mississippi Delta for over thirty years. One November morning I went with my colleague on a research trip, to retrieve data from recorders he has anchored throughout the delta. As we cruised among the stands of reeds and floating islands of water hyacinths, I felt overwhelmed by the scale of the place—the river a quarter of a mile wide in the main channel, disgorging itself out into the gulf which in turn pushes its salty currents back into the river. The delta is a porous landscape par excellence, a place that can hardly be called a place at all—and yet here it is giving my colleague incontrovertible evidence of climate change in the form of shifting plant species due to changes in water temperature. Landscape ecology has this utility: accepting the limited, fragmented quality of place—as region, home, or wilderness—while gesturing toward a far more suffused and ever-receding sense of place, planetary and geological.

This is a place we can only intuit in the form of a crushing humility, a realization of all that beyond what we can only wait for, knowing it must always exceed us.

3. Mary Gaitskill's "The Other Place" tells its disturbing story through a narrator who admits early on an interior life of violent, misogynistic fantasies—a realm he labels "the other place" so as to distinguish it from the real world, where he seems normal enough. Part of the story involves our narrator's relationship with his son and a wavering worry that he has passed on "the other place" to him. But weirdly, our narrator—and he is made ours by his confessions, which insinuate the reader into the story—is also at pains to delineate how his flirtations with "the other place" indicate a kind of ecological awareness of objects, place, and space. After describing his arguably creepy childhood routine of walking around his neighborhood at night and basically spying on the neighbors, our narrator brushes up against deviancy only to disavow it, and indeed even embrace it: "I just wanted to sit and watch, to touch other people's things, to drink in their lives. I suspect that it's some version of these impulses that makes me the most successful real-estate agent in the Hudson Valley now: the ability to know what physical objects and surroundings will most please a person's sense of identity and make him feel at home."[6] Gaitskill here is both sending up the American suburban idyll as a charade and pointing at something undeniably real: how physical objects make us feel at home. Directly after this passage, our narrator turns to thinking about his son: "I wish that Doug had this sensitivity to the physical world, and the ability to drink from it. I've tried different things with him: I used to throw the ball with him out in the yard, but he got tired of that; he hates hiking and likes biking only if he has to get someplace. What's working now a little bit is fishing, fly-fishing hip deep in the Hudson. An ideal picture of normal childhood." In this story Gaitskill slyly deploys images of fishing to highlight not only the tenuous connection between the father and son but also the tenuous boundaries between the normal and the abnormal, the two "places" that gradually blur into the same place—namely, life on Earth, including the inescapable and at times uncomfortably awkward realities of coexistence. "The Other Place" explores habitation as something spatial and physical, something commoditized and artificial (real estate), and something processual and connective, as with fly-fishing but also as in reading, as the reader is

drawn into the story, to wait for—and with—this (unlikable, or all-too-familiar?) narrator as he tells his story. Fishing serves as a figure and a ground, a foil for dysfunction and a possible escape, but also, importantly, an ironic signifier as an "ideal picture of normal"—hardly a certain place at all, and certainly not anything one can wait for with confidence that it will ever arrive.

2. What is the place of fishing? In the fiction of Ernest Hemingway, fishing scenes often open up to profound awareness—for instance, Bill Gorton in *The Sun Also Rises* remarks, on the way to a trout stream, "this is country."[7] What a ridiculous and yet accurate statement this is, as the two characters hike through a forest on their way to fish. This utterance occurs during a pastoral exception to the rule of this novel, which is overwhelmingly rife with tension, aggression, and pent up energy in the postwar landscape of France and Spain—all sieved through Hemingway's understated narration and the ruse of one party after another. The fishing scene in the middle of the book works like Gaitskill's use of fishing in "The Other Place," to normalize while simultaneously undermining any clear basis for determining normality and abnormality: these are messy worlds, unified by history, by the requirement to wait for whatever is to come. In Hemingway's story "Big Two Hearted River," fishing is something that not only conjures in Nick Adams "all the old feeling" (connectivity, familiarity with the natural—perhaps in Henry David Thoreau's word, "wildness"); fishing also operates as an exception to all that waits beyond, things dynamically repressed by the war-scarred Nick. Violence, trauma, the horror of artillery fire: all those things that cannot be talked about directly in Hemingway's fictions are at once evoked and pushed away as his characters try to focus on the act of fishing. Turning back to *The Sun Also Rises,* what delivered Jake Barnes his unutterable accident? An airplane—he was a pilot in the war, a fact only alluded to elliptically. The act of fishing, Jake's wait for a fish to tug on the end of his line, is haunted by the specter of airborne violence. The place of the fishing retreat is encased within a larger context of global tensions, just as it is shaped by Jake's own anxiety and the indifferent water coursing through the Irati River.

1. In Istvan Banyai's picture book *Zoom,* each recto page shows an image that is revealed on the following recto page to be part of a larger scene.[8] Each place in the book is shown to be relative to scale and

context—aspects can be expanded or contracted infinitely (or as far as we can imagine, anyway). As the reader turns each page, the mind must adjust for these aspects to figure out what the image *is*, and then what it is in relation to things larger and smaller. The opening page is a blot of color that turns out to be the hackle of a rooster; the final page is a white dot that a page earlier was the blue marble of Earth floating distantly in space. I am not ruining anything in summing up the book this way, because each page is fascinating in its fractal poignancy. Like fishing for Mary Gaitskill and Ernest Hemingway, like the vacation for Joanna Walsh, like Timothy Morton's ecological thought, the places where we wait peel away layers of existence and coexistence, incommensurable levels of life at once contained and yet spilling over and into each other. A proper description for what it feels like to read Banyai's book *Zoom*—particularly since there are few words, and none in the service of narrative—might be that the reader at once waits and *drinks it in*.

.5. Conflicted, I am still drawn to fish, to wait for the odd sensation on the tip of my rod that tells me a fish is there, commingling with me, and me with it. Lately I have been fishing on the Mississippi River near my home in New Orleans. Often I fly-fish on the river, but it is haphazard, totally unlike fly-fishing in Montana or Michigan. I cast shimmering subsurface streamers and catch fish infrequently: ladyfish, needlefish, and white bass. The water is murky brown and the bottom of the river dangerous with intermittent shards of broken glass, rusty metal reefs, and concrete slab embedded in worn channels of mud. I don't dare wade deeper than my knees, and after fishing my feet remain stained for days. Sometimes my Mississippi River fishing is safer, if less elegant, using chunky one-ounce weights and gobs of worms cast from higher up the riprap, then to wait as the bait bounces along the bottom, waiting for catfish to signal with their characteristic tap-tap-taps. The catfish might be the size of my hand or the size of my body entire, you never know until the hook is set. Of course it may also be a log, a piece of rebar, or any number of other wrecked items lodged in the depthless river bottom. One time my friend hooked something that gave his rod an encouraging bend; he reeled it in to discover that it he had hooked another fishing rod, long lost and encased in muck. This recursive catch was an uncanny reminder of other people who have fished this river, and more to come, even as we wait for fish in the present. To fish on the Mississippi is not only to wait for actual fish but also to glimpse scales of place flow past and unfold. The landscape ecology here includes the invasive *Triadica sebifera* trees with their long-delay seedpods and also the freighters and tugs moving cargo and waste up and down river. It is the interminable churn of the river and the gradual changing of climate and coastal erosion, forces that will eventually wipe the place of New Orleans off the map, if we wait long enough. In the *Arcades Project* Benjamin muses, "As life becomes more subject to administrative norms, people must learn to wait more. Games of chance possess the great charm of freeing people from having to wait."[9] At this drainage point of late capitalist North America I wait. Fishing is a game of chance, perhaps, that cuts through my having to wait. But it is also something that invites me to wait, to see the layers of life teeming around me in a nonsimple amalgam along this riparian corridor. I know the norms are not norms at all, hurtling humanity toward disaster; but we also know that on another scale, this hardly matters. Ecology waits somewhere between these realizations, and somehow acts from a point of suspension.

.25. When I worked at the airport outside Bozeman, Montana, a familiar scene would occasionally play out: fly-fishers at the airport, at the end of their trips but suddenly faced with a delay or cancellation—mechanical, weather, or unexplained. Angry, frustrated, fuming, they punched numbers into their phones, leveraging their elite statuses, on hold with other airline representatives. Rod cases leaned up against the granite columns, fishing shirts still rumpled from days of adventure, their equipment neatly packed in clever padded boxes and bags. And just think: a day or two prior, these fly-fishers had been presumably enjoying the place; they were "in the zone," as we say, tuned in to such things as water temperature, the hatch, wind conditions. I was often struck by this paradox of place: how the landscape could so readily diminish within the constraints of the terminal. How the romantic feelings of inspiration, suspense, chance, and wildness were nowhere to be found around the check-in counter, while they had most certainly (presumably) been happening out on the river, and in fact some variant of these same forces were playing out in the logistics of flight: uncertainty, calculation, attempts to overcome a challenge by technological means, fluctuating weather conditions—all the things that make fishing fun also, weirdly, inform the practice of everyday air travel. But while the one becomes a destination activity to be savored, the other becomes a dreaded ordeal to be dealt with as quickly as possible and to be done with. The drawing out of the airport delay could be embraced as a challenge; to wait for a plane might not be all that different from casting a line and all the material contingencies involved therein.

.125. Wait, it's almost here, ground zero, an ecological thought. I have been writing this essay in autobiographical spirals, while sitting in the winter landscape of northern Michigan, this place I know so well but which strikes me increasingly as an utterly alien environment. The black squirrels are catching my attention this season, the high contrast of their plump jumping bodies against the eerie orange of fallen maple leaves and dead bracken fern fronds laying on the ground. This winter it is unusually warm, pleasant enough to be outside taking long walks in the woods, or doing one of my favorite things: with my nieces and nephews and my son and daughter in pursuit, I run into the forest or the meadow and find a pine tree to hide beneath, or an oak tree to climb up. The children try to find me, and when they get off track I give them hints with exaggerated animal calls—crow caws, ungulate grunts, owl hoots, the shrill shrieks of attacked rabbits. I wait as they get closer, and as soon as one of them spots me, I bolt for another redoubt, to repeat the game. This play will go on for hours sometimes. Writing this, it sounds infantile and silly. But to be there in the moment, to wait for children in hot pursuit with my face pressed into pine needles, lichen, or moss—to wait like this is to be confronted with animal inhabitance on this planet, and more. This is coexistence, dynamic and resplendent, but at the same time low-key, ordinary. This is being with my family but also being in an exploded view: as I wait in those spontaneous redoubts, I feel my heart beat inside me, and I become aware of the creaturely life, floral elaborations, and lithic substrate . . . surfaces, textures, and actants intersecting and brimming all around. I feel life moving inward and outward, at once. I lie there poised, ready, ready to continue to wait. Whitman put it this way: "Space and Time! now I see it is true, what I guess'd at, / What I guess'd when I loaf'd on the grass."[10] I give into it, fully, for a few protracted minutes at least. To be ecological might be to wait, knowing at the same time that this is it.

Notes

1. Christopher Schaberg, *The Textual Life of Airports* (New York: Bloomsbury 2013), 99–115.

2. Joanna Walsh, *Vertigo* (St. Louis: Dorothy Project, 2015), 88.

3. Dr. Seuss, *Oh, the Places You'll Go!* (New York: Random House, 1990).

4. Timothy Morton, *The Ecological Thought* (Cambridge, Mass.: Harvard University Press, 2010), 125.

5. Walter Benjamin, *The Arcades Project* (Cambridge, Mass.: Harvard University Press, 1999), 107.

6. Mary Gaitskill, "The Other Place," *New Yorker,* http://www.newyorker.com/magazine/2011/02/14/the-other-place.

7. Ernest Hemingway, *The Sun Also Rises* (New York: Scribner, 2006), 122.

8. Istvan Banyai, *Zoom* (New York: Viking Penguin, 1998).

9. Benjamin, *The Arcades Project,* 119.

10. "Song of Myself," *Walt Whitman Poetry and Prose* (New York: Library of America, 1996), 219.

Play

J. ALLAN MITCHELL

P*lay* recreates itself moment-to-moment as a volatile field of interplay, veering among the elements. No reified concept quite captures
the fugitive agencies and energies set loose. In the event, the meaning of
play remains elusive even when sensed as a persistent stimulus, agitating
and captivating those involved. Eliciting successive turns, moves, throws,
rolls, feints, leaps, kicks, and blows, playing subsists in a dynamic negotiation. I hazard that the palpable, permutable, and relational aspects
of playing open possibilities for future ecocritical endeavors, supplying
motive and means for an immanent critique of possessive individualism and human triumphalism. Prompted by those who have championed "critical play,"[1] I want to venture an associated notion of *crisis play*:
an avid, vigorously embodied rapport with the immense play-of-world.
Play theory has yet to address the issue of playing in the face of imminent
ecocatastrophe and, if it did, might risk appearing frivolous for invoking recreational activity alongside planetary disaster. And yet it should
actually be well placed to advance environmental efforts against excep-
tionalist claims on behalf of this or that species or social order. All leading play concepts seem to be after what is most essential, elemental, and
creatural—with puzzling results so far. For a long time, in fact, play has
been enlisted to denote profound human acts and achievements, civili-
zational processes, and species superiority, and it is precisely those grand
claims that need to be challenged in the current epoch. Let's now turn
the tables to consider the alternatives, notably the fact that playing is
scarcely human and colludes ("playing together" from *col + ludere*) with

a welter of things. In the combustible event of playing, heteronomous relations take primacy over autonomous subjects and substances, rendering play an emergent and protean activity pitched to a phenomenal world beyond the merely human.

Here is a chance to rethink play outside anthropocentric or species-bound frames, pursuing more hopeful, reparative possibilities of a play ecology—call it "playing house" (recalling the *oikos* in *ecology*), if you will—identifying habitable conditions for a ludic environmentality. It will require more than saying some favored species of plant or animal, in the bland journalistic formulation, "play an important role" in the ecosystem. We need a better sense of how play re-creates roles and systems in crisis; we need to see how play can become radicalized in the event. The aim in this brief essay is to address pressing ecological questions, earnest as any game, about the possible ends of playing in the ruins to re-create conditions on the ground. In which situated environments do we find people, animals, and things at play? Is playing ever more than a hedonic interlude? How is it possible to play in and through serious emergencies? Consider what follows as a set of rough field notes toward a play ecology. Since a normative theory of play would soon lose a feel for contingencies, I have tended to favor observation and experimentation of the sort encouraged by the critical essay, preferring examples over precepts. The essay is a critical test to the extent that it plays—playing around with matters whose allure and allusiveness (i.e., *playfulness*) point to broader ecologies in which everything is caught up in the field. In the spirit of returning play to an elemental modality—excitable, interactive, recombinatory—my piece is offered without a certain end game.

Wordplay to Worldplay

Noticing how language plays is one way to enlarge the field of play and intensify the dynamic, underscoring the exigency of a world at play. *Playing* avoids capture almost by definition, the exact origins of the word remaining enigmatic. Various cognates suggest themselves: Old English *plegan* (move rapidly, frolic; perform music) and *pleon* (to risk, venture, expose to danger), Middle Dutch *pleyen* (dance, leap, rejoice), among other words derived from Indo-Germanic terms for lively activity. In English, brisk bodily movement or exercise belongs to the earliest and primary senses of the verb.[2] Intransitively, play refers to mental or physical acts and irregular or rhythmic motions. Workers handling tools and all manner of other living creatures—surfacing fish, gamboling deer, flying bees, prancing birds—are playing. The play of inanimate things (wheels, dials, watercraft) and alternating appearances and phase-states (flowing or boiling liquid, shimmering sunlight, flickering mental images) yield senses of phenomenal rapidity and unimpeded movement of whatever kind. Transitive senses include shifting game pieces about on a board, manipulating tools, exhausting a fish on a line, and seducing or scheming. Activities that are pleasurable rather than practical are perhaps foremost in common parlance: striking a ball, performing music, flirting. You can tease, joke, or goof off in ways typically associated with children at play. Whatever else the word means, *playing* seems less a matter of stable reference than of performance (acting, reacting, advancing, animating). A sense of sheer engagement and energy expenditure remains common to play activity in the language.

Verbally then, *playing* is highly *elusive* and *allusive* (both happen to be play words whose etymological origins are *ludic*), crossing thresholds that are by convention supposed to divide incommensurable species and substances. And that may be just how words mean in one or another language game, following Ludwig Wittgenstein: "The meaning of a word is its use in the language."[3] There is no single generalizable sense; there's only a play of senses in this or that utterance. There are just so many specific examples including the very notion of game: "Look for example at board-games, with their multifarious relationships. Now pass to card-games; here you find many correspondences with the first group, but many common features drop out, and others appear. When we pass next to ball-games, much that is common is retained, but much is lost.— Are they all 'amusing'? Compare chess with naughts and crosses. Or is there always winning and losing, or competition between players?" It is not that nothing is shared in common, but that commonality is taken for granted where a naming game is in fact taking place in the language: "We see a complicated network of similarities overlapping and criss-crossing: sometimes overall similarities, sometimes similarities in detail."[4] Jacques Derrida's notion of the play of the signifier has also served well to underscore the continual supplementation of meaning in a field of infinite substitutions, exposing the ceaseless movement of word-senses in game-like systems.[5] The play of words is one element among others, though it is a good index of the way something significant eludes human mastery in a volatile field. Language takes so many turns.

Yet if those theoretical formulations are primarily devoted to actions of semiotic tokens or texts, they should not be taken to apply too narrowly to artifacts of speech. In practice, the things played (cards, sports, drama, politics) are heterogeneous, involving so many material and practical differences besides lingual ones. Nor are the rules of individual games always articulable. For as Wittgenstein also says, one can be a proficient player and not know the rules just as one can know the rules without ever having played.[6] Playing involves unregulated actions of one sort or another even in rule-bound environments. What Brian Upton calls the mutable "phase space" of gameplay consists of multidimensional and aleatory game states.[7] Ping-pong has only a few unstable dimensions (e.g., speed, position, and spin of the ball), while a game of make-believe zombie pirates has many times more. A complex game generates moves and countermoves unforeseen. Playing with rules is also not the same as playing by rules. Cheats, bluffs, loopholes, and do-overs are sometimes tolerated and enjoyed; unruliness may be encouraged. Fantasy role-play expands itself in the process of inventing and destroying worlds, generating unpredictable social situations that are far more than just verbal constructs. Examples soon proliferate of "overlapping and criss-crossing" elements that are irreducible to a single articulable sense or regulated system. Play occurs in an ongoing dialectic between choices, challenges, and unknown eventualities that can be hard to name.

Playing is sometimes misrecognized or mislabeled; at other times we find things at play where we least expect them. As Greg Costikyan observes, simple games such as tic-tac-toe, rock-paper-scissors, and *Monopoly* all have optimal strategies for winning that can be learned as "plays." But because they are "solved" games, they allow for little play in the event.[8] Here the word *play* may be a poor guide to the actual performance. Uncertainty extends to the question of whether many complex acts are play or games, as Gregory Bateson once noted: the question "Is this play?" is often held in suspension in fantasy, art, magic, and practical jokes.[9] So, if solved games are some of the least playful, other activities of grave consequence can be the most. Strategic and high-risk gameplay seem to be essential to aspects of public life, where everyday journalistic and bureaucratic usage yields senses of manipulation ("play at" or "play a person false"), cowardliness ("play safe"), and dangerous ambition ("to play politics"). Playing for advantage is to play the angles. Playing into the hands of others is imperiling one's position to aid the enemy. Playing describes diplomatic activity carried out between officials at the highest levels (just as stock pickers play the market; hedge fund managers hedge bets), but it is also a serious tactic of the relatively powerless, as in a variety of recent artistic and activist games. Some present creative strategies to cope with repressive state actors and security apparatus, often exposing the violence within larger gamed systems. For example, in *Camover*, "Players are encouraged to destroy CCTV cameras in a specific urban environment and are awarded points for doing so." *Metakettle* is a protest game to thwart the kettling tactics of riot police during demonstrations. *September 12th* presents a Middle Eastern market town on a screen for players to attack terrorists while mostly killing civilians, and *Darfur Is Dying* sets players the task of managing a refugee camp that is under threat of attack.[10] Playing with transnational corporations and geopolitical entities that do not authorize the game has proven ethically engaging and adventurous. The 2015 Climate Games initiated a series of interventions ("more than 120 teams organising 140 beautiful, quirky, funny, courageous, disobedient actions") during the Paris Climate Conference to oppose government and corporate complicity.[11] They pitted themselves against Team Blue, or the police. Protest play is one way to play hard and for keeps, as in the extraordinary social-impact game *Survivance*: "Players choose from non-linear quests that are structured in the phases of the Indigenous life journey. . . . Survivance merges survival and endurance in asserting Indigenous presence in contemporary media."[12]

Play is so capacious and potentially ambiguous that it can be hard to tell apart from work, whether that of protestors or plutocrats. Intramural games are employed in modern workspaces to optimize behaviors or explain them organizationally. Even as play is idealized and segregated—Upton asserts that "work isn't play" on the commonsensical grounds that play would otherwise not seem like an escape from ordinary obligations[13]—exceptions are readily found. Play has been treated as formative (educational psychology touts the maxim "Play is children's work") and functional, serving species fitness (as when evolutionary biology identifies adaptive traits that stem from exploratory play) or mental health (as in some therapeutic techniques). Corporate team-building exercises employ play to strengthen morale, or—more cynically—to mollify or distract bored workers. Tradeshows are abuzz with the promise of gamifying classrooms, boardrooms, and online showrooms, motivating consumption and underwriting capitalism at every level. Economic game theory proffers a sophisticated neoliberal model of rules, moves, and goals in competitive marketplaces. Fortunately, powerful critiques of corporatized gaming and rationalized play are not far to seek. Ian Bogost calls out recent gamification initiatives as fraudulent and exploitive.[14] Wendy Chun faults economic game theory for turning participants into automata.[15] Andrew Galloway offers a subtle account of the coercions of "ludic capitalism."[16] It is right to be wary of claims made on behalf of managed play, because no doubt playing becomes something less than playful as a result. And yet for that very reason, play may be worth crediting for interfering with business as usual. Playing may introduce inutile pleasures into a workplace that is otherwise far from playful, exposing faults in the system; playing may collude with corporate systems to propagate new, extramural relations and thwart the transactional efficiencies preferred by global financiers, college presidents, and state officials. The intensity and spontaneity of play will often as not result in profitless expenditure whose excesses are aimless mischief, masquerade, and folly, precipitating a crisis for those who seek to control all the resources. Robert Fagen has noted that many nonhuman animals "risk time, energy, and injury to play,"[17] suggesting that playing—for all the hypothetical good it may do for biological development, stimulation, and adaptation—is a manner of thriving that exceeds the requirements of just surviving. Play is irreducible to mere utility, as observed by Roger Caillois long ago when he spoke of playing as "an occasion of pure waste; waste of time, energy, ingenuity, skill, and often money."[18] Playing exemplifies a critical distinction between what Michel Serres playfully calls "use value" and "abuse value," the latter designating a preeconomic relation that resists economism.[19] If anything, playing with zeal is highly sociable and, moreover, liable to stimulate a false economy or free up resources for other ends, exhibiting powerful disequilibrial effects. Playing presents an occasion to start noticing what is missing from established systems and recognized utilities. Asymmetrical to predatory or proprietary capital, playing can be a radical mode of encounter with contingencies and externalities.

Multispecies Play

Play theory over the past half-century has singled out one or more species as particularly playful, claiming that play belongs to higher mammals and, more specifically, counts as a distinctly humanizing activity. Johann Huizinga's thesis about *homo ludens* is often reiterated and informs recent and otherwise very sophisticated treatments. It appears in the guise of Costikyan's notion that only humans "create culture out of play" and Miguel Sicart's initial assertion: "Play is what we do when we are human."[20] Such statements should be treated as highly equivocal given the notes amassed so far. If play is, as Costikyan also contends, "a mode of being human,"[21] surely that does not make play just a human mode. Playthings and playscapes are hardly restricted to specific natures or cultures, reserved only for a splendid few. Play is manifold—immanent to host environments we may think the least playful—and multispecies.

Observing romping monkeys, Bateson famously witnessed signaling play, which he attributed to metacommunication processes required to frame an activity as distinctly playful: "The playful nip denotes the bite, but it does not denote what would be denoted by the bite."[22] Since then the ethology of animal play has been extensively documented. Observing the "rules of engagement" of animals at play, Marc Berkoff has identified complex play signals in minute changes in behavior—for example, invitations to play through "bowing," enforcement of "fair play," and "role-reversing and self-handicapping."[23] The known variety of social and solo animal play includes horse foals dancing, play kicking, and neck wrestling; squirrels pouncing, somersaulting, and leapfrogging; bears pawing, play biting, and manipulating objects; cats stalking, rushing, and dribbling and chasing objects; harbor seals play biting and wrestling; birds fetching, dropping, and midair catching; and nonhuman primates wrestling, vaulting, and manipulating objects. For some, play fighting is a way of playing around sexually. Primates among other mammals engage in same-sex and erotically charged bouts of wrestling and mounting.[24] Animals play with themselves, masturbating. Some play joyously and inventively across species divisions too, enlisting humans among others in their sports and games. Drawing on her experience with agility sports, Donna Haraway has focused attention on how interspecies play catalyzes strange intimacies—what she calls the "play of strangers"—between woman and dog.[25] As she writes, bonding with her dog in "doggishly appropriate ways"[26] is just one exemplary case of fortifying attachments between companion species.

Huizinga's other notable contribution is the idea that playing occurs in a safe, restricted zone such as a purpose-built playground, board, rink, or field—an enclosed playspace he calls the "magic circle." Playing is circumscribed within a particular space or environment reserved for some enchanting game or sport. Yet his description seems to be at cross-purposes with *homo ludens,* given the supposed "civilizing functions" of play throughout history. For while play is apparently "executed within certain fixed limits of time and place" and possessed of a "consciousness that it is 'different' from 'ordinary life,'" Huizinga posits a primitive anthropology according to which play is foundational and practically everywhere. "Culture arises in the form of play . . . [and] is played from the very beginning."[27] That play is coeval with a primary sociality suggests that playing cuts across artificial divisions (nature and culture, play and work, reality and fantasy, here and there). The magic circle spirals to include other elements as coplayers, and they are scarcely fixed by and for humans within a bounded playspace. It is perhaps as simple as noticing how play ever depends on the agencies and affordance of surrounding objects and areas, as in skateboarding, urban parkour, and outdoor races and games, incorporating whatever happens to be around. A preexistent space becomes newly configured as a playscape. Flanagan documents examples of more sophisticated types of "locational gaming" that deliberately involve myriad things found in a particular streetscape: *Transition Algorithm* and *Chain Reaction* set players tasks that identify changes in the urban environments, attuning themselves to mundane objects and events, recomposed as avatars and game tokens, found in public spaces.[28] Other, less structured kinds of playing require a degree of responsive to host environments and makeshift assemblages. Sliding down a snow bank; playing conkers under a horse chestnut tree; building with driftwood and sand on a beach; matching movements to those of an avatar on screen; delivering a ball down a field—these are events not determined by human designs alone. Likewise, sculpted Play-Doh realizes the shape of a relation between mind and matter at hand. "Play values experimentation," as L. O. Aranye Fradenburg says.[29] It is a style of observation and invention through the recombination of whatever happens to be lying about. Lego, like Play-Doh, creates the conditions for play out of blocks of plastic matter that do not predetermine the resulting model. (Such play models *modeling* as such.) It would be misleading to say playing is ever completely free of time, space, and matter; it is rather more like the discovery of freedoms within given time, space, and matter. Playing is a disciplined attentiveness to empirical elements. "Play doesn't screen out reality; it plays with it."[30]

What I am noticing here is that playing redistributes agencies outside of the usual anthropocentricity of human habits of thought and action even while appealing to the individual senses. What I am calling *multispecies playscapes* host persons, animals, and things at play, extending a set of captivating challenges that are *playable* rather than merely *manipulable*. Nor are they "magic." In this augmented sense of playing, interacting agencies recreate themselves in relation to the available elements; and it is perhaps a most ordinary, enworlding effect of recreational play. When, for example, Serres speaks of players becoming "the attribute of the ball as substance," the field of play is construed as expressive action coproduced by persons and things.[31] To play is to enter into a field of immanence where, as Brian Massumi elaborates, the ball and not the player becomes the subject. "The ball moves the players."[32] Play travels along the virtual vectors of substances to reveal the play *of* substances within situated environments. In other words, one never plays alone. Play is interplay. Play yields to the constraints and cunning capabilities of others to discover what they can become in the event. In a comparable view, Sicart observes that playthings *matter* in the verbal sense of manifesting themselves: "Dolls, toy cars, construction sets: they foster the creation of an intrinsic, object-centric context that emanates from the toy itself."[33] Repeated handling of play objects draws them into new patterns of interaction and mutual inhabitance, something that holds just as much for bodies at play—playfighting, leaping, chasing, or caressing—as for the play of senses across bodies. The modulating styles of everything from blocks, gears, limbs, clay, branches, musical instruments, and the ocean surf is disclosed in relation. In the sensuous rhythms of play you probe an uncertain materiality and circumstantiality that may be indifferent to you even as it relays a strong affect. You encounter an abyssal object, what Francis Ponge once called *l'objeu* (from *objet* and *jeu* for "play") to describe the phenomenal depth of things that escape full cognition while yielding to *objoie*, the joy of things.[34] Ponge's example is the sun. You play in the radiance of such things. To play is to enter a more-than-human cosmos, what Georges Bataille calls "a play of *living matter in general*."[35] It is the mundane condition of a marvelous world.

Crisis Play

But what *objoie* remains under the sun? In an age of rising temperatures and environmental degradation, when lively matter is scarce and imperiled, what becomes of the play-of-world? Has play itself been wrecked along with all the other things we may cherish in the environment? A practical way forward may be to urge that playing is often a mode of reclaiming neglected, useless, or wasted items junked by international capital and corporate systems. It is precisely what survives within—and thrives on—the ruins. What we need now more than ever is an ardent play ecology—playing on the *political ecology* of the likes of Bruno Latour and Jane Bennett—that ventures a ludic critique. Given the foregoing observations, let me propose among other strategies gamesome multispecies engagements and enthusiastic salvage: more intensive play.

One place to look is child's play. Rummaging for playthings, children regularly engage in the arts of assembling and repurposing obsolescent goods. Playthings are reclaimed from waste, recalling a likely etymological connection between toy and trash. Walter Benjamin was not the first to observe a secondary economy, or rather play ecology, established and sustained at a young age: "No one is more chaste in the use of materials than children: a bit of wood, a pinecone, a small stone—however unified and ambiguous the material is, the more it seems to embrace the possibility of a multitude of figures of the most varied sort."[36] Playthings manifest possibilities inherent in them but ignored by domestic regimens, recalling forgotten histories and possible futures. Playing with things *as* things is an opening to the sensuousness and surprises of objects as such, disclosing the latency of play within them. As Giorgio Agamben writes of the toy: "The toy is a materialization of the historicity contained in objects."[37] Items that may have appeared to be single-use afford unexpected acts and alternative futures in the event, as exemplified in miniature by some forms of doll play. As Brian Sutton-Smith asks: "Will the plastic doll be used for mothering or to make a mock of mothers? The toy itself cannot tell us."[38] Benjamin likewise notes: "Even the most princely doll becomes a capable proletarian comrade in the children's play commune."[39] Human simulacra present other occasions to transgress conventional interdictions and experiment with received matters, as in the case of *Sims* players who alternate gender presentations, sexual orientations, domestic arrangements, and somatic pleasures.[40] Granted, playing house has not often conjured radical and unsentimental alternatives to conventional domesticity, but that is to overlook powerful contingencies inherent in everyday minutiae. The queer spectacle of something so familiar as dollhouse play can open up ecumenical possibilities latent in matter; the very act of miniaturizing or virtualizing paradoxically enlarges the field. Children at play with things alter or ignore assigned scripts and scenography to discover a great variety of alternative plots available within the given materials. Flanagan describes three emancipatory kinds of critical play: "unplaying" (secret or forbidden play), "re-dressing or reskinning" (altering or effacing received appearances), and "rewriting" (subverting domestic scripts and fictions of domesticity).[41] All are types of improvised "modding," or let us say means of "hacking" systems, no less. In this way, playing can be a conspicuous means of radicalizing agency and founding new material conditions and relations that may be more inclusive and sustainable. Adapting Latour's notion of political ecology, play ecology may proliferate risky attachments to objects whose ontologies are unsettled and demand reengagement.[42] Playing is the chance to release and mobilize untold matters and energies beyond the human, suspending all-too-human concepts and compartmentalizations.

Here then is a renegade model of ecoplay, easy to trivialize and stigmatize as childish but all the more critical for escaping the adults much of the time. A ludic critique may then be ventured in the process of modding, hacking, fidgeting, animating, or toying, activities that are always liable to look immature, ignorant, inefficient, or dysfunctional. Perhaps to play well is by definition to become impertinent, irresponsible, or simply foolish children, if we take a cue from recent thinkers. When children play, "creating miniature rebellions" within the anarchic space of a game,[43] as Flanagan writes, they surrender to a lavish and profitless flow of things. Pursuing a similar intuition about childishness, J. Jack Halberstam's *Gaga Feminism* urges a queer politics modeled on the "anarchic child" who "comes at the world a little differently than the post-shame, post-guilt, post-recognition, disciplined adult."[44] Bennett begins her book on vital materialism with the remark that "childhood experiences [are] of a world populated by animate things rather than passive objects,"[45] again drawing attention to the vitality of childish play and playthings against the comparatively disenchanted lifeworlds of mature society.

Yet play ecology can be still more radical than a puerile figure allows, refusing to be identified with phases of human development. Part of the critical energy, indeed the felt emergency, of playing is losing oneself in the process, leading to what Hans-Georg Gadamer calls the "primacy of the play over the consciousness of the player." Playing is potentially so absorbing that individuals are practically beside themselves: "The players are merely the way the play comes into presentation."[46] It is as though the comportment of mind and body at play levels the field, engrossing subjects and objects in elemental matter and shared environmentality. Berkoff observes other species so involved in playing that "they *are* the play."[47] Massumi finds enthrallment indispensable too: "The players, in the heat of the game, are drawn out of themselves."[48] Play ecology must include the possibility of an enlarged, ecstatic, immersive, and ecocentric orientation without recurring to mere human verve (childish or otherwise). Playing hard refuses to attribute all the moves to individual style and strategy that otherwise adheres to creative praxis, and this is all the more urgent to notice in an era of planetary crisis. A ludic critique would expose the involution of subjects and objects in a common medium, one that shows up as much more than a passing condition. Play ecology acknowledges the exuberant and mercurial veer of environments (fields in which so many others besides the human are brought into play) and tends toward what we might call an identity crisis. Fortunately, playing is one of the most desirous and enthusiastic ways we have of dispensing with the presumption of human autonomy, opening up heteronomous zones of encounter where subjects and objects are in flux. Playing, in this sense, is the eternal rapport of interplay. Who can resist the tautological to-and-fro? A typical impulse of a game is to play on, seeking to sustain a convivial comingling despite divisions. *Crisis play* is an affective recognition of companionable acts required to thrive collectively—as in *Climate Games* and *Survivance*, to recall two hopeful playscapes.

Let me end by underlining how I am coming to think of critical and creative acts together, including those written and bound in the present volume, as profoundly playful. Talk of ecstatic play and enjoyment may seem to align my observations with a recent turn against a brooding "hermeneutics of suspicion," now regularly impeached for a supposedly aloof, joyless, and disenchanted orientation. Critique has appeared to some, of late, as a stern reprimand ("Stop playing and be serious!") and smug withdrawal ("The game is up!"). Depending on the daily news cycle, I may be more or less inclined to pack it all in. Yet even negative criticism gamely "combines rules and expectations with the possibility of unexpected moves and inventive calculations, enabling a form of carefully controlled play."[49] To play is not to relinquish the agonistic energies of critique any more than it is to be completely disarmed and delighted, merely passive. Critical acts always have a head in the game. No one really plays in a neutral position. I like to think that, taken together, critical engagements inside and outside academe express the erratic energy and exuberance of playing. As suggested by the many recent turns across the disciplines (from "material turn" to "animal turn" and beyond), critics tend to play the long game. An exigent critique depends on nothing less than sustaining enthusiasm for the veer. To designate the university as one among other fields of play is not to reduce intellectual activity to trifling; it is to describe serious recreative acts. Playing has an urgent ecopolitical aspect precisely to the extent that it describes the tensile pleasures and pressures that inform committed acts of criticism.

Notes

1. See Brian Upton, *The Aesthetic of Play* (Cambridge, Mass.: MIT Press, 2015); Mary Flanagan, *Critical Play* (Cambridge, Mass.: MIT Press, 2009).

2. *Oxford English Dictionary*, s.v., *play*, v.

3. Ludwig Wittgenstein, *Philosophical Investigations*, trans. G. E. M. Anscombe (Oxford: Basil Blackwell, 1958), part 1, section 43.

4. Ibid., part 1, section 66.

5. Jacques Derrida, *Writing and Difference*, trans. Alan Bass (Chicago: University of Chicago Press, 1978), 289.

6. Wittgenstein, *Philosophical Investigations*, part 1, section 31.

7. Upton, *Aesthetic*, 38.

8. Greg Costikyan, *Uncertainty in Games* (Cambridge, Mass.: MIT Press, 2013), 10.

9. Gregory Bateson, "A Theory of Play and Fantasy," in *Steps to an Ecology of Mind: Collected Essays in Anthropology, Psychiatry, Evolution, and Epistemology* (Northvale, N.J.: Jason Aronson, 1972), 188.

10. For these and similar games, see Miguel Sicart, *Play Matters* (Cambridge, Mass.: MIT Press, 2014), 15, 74; Flanagan, *Critical Play*, 236–37, 239–40, 245.

11. *Climate Games,* https://www.climategames.net/en/home.

12. *Survivance,* http://survivance.org/about. For a detailed study of the game design process and effects of community gameplay, see Elizabeth LaPensée, "Survivance: An Indigenous Social Impact Game" (PhD diss., Simon Fraser University, BC, 2014).

13. Upton, *Aesthetic,* 19.

14. Ian Bogost, "Persuasive Games: Exploitationware," Gamasutra, May 3, 2011, http://www.gamasutra.com/view/feature/6366/persuasive_games_exploita tionware.php.

15. Wendy Hui Kyong Chun, *Programmed Visions* (Cambridge, Mass.: MIT Press, 2011), 166.

16. Alexander R. Galloway, *The Interface Effect* (Cambridge: Polity Press, 2012), 27.

17. Robert Fagen, *Animal Play Behavior* (New York: Oxford University Press, 1981), 19.

18. Roger Caillois, *Man, Play, and Games,* trans. Meyer Barash (Urbana: University of Illinois Press, 2001), 5.

19. Michael Serres, *The Parasite,* trans. Lawrence R. Schehr (Baltimore: Johns Hopkins University Press, 1982), 80.

20. Sicart, *Play Matters,* 1; Costikyan, *Uncertainty in Games,* 6.

21. Costikyan, *Uncertainty in Games,* 6.

22. Bateson, "Theory," 186.

23. Marc Berkoff, *The Emotional Lives of Animals* (Novato, Calif.: New World Library, 2007), 85–109.

24. Bruce Bagemihl, *Biological Exuberance: Animal Homosexuality and Natural Diversity* (New York: St. Martin's Press, 1999), 17.

25. Donna J. Haraway, *When Species Meet* (Minneapolis: University of Minnesota Press, 2008), 241.

26. Donna J. Haraway, *The Companion Species Manifesto: Dogs, People, and Significant Others* (Chicago: Prickly Paradigm Press, 2003), 24.

27. Johan Huizinga, *Homo Ludens: A Study of the Play Element in Culture* (London: Routledge and Kegan Paul, 1950), 28, 46.

28. Flanagan, *Critical Play,* 210–11.

29. L. O. Aranye Fradenburg, "Living Chaucer," *Studies in the Age of Chaucer* 33 (2011): 57.

30. Ibid., 61.

31. Serres, *The Parasite,* 226.

32. Brian Massumi, *Parables for the Virtual: Movement, Affect, Sensation* (Durham: Duke University Press, 2002), 73.

33. Sicart, *Play Matters,* 38.

34. Shirley Ann Jordon, *The Art Criticism of Francis Ponge* (Leeds: W. S. Maney, 1994), 21.

35. Georges Bataille, *The Accursed Share*, vol. 1, trans. Robert Hurley (Brooklyn, N.Y.: Zone Books, 1989), 23.

36. Walter Benjamin, "The Cultural History of Toys," in *Walter Benjamin: Selected Writings*, vol. 2, part 2, ed. Michael W. Jennings et al. (Cambridge, Mass.: Harvard University Press, 2005), 115.

37. Giorgio Agamben, *Infancy and History: On the Destruction of Experience*, trans. Liz Heron (London: Verso, 1993), 80.

38. Brian Sutton-Smith, *Toys as Culture* (New York: Gardner Press, 1986), 251.

39. Benjamin, "Old Toys," in Jennings et al., *Walter Benjamin*, 101.

40. Mia Consalvo, "It's a Queer World After All: Studying *The Sims* and Sexuality," *GLAAD* (2003): 1–49; Flanagan, *Critical Play*, 48–62.

41. Flanagan, *Critical Play*, 33–34.

42. Cf. Bruno Latour, *Politics of Nature*, trans. Catherine Porter (Cambridge, Mass.: Harvard University Press, 2004), 22.

43. Flanagan, *Critical Play*, 57.

44. J. Jack Halberstam, *Gaga Feminism* (Boston: Beacon Press, 2012), xxiv.

45. Jane Bennett, *Vibrant Matter: A Political Ecology of Things* (Durham: Duke University Press, 2010), vii.

46. Hans-Georg Gadamer, *Truth and Method*, trans. Joel Weinsheimer and Donald G. Marshall, 2nd ed. (New York: Crossroad, 1991), 92, 98.

47. Berkoff, *Emotional Lives*, 94.

48. Massumi, *Parables*, 74.

49. Rita Felski, "Suspicious Minds," *Poetics Today* 32, no. 2 (2011): 229.

Ape

HOLLY DUGAN AND SCOTT MAISANO

When the matrices are strong, persistent and reinforce each other, we see
an actor, no matter how ordinary the behavior

—MICHAEL KIRBY, "On Acting and Not-Acting"

Until the nineteenth century . . . anthropomorphism was integral to the
relation between man and animal and was an expression of their
proximity. . . . In the last two centuries, animals have gradually
disappeared. . . . And in this new solitude, anthropomorphism makes us
doubly uneasy.

—JOHN BERGER, "Why Look at Animals?"

Film adaptations of *Romeo and Juliet* frequently veer far from Shakespeare's script. Both Franco Zeffirelli (1968) and Baz Luhrmann (1996),
for example, eliminate the character of Paris from the final scene in the
Capulet's tomb.[1] But a more radical reimagining of the play's end comes
in Karina Holden's *Romeo & Juliet A Monkey's Tale*, a television documentary produced by Animal Planet in 2006 about Macaque monkeys
in Lopburi, Thailand.[2] From the opening iambs of its modified prologue
("In Thailand's town . . ."), Holden's urban wildlife film pulls two seemingly far-removed worlds—Shakespeare's Globe and Animal Planet—into
a potentially catastrophic orbit with one another. In its final moments,
Tybalt, who ought to be dead halfway through the film, is alive to explain
in voiceover narration: "Juliet had stolen away. One of the guards had

spied her heading out of town on the back of a pickup truck." While Tybalt speaks, we see the macaque Juliet astride the bed of a Toyota, racing against a background that shifts from suburban traffic to rice fields and back again, to the soundtrack of Sam and Dave singing "Hold On, I'm Comin." These monkeys never knew that they had been cast in the roles of Romeo, Juliet, Tybalt, and the rest. Instead, Shakespeare's narrative was imposed on them in postproduction—as raw footage was cut, edited, and dubbed—only after filming had wrapped. Admittedly, the documentary avails itself of virtually every problematic convention associated with popular wildlife films: a "market-driven, formulaic emphasis on dramatic narrative and ever-present danger," giving the animals names, use of a musical soundtrack, "point-of-view shots, reaction shots, and so forth."[3] And Holden herself acknowledges resorting to some tricks of the trade— such as continuity editing, using multiple (similar-looking) monkeys to portray individual characters, and cuing action and attention by means of peanut butter and molasses—while filming in Lopburi.[4] These built-in controls notwithstanding, no auteur has ever veered as far off script as Holden and her macaque "actors" (see Figure 1).

In a recent essay responding to the uproar surrounding the Oregon Shakespeare Festival's decision to commission playwrights to translate Shakespeare's texts into twenty-first-century parlance, Sheila Cavanagh concludes: "The best adaptations—*West Side Story*, the musical *Kiss Me, Kate*, and the Japanese film *Throne of Blood*—thrive. The bad, silly and unfortunate—*Romeo and Juliet Sealed with a Kiss* and Animal Planet's *Romeo and Juliet A Monkey's Tale*—fall by the wayside."[5] We wish to rephrase this assessment by suggesting that Animal Planet's *Romeo & Juliet*, like its leading lady, ends up going off the beaten path—and "by the wayside"—in part because instead of *adapting* or *appropriating* Shakespeare's star-crossed lovers, this documentary "apes" them. *To ape*, of course, is to imitate in a way that is, to borrow Cavanagh's terms, "bad, silly, and unfortunate", or, to quote the *Oxford English Dictionary*, to mimic "pretentiously, irrationally, or absurdly." And who could more absurdly impersonate—*what* could be a sillier substitute for—Shakespeare's legendary lovers than macaque monkeys? And yet, in what is now regarded as a classic study in the history of modern psychology, Harry Harlow selected neonate and infant rhesus macaque monkeys for his pioneering investigations into "The Nature of Love," the title of his 1958 address to

On the clapperboard:

HOLLYWOOD

PRODUCTION ROMEO + JULIET

DIRECTOR KARINA HOLDEN

CAMERA PETER NEARHOS

DATE 22/11 SCENE 12 TAKE 5

Figure 1. Macaque actor on the set of *Romeo & Juliet A Monkey's Tale*.
Image courtesy of Karina Holden.

the annual convention of the American Psychological Association, because in his words

> the macaque infant differs from the human infant in that the monkey is more mature at birth and grows more rapidly; but the basic responses relating to affection, including nursing, contact, clinging, and even visual and auditory exploration, exhibit *no fundamental differences in the two species.*[6]

Whereas macaque monkeys are routinely used *as if they were human* in laboratory experiments, including but hardly limited to Harlow's gruesome quest to discover the nature of love by starving and by depriving infant macaques of "contact comfort" (see Figure 2), the appearance of macaques in a performance of *Romeo and Juliet* strikes most viewers, not just Cavanagh, as awkward, inappropriate, inadequate, and foolish: in short, an embarrassment.[7]

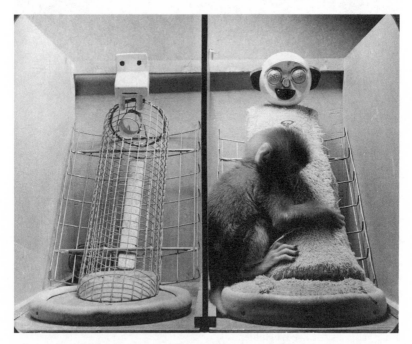

Figure 2. Harlow's famous experiment on the nature of love isolated infant macaques from their mothers in order to study their responses to wire and cloth surrogate mothers. Image from Science Source/Getty Images.

Why are the same animals viewed as ideal impersonators of humanity in one context and yet off-kilter in the other? In this essay we analyze aping—specifically, ethical cross-species casting of Shakespeare's dramatis personae—not as a way of experimenting with animals in the name of science but rather as a way of experimenting with the humanities in the name of nonhuman animals. Taking what Anna Blackwell calls an "actor-based approach to adaptation studies," in which "the actor shapes the adaptation, not only in their performance but through their intertextual physicality," we consider how casting a macaque Juliet (or a Shetland sheep Cordelia in *King Lear*) can change the context of mimesis and unexpectedly shift the significance of their performances into realms of posthuman ethics and environmental politics.[8]

The casting of animals in such roles may seem arbitrary: that is, one could easily suppose that macaques have no more relevant connection to Shakespeare's *Romeo and Juliet* than any other animal. In *CIFERAE: A Bestiary in Five Fingers,* Tom Tyler describes two ways philosophers (and, we would add, filmmakers) have found to employ animals: either as "cipher" or as "index." Tyler explains how the "cipherous use, in which the choice of creature is entirely arbitrary, stands in direct contrast to the indexical use, in which specific traits are especially selected."[9] Holden's macaques had to be macaques: few other nonhuman animals could convincingly play even the opening scene of *Romeo and Juliet*, where an erstwhile "ancient grudge" erupts into "new mutiny" over a simple nonverbal gesture, "the biting of thumbs" (prologue: line 3; 1.2.40–41).[10] Indeed, as Tyler makes clear in his book, the opposable thumb has long served as readymade proof of human exceptionalism, what sets us apart from all other animals. But this anatomical shibboleth is itself easily opposed by a quick glance at macaque phalanges, which Holden's camera provides us, as Tybalt, who identifies himself as a "temple monkey," recounts how the rival clan of "market monkeys" did "bite their thumbs at us."

We also contend that such casting reveals implicit categorical relationships between the human domain of acting and the animal domain of aping as well as the continued role Shakespeare plays in drawing such lines. There are many who argue precisely that: for instance, Jeremy Taylor, a documentary filmmaker and author of the book *Not a Chimp: The Hunt to Find the Genes That Make Us Human*, argues that, despite primatologists' and evolutionary psychologists' claims to have located "the foundations for human empathy, social intelligence, technology, mathematical skills, and moral rectitude in chimpanzees," it is simply "an insult to human ingenuity and culture" for any self-respecting scientist to make comparisons "between alarm calls, food-specific grunts, whoops, and Shakespeare."[11] Taylor concedes that humans and chimpanzees share 96 percent of their genetic material but insists there is, nonetheless, a difference of kind, not degree, separating chimpanzees' pant-hoots and lexigrams from Shakespeare's autotelic poetry and drama.

Not only does Shakespeare make us "not a chimp" but, according to Taylor, sentence structure is foundational for what is called "theory of mind," the cognitive capacity to understand that others have beliefs, desires, and motivations different from one's own. Whereas Taylor argues that the ability to form complex sentences predates and in some sense precipitates theory of mind in humans, many primatologists have argued, contrarily, that apes, like humans, possess theory of mind. Robert Sapolsky, a prominent neuroscientist best known for his studies of nonhuman primates, clarifies in his TED talk, "The Uniqueness of Humans," that chimpanzees do possess theory of mind but emphasizes that only humans possess a "secondary theory of mind," which, he says,

> is when you understand that *that* [second] individual has information that *that* [third] one doesn't; and thus *that* one thinks that *this* one is doing *this* but it's actually doing *that*; and [so] no other animal can sit through something like a performance of *A Midsummer Night's Dream* and understand what's going on there, with whom. We [humans] are the only ones who would be willing to spend an evening doing that and have a clue what was happening there. So, we are alone in that realm.[12]

Whether advancing or opposing the idea that macaques, baboons, bonobos, and other primates are "almost human," Shakespeare is invariably assumed to be the furthest thing imaginable from nonhuman primates. Western philosophy and science have always employed some form of "anthropological machine" to separate humans not only from other animals but from their own animality, a point that Giorgio Agamben makes in *The Open*; but what this more recent history shows is that Shakespeare has increasingly *become* the anthropological machine.[13]

This is a surprising turn of events, particularly if one revisits the long history of aping in the Renaissance or even just recalls how Shakespeare begins act 2 of *Romeo and Juliet* with Mercutio calling Romeo an "ape." Immediately after Romeo scales the wall surrounding the Capulets' orchard, Benvolio and Mercutio enter in search of their smitten chum. As he calls Romeo's name, Benvolio explains to Mercutio what has just happened: "He ran this way, and leaped this orchard wall: / Call, good Mercutio." Mercutio replies: "Nay, I'll conjure too. / Romeo! humours! madman! passion! lover! / Appear thou in the likeness of a sigh: / Speak but one rhyme, and I am satisfied." When Romeo fails to appear or to speak, Mercutio jests: "He heareth not, he stirreth not, he moveth not; / The ape is dead and I must conjure him" (2.1.9–16). The Signet Classic edition of the play glosses Mercutio's cryptic phrase—"The ape is dead"—as "Romeo plays dead, like a performing ape," conjugating Romeo's action from Mercutio's noun. To conjure is to "swear together," but here the ape is silent and still, forcing meaning to career around it.[14] The Bedford Shakespeare offers a similar explanation: "i.e., like a performing monkey playing dead."[15] Mercutio calls Romeo an "ape" at this point precisely because he knows that Romeo can hear him and that Romeo is only *performing* deafness, *pretending* not to hear his name called, just as monkeys in Elizabethan England were trained to *act* dead in paratheatrical entertainments (though many were indeed beaten to death as part of their training).

Part of our contention here is that aping has a long history, one that culminates in (rather than emerges from) recent posthuman and ecocritical investments in Shakespeare. Students and scholars of Shakespeare have long been familiar with the proximity of Shakespeare's Globe Theatre to the Bear Garden in Southwark: in John Norden's 1593 map of London the two edifices sit right next to each other on London's Bankside (see Figure 3).[16] More recently, Andreas Höfele has argued "that the theatre's family resemblance to animal baiting . . . bred an ever-ready potential for a transfer of powerfully affective images and meanings. The staging of one of these kinds of performances is always framed by, always grounded in, an awareness of the other."[17] Höfele adds that "the effect could be described as double vision or synopsis, in the literal sense of 'seeing together,' of superimposing one image upon the other. What spectators perceived as human or as animal no longer exists in clear-cut separation; it occupies a border zone of blurring distinctions where the animal becomes uncannily familiar and the human disturbingly strange."[18] Höfele effectively shows that "the Elizabethan theatregoer . . . saw plays differently . . . from someone living in a world without bear-baiting."[19]

Yet aping also colors Shakespeare's works. Over a hundred years ago, around the time he was coauthoring *The Elements of Style* with his former student E. B. White, William Strunk Jr. identified "the Elizabethan Showman's Ape" as a performer who competed with Shakespeare for audiences, if not for eminence, in early modern England.[20] Though Höfele, Erica Fudge, and others have focused on bear-baiting's proximity to and influence on Shakespeare's Globe, including the important role of monkey-baiting in those arenas, we would note that Shakespeare's plays are shot through with the idea of primate performances—allusions to "apes," "baboons," and "monkeys" appear in twenty-four, or roughly two-thirds, of his plays—even more so than bear-baiting.[21] Though Holden's decision

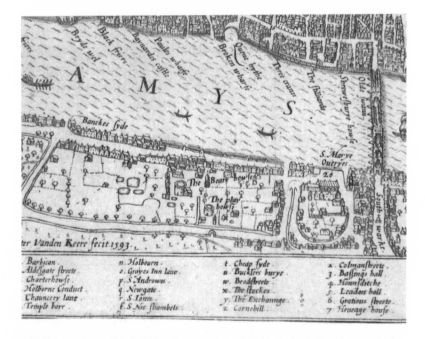

Figure 3. Detail of John Norden's Map of London, 1593, depicting the proximity of baiting arenas with public playhouses. Courtesy of the British Library and Wikimedia Commons, Maps. Crace.1.21.

to edit footage of macaques into a Shakespearean plot line could be said to belong to this longstanding performance tradition, most examples of monkey performers, then and now, are far less savory.

While bear-baiting has gone the way of the ruff and the codpiece, monkey-baiting has not. We still have an appetite for simian performance that would not be out of place in the Elizabethan Paris Garden. In November of last year, the NFL, for instance, thought it wise to entertain Cincinnati Bengals fans during a particularly "brutal" game with the Texans with an "exciting" halftime performance by Team Ghost Riders, a company that aims to bring the jungle and the rodeo arena together: three monkeys, riding sheepdogs, herded rams on the football field.[22] The only difference between this performance and Elizabethan ones was the rodeo costume and a GoPro camera, which provided fans with an up close shot of the terrified monkeys' faces.[23]While this halftime entertainment was not intended to raise the issue of nonhuman subjectivity, the inclusion of the GoPro unexpectedly offered human observers the far-less-amusing "point of view" of a nonhuman performer.

The long history of the performing monkey also received some recognition in the BBC's television adaptation of Shakespeare's *Richard II* for its *Hollow Crown* miniseries.[24] The eponymous monarch, played by Ben Whishaw, is seen early and often holding a pet monkey in his arms. Indeed, even as the opening credits are still rolling, we see this monkey, quite literally the king's right-hand man, sitting close enough to touch King Richard's sleeve. For most of the film, Rosey, the monkey, has neither agency nor a point of view; she is merely a living symbol, cipher, ornament, or accoutrement: "For a long time," Rupert Goold, the director, explained in an interview, "I was interested in Richard II as a Michael Jackson figure." In terms of performance, then, Rosie's body—its expressiveness—is primarily there to realize and to build Richard's character, not her own.

But more than two hours into the film, following Richard's abdication, there is a shot of Bolingbroke sitting alone in profile on his throne at Windsor Castle. We watch as his eyes look upward and take note of something overhead; the film then cuts from the shot of Bolingbroke to a low-angle close-up of the monkey, perched aloft and staring down at the newly crowned king, before cutting back to Bolingbroke whose face now registers the effect of realizing there is another set of eyes—another point of view—still with him in the room. Known as a "shot reverse shot," this ubiquitous technique of continuity editing gives viewers the sense of two characters looking at each other: it first shows one character looking at another character (offscreen), then shows the other character, facing the opposite direction, looking back at the first character (offscreen). Here, for a moment, the monkey ceases to be merely an object for humans to look at. The shot reverse shot credits the monkey with a gaze equal to the monarch's: no longer tucked marsupial-like into the folds of the old king's clothing, the monkey now fills the screen and directs its own estranging gaze downward at the new monarch.

Team Ghost Riders and the Hollow Crown remind us that this history of performing monkeys is long and complex, even as it is recent; it is almost impossible to imagine any kind of macaque performance existing outside of it. Though the monkeys of Lopburi were not necessarily acting for a camera, they are skilled performers: most have been trained to busk both in the temple and in nearby markets, a skill that makes them effective petty thieves. Elsewhere, macaques are defined as a nuisance pest; in 2013, 91 percent of imported primates into the United States were crab-eating (also known as long-tailed) macaques.[25] Over seventeen thousand of these animals were used in toxicology and biomedical research, a trend that was repeated in previous years.[26] The United States is one of the top three importers of long-tailed macaques, and though most are officially listed as bred in captivity, researchers hypothesize that these records are inaccurate, due to "laundering" in the industry.[27]

Defined as both a nuisance pest as well as a tourist headline, the crab-eating or long-tailed macaques of Lopburi are in high demand as exports as biomedical research subjects, food, and pets. Though tourism is the main economy of the town—all centered on the monkey troops who live in its center—very little of that income makes its way to local residents. The human residents of Lopburi adopt a mostly positive approach to their long-tailed neighbors, choosing conservation (like birth control) over pest control options, sharing water resources and willingly providing food, culminating in the yearly celebration shown in the film—the monkey buffet.[28]

While *Romeo & Juliet A Monkey's Tale* was neither intended nor marketed as avant-garde performance art, the film would not be out of place in Una Chaudhuri's introduction to *Animal Acts: Performing Species*

Today, where she stresses that a "primary mode of [twenty-first-century] interspecies performance . . . is literalization, a steady focus on—or regular return to—the animal or animals around whom the performance revolves," and adds that "the animal acts being forged today are committed to never forgetting the animal and to always asking: 'Where are the real animals in all this?'"[29] Holden's documentary does indeed maintain a steady focus on and regularly returns to its pair of macaque lovers, Romeo and Juliet. But sometimes Holden substitutes one mohawked male macaque for another in the role of Romeo, and one young female macaque for another as Juliet. These substitutions went unnoticed by us, even after multiple viewings, and were only revealed by Holden herself in conversation. This practice of casting multiple actors to play a single role not only monkeys with the more familiar practice of doubling (where one actor plays two or more roles, as in Shakespeare's plays) but, more importantly, wreaks havoc with the very theater-watching abilities that Sapolsky singles out as unique to our species: as it turns out, humans watching macaque actors frequently do *not* understand "which individual is which" or "what's going on there, with whom" any better than macaques watching human actors. Moreover, for the first thirty-seven of its forty-four total minutes, as we watch "street-smart" macaques scavenge food from human vendors and patiently wait curbside to cross the road while a uniformed officer directs traffic, we're reminded that "the real animals" are *both* macaques and humans. The two species prove equally adapted to the anthropogenic environment of the city. The attribution of human voices to characters with simian bodies further blurs any clear-cut lines between "animal" and "human" behavior. Whether looking at man or macaque, the same performance-related question arises: Is this acting or aping? It hangs by a tail.

The monkey's tale veers in other ways too: there are good and bad monkeys. The "peaceable" temple troop is at war with the "market" monkeys, or at least, so says Tybalt, the narrator of Animal Planet's footage. That isn't quite the same as two households "alike" in dignity; nor is their "traffic" the same. Indeed, the preamble drops all mention of the stage or its history, replacing it with that of Lopburi's Buddhist temple and its warring troupes of macaques. But this is where, per Tybalt's narration, another unexpected swerve takes place: "To bring our ancient grudge to a close, the humans hatched a cunning plan . . . an old, accustom'd feast."

Readers familiar with Shakespeare's play will recognize in those final few words a direct quote from Shakespeare, spoken by Juliet's father, Capulet, in act 1, scene 2. In Shakespeare's text, this "old, accustom'd feast" supplies the setting for Romeo and Juliet's love-at-first-sight encounter, a scene where all is well, Capulet himself describes Romeo as "a virtuous and well-govern'd youth," and only Tybalt is spoiling for a fight. So what is this joyous feast doing, in place of the tomb scene, at the end of *Romeo & Juliet A Monkey's Tale*? This "party," the macaque Tybalt explains, "would end once and for all the bitter quarrel of the monkey clans." When the humans appear at the annual Lopburi Monkey Buffet Festival, some are carrying larger-than-life effigies of monkeys, others have their faces painted to resemble monkeys, others wear monkey masks over their faces, and still others dance in ways meant to simulate monkeys' movements. In the documentary, where Tybalt does all the talking, we never hear the humans speak. The effect renders Animal Planet into a Shakespearean *Planet of the Apes,* if the latter were a nature documentary.

Indeed, in their travel reviews to Phra Prang Sam Yot, tourists (mostly white and Western) repeatedly describe their experience of visiting the temple space in Lopburi through references to *Planet of the Apes*: "At times you feel surrounded, like you're in a homemade Land of the Apes flick"; "The backdrop is quite cinematic. I wonder if they're waiting for the next Planet of the Apes movie."[30] In doing so, the tourists conflate monkeys as apes and define themselves outside of both categories, narrating their experience as one outside of time or history but not necessarily science or fiction: the "monkey" temple of Lopburi becomes an otherworldly space of (white, Western) imagination, where apes reign (not a present-day space where humans choose to live with animals in radically different ways).

That double vision is, of course, embedded in the text that they cite. Pierre Boulle's *La Planète des Singes* (1963) and its subsequent film versions beg the question, who is the greatest of the great apes: humans or apes?[31] In it, "astronauts from the future" crash-land on a planet "where intelligent talking apes are the dominant species"—that is, earth.[32] The film's strange logic makes sense only when we imagine chimpanzees, gorillas, and other hominids as human—that is, as performing intelligent, *talking* apes (not intelligent aping apes). That we trip over the very words that we use to define such premises makes the irony all the sweeter.

Lambasting both the original premise, as well as the numerous times the 2011 film was described by reporters as involving monkeys rather than apes, science writer Martin Robbins emphasizes the *lack* of logic in such errors of diction: "People think that they *aren't* apes but that apes *are* monkeys."[33] Robbins gleefully seizes on headlines like this one from the *Daily Mail* as evidence of his point: "Fans go bananas for new *Planet of the Apes* trailer which takes humanized monkey effects to a whole new level."[34]

Semantics matter. Robbins emphasizes the lack of specificity at work in the end of the sentence, but we're struck by its start: Who or what are the fans aping when they "go bananas" for this film about humanized (and computer-generated) monkey effects, effects that included, for instance, mixing human voices with chimpanzee growling? What is it that audiences want to see reflected through performance, and does participating in such a structure of imitation change them? What happens when we visit the monkey temple of Lopburi and imagine it as an inverted evolutionary space, or when we elevate it to a stage worthy of Shakespearean adaptation?

Romeo & Juliet A Monkey's Tale, for instance, begins with its title spelled out in a sparkling Disneyesque font. There are no punctuation marks in the opening credits, creating semantic confusion (see Figure 4). The juxtaposition of the two phrases suggests coauthorship: this tale is both Shakespearean and simian. The opening credits set the stage for this shared act of storytelling, signaled by the shimmering, vibrant tail of the ampersand between *Romeo & Juliet.* It emphasizes the separation between Romeo and Juliet, rather than one between Shakespeare's title and the macaques of Lopburi. Yet critics quietly corrected this in their reviews by adding a colon. Indeed, we did, too, in initial drafts of this essay. Though this correction makes grammatical sense, it signals the relationship between Shakespeare's tale and those of the macaques as one of analogous, if unequal, equivalents. *Romeo & Juliet: A Monkey's Tale* emphasizes a grammatical logic that begins with Shakespeare and ends with a homophonic pun about animals. To quote the language of the trailer, "the world's greatest love story was to become a monkey's tale."

The title *Romeo & Juliet A Monkey's Tale* creates a useful run-on phrase, one that refuses to separate the animal from Shakespeare into hierarchies or syntactical sense. This relationship may be grammatically

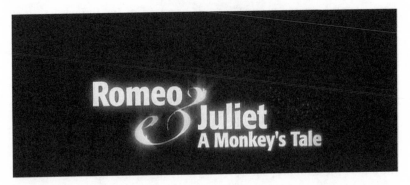

Figure 4. Film still from the opening credits of Animal Planet's *Romeo & Juliet A Monkey's Tale*.

incorrect, but it is ethically sound, inviting us to view monkeys as actors, no matter how animal the behavior. It crafts a relationship between animal studies, performance studies, and Shakespeare that *might* produce something different from what we already know, particularly dominant, scientific accounts of who or what is an ape and who or what can ape. Aping, we argue, refuses to choose a side, swinging in and out of focus between such structures of meaning.

Ape as a verb, rather than as a noun, allows us to ask such questions and transforms the smooth logic of grammar into a rougher plane of engagement. This logic unsettles hierarchical taxonomies, especially a grammar of species difference based on nouns, embracing other kinds of verbs than *to be*. Macaques, after all, have been trained to "ape"—that is, to mimic and imitate humans in ridiculous fashion—from before Shakespeare's time to our own. Giving priority to doing in place of being, the verb *ape* in place of the noun, one could argue that "if it 'apes,' then it is 'an ape.'" Ben Jonson's satirical sonnet, "On Poet-Ape," for instance, tartly derides Shakespeare as the eponymous animal, one whose literary gifts are more imitation than inspiration. Shakespeare's just a "bold thief," dressed up in a poor man's wool, one whose "brokerage" may fool "sluggish, gaping auditors" who can't tell "a fleece from locks of wool or shreds of the whole piece." Jonson's distinction between poets and poet-apes is wooly, but he insists that the difference will out in performance. As the character Envy in his play *Poetaster* makes clear, "Apes are apes," even if dressed up and made to perform.

We screech instead: Apes *ape* apes. It's a sound that mimics Lear's epizeuxis of grief—"never, never, never, never, never!"—even as it raises the question: Is there any limit to aping? If, as Jonson suggests, Shakespeare was an ape in sheep's clothing (dressed in "locks of wool" "fleeced" from other playwrights), then can even sheep ape? We ask this question in all seriousness, given the recent success of Missouri Williams's *King Lear with Sheep*. Williams's theatrical experiment shares a willingness to see what happens when we stage an animal Shakespeare: in the play chaos ensues when "a stubborn director" tries to put on a production of the play cast with sheep.[35] The result is an ovine fiasco. Williams's sees her play as taking aim at our theater conventions, using animals to do so: "I wanted to do something extreme that would challenge our ideas about both Shakespeare and the theatre as a whole—to re-establish this sense of presence, a violent, unpredictable animal presence (sheep), that people might feel uneasy, scared or embarrassed by. Expose the performance as a performance, if you will."[36]

But the play also takes aim at our habitual way of looking at animals, using theater to do so: sheep who won't follow, be led, or take direction, of course, goes against all of our assumptions about sheep. The conceit of the play—that the sheep's refusal to perform is limited to this particular night—invites us to suspend our disbelief and to imagine the sheep playing against type, as rebellious and unpredictable (see Figure 5). Williams primes spectators to watch sheep in ways previously reserved for trained wild animals—horses, tigers, bears—performing with human handlers as part of circuses and magic acts. Donna Haraway, noting how "most sheep farmers . . . rob sheep of virtually every decision until whole breeds may well have lost the capacity to find their way in life without overweening human supervision," praises British primatologist Thelma Rowell's "practice of setting out a twenty-third bowl in her farmyard in Lancashire when she has only twenty-two sheep to feed."[37] As Haraway puts it, this "homely twenty-third bowl . . . is the space of what is not yet and may or may not ever be; it is a making available to events; [and] making it possible for something unexpected to happen."[38] In Williams's play that twenty-third bowl expands to the size of the theatrical stage.

Is *King Lear* a sheep's tale? Williams's play certainly suggests that it could be. But despite Rowell's work to elevate sheep to the status of "honorary primates," that privileged class of charismatically social creatures,

Figure 5. Alasdair Saksena as the director in the Courtyard Theater's 2015 production of *King Lear with Sheep*. Image from Nick Morris.

sheep are not afforded the right to ape or act. To sheep with Shakespeare is to pasture the field, a point made dramatically literal in this case. Though some reviewers, for instance, described the play as both "odd and wondrous," others dismissed it as just a one-man production of Lear "surrounded by pooping sheep."[39] The critical success of Williams's play, which includes three separate stagings so far, culminating in an extended run in London in 2015, suggests that we much prefer to see an animal fail at acting human, even as the material conditions reveal their capacity to ape us well.[40]

King Lear with Sheep offers a chance to see not just how our perception of the natural world is shaped through linguistic categories but also the lengths that we'll go to in order to have them performed for us. That we refuse that relationship with some animals (sheep) but require it for others (apes) reveals the importance of that performance. But aping—as a shared practice of performance and as an acknowledged intimate entanglement—also provides us with an opportunity to play against type. What are we willing to acknowledge about ourselves through our relationships with animals? What might we learn from aping macaques, rather than subjecting them to scientific or Shakespearean tests about the nature of love?

In *Shakespeare's Universality* Kiernan Ryan criticizes the historicist approach to Shakespeare's drama as focusing on "what it once meant to people long since dead in a world that no longer exists, or with what it has been made to mean in subsequent eras by diverse cultures."[41] Raising the "question of what sets Shakespeare apart" from other Elizabethan playwrights, Ryan answers that Shakespeare owes his unique "universality," the continued success of his plays in countless adaptations across all media platforms for more than four centuries, to the fact that his plays are "dramatized from" (and ceaselessly gesture toward) a utopian future perspective, what Ryan calls "an egalitarian standpoint that is still in advance of our own time."[42] To elaborate, Ryan cites a line from *Troilus and Cressida* as evidence of Shakespeare's "pervasive alertness to every irrefutable fact that 'makes the whole world kin.'"[43]

For Ryan, this offers human potential: Differences in rank, gender, race, creed, and nationality are all transcended by an appeal to a human "equality," to our "belonging to the same species."[44] If we're invested in a Shakespearean universality, one veering toward the future, why limit it to human differences? Ryan himself notes that "kindness" derives from "kin"—"A group of persons descended from a common ancestor, and so connected by blood-relationship," according to the *OED*, "a family, stock, clan." Why not biological clades, too?

Making the whole world kin can be a dangerous proposition: *Romeo and Juliet* as a play demonstrates the risks of approaching the world in this way, of falling in love without paying close attention to the grammar of difference or of kind. Juliet's famous refusal of the proper nouns of identity—"'Tis but thy name that is my enemy," and "What's in a name?"—is matched by Romeo's response, "By a name I know not how to tell thee who I am" (2.2.38–39; 2.2.43; 2.2.53–55). Their shared capacity for love (whether imagined as human or simian) violates Verona's rules of order, and they die as a result, a violent sacrifice that reminds witnesses of their own capacity to love.

Described in this way, the play echoes Harlow's own tests about the nature of love. His description of human and macaque infants emphasizes their shared capacities without extending any material consequences as a result: "The basic responses relating to affection, including nursing, contact, clinging, and even visual and auditory exploration, exhibit *no fundamental differences in the two species.*[45] As infants, macaques and humans love alike. Might not human and simian Romeos and Juliets love alike as well?

Like Romeo and Juliet, the end for Harlow's test subjects was tragic: violent, asocial, and isolated from both their biological kin and from their human handlers, the animals eventually hurt themselves and others. The solution, Harlow went on to prove, was playful interaction. Indeed, very little play was needed to correct the horrors of such deprivations. As one reporter concluded, "A little jiggle, a soft sweater, and only 30 minutes of actual primate interaction . . . any mother can do this: lazy, working, wired, iron."[46] Played in this way, by macaques who share our capacity to love and, more importantly, our capacity to suffer for it, Shakespeare's tale takes on new meaning.

Harlow's research on macaques had long-ranging, positive impact for human children. So why are we hesitant to interpret the macaques' Shakespearean play as worthy as ours? Animal advocate Kathryn Dawn argues that macaques who undergo experimentation for human benefits should earn luxurious retirements; Karen Rudy argues that they should be deemed public heroes.[47] For Rudy, loving animals should be an important part of our advocacy for them: "affective connection is the basis for mass change."[48] And though we agree, we note that Shakespeare's plays offer an important footnote about the shape that this love may take. It's hard not to ape Lady Macduff, whose final words to her precious child before he is violently killed by Macbeth's murderers, are, "Now God help thee, poor monkey!"[49]

Our hope is instead that through creative performances, adaptations, and appropriations, "Shakespeare's universality" might, like Rowell's primatology, veer far enough from these harrowing histories toward a sense of "kindness" that extends beyond our own species, genus, family, order, and kingdom, ultimately including any animal having the capacity under the right circumstances *to ape.* Incongruous as these "apes" may be— humans, long-tailed macaques, Shetland sheep—they would all be alike in dignity. For now, Holden's *Romeo & Juliet A Monkey's Tale* and Missouri Williams's *King Lear with Sheep* will point the way toward this expanded universality of a nonhuman Shakespeare, toward, as it were, a ShakesVeer.

Notes

1. Franco Zeffirelli, dir., *Romeo and Juliet* (Paramount Home Video, 1968); Baz Luhrmann, dir., *Romeo + Juliet* (Twentieth Century Fox, 1996).

2. Karina Holden, dir., *Romeo and Juliet A Monkey's Tale* (Becker Entertainment, 2005).

3. Derek Bousé, *Wildlife Films* (Philadelphia: University of Pennsylvania Press, 2000), 5, 28.

4. It is equally, if not more important, to underscore Holden's training in conservation biology, her closely read and heavily marked copy of *Romeo and Juliet* (which she had at hand while scripting the film), and the fact that her editor, Jonathan Redmond, had also worked with fellow Australian filmmaker and adapter of *Romeo and Juliet*, Baz Luhrmann. Karina Holden, phone interview by Scott Maisano, February 2, 2016.

5. Sheila Cavanaugh, "Does 'Translating' Shakespeare into Modern English Diminish Its Greatness?" The Conversation, October 28, 2015, http://thecon versation.com/does-translating-shakespeare-into-modern-english-diminish-its -greatness-48297.

6. Harry Harlow, "The Nature of Love," in *The Macaque Connection: Cooperation and Conflict between Humans and Macaques*, ed. Sindhu Radnakrishna, Michael A. Huffman, and Anindya Sinha (New York: Springer, 2013), 21, emphasis added.

7. The etymological history of the word *embarass*, too, is linked to our argument: "to embarrass" someone in the sixteenth century included "hamper[ing] or imped[ing] a person, action, or process," or veering them off course. See *OED Online*, s.v. *embarrass*, v. Thanks to Jeffrey Cohen for pointing this connection out to us.

8. Anna Blackwell, "Adapting *Coriolanus*: Tom Hiddleston's Body and Action Cinema," *Adaptation* 7, no. 3 (2014): 344–52 (351).

9. Tom Tyler, *CIFERAE: A Bestiary in Five Fingers* (Minneapolis: University of Minnesota Press, 2012), 32–33.

10. In-text references are from William Shakespeare, *Romeo and Juliet*, in *The Bedford Shakespeare: Based on the New Cambridge Shakespeare Edition*, ed. Russ McDonald and Lena Cowen Orlin (New York: Bedford St. Martin, 2015), 272–342 (276).

11. Jeremy Taylor, *Not a Chimp: The Hunt to Find the Genes That Makes Us Human* (New York: Oxford University Press, 2010), 74.

12. Robert Sapolsky, "The Uniqueness of Humans," September 2009, http://www.ted.com/talks/robert_sapolsky_the_uniqueness_of_humans.

13. Giorgio Agamben, "Taxonomies," in *The Open*, trans. Kevin Attell (Stanford: Stanford University Press, 2004), 27.

14. J. A. Bryant Jr., ed., *The Tragedy of Romeo and Juliet* (New York: Signet Classics, 1986), 34.

15. Shakespeare, *Romeo and Juliet*, 292.

16. C. J. Visscher, *[Londi]num florentissima Britanniae [ur]bs emporiumque toto orbe celeberrimu* (Amsterdam, 1650).

17. Andreas Hofele, *Stage, Stake, and Scaffold: Humans and Animals in Shakespeare's Theatre* (New York: Oxford University Press, 2014), 12.

18. Ibid, 15.

19. Ibid, 14. Indeed, the history of the Hope Theatre suggests that we might inverse this relationship: that early modern bear-baiting audiences understood the violence of baiting differently from someone living in a world without public playhouses. Built to host both baiting events and plays, the Hope Theatre was a mixed-use venue, though one that was never very quite as successful as a playhouse as it was as a baiting arena. And as Lowell Duckert reminds us, bears appeared on other kinds of Renaissance stages. Their presence on early modern stages reminds us of the dangers of staking ourselves to historical or ontological categories of difference: with such stakes, "the end is violence." Bear baiting (at least at the Hope) ended in the final years of the Interregnum after parliamentary soldiers shot all of the bears, save one. Lowell Duckert, "Exit, Pursued by a Bear (More to Follow)," June 24, 2013, http://www.clemson.edu/upstart/Essays/exit-pursued-by-a-polar-bear/exit-pursued-by-a-polar-bear.xhtml#t19.

20. W. Strunk, "The Elizabethan Showman's Ape," *Modern Language Notes* 32, no. 4 (1917): 215–21.

21. See Andreas Höfele, *Stage, Stake, and Scaffold: Humans and Animals in Shakespeare's Theatres* (New York: Oxford University Press, 2011); see also Erica Fudge, *Perceiving Animals: Human and Beasts in Early Modern English Culture* (Urbana: University of Illinois Press, 2002); Karen Raber, *Animal Bodies, Renaissance Culture* (Philadelphia: University of Pennsylvania Press, 2014); Laurie Shannon, *The Accommodated Animal: Cosmopolity in Shakespearean Locales* (Chicago: University of Chicago Press, 2013). On bears as performers, see Duckert, "Exit."

22. See Will Brinson, "Monkey in Bengals Jersey Rides a Dog While Holding a GoPro at Halftime," CBSSports, November 16, 2015, http://www.cbssports.com/nfl/news/monkey-in-bengals-jersey-rides-a-dog-while-holding-a-gopro-at-halftime/; for more on "Team Ghost Riders," see http://www.humanesociety.org/assets/pdfs/abuse/cowboy_monkeys_fact_sheet.pdf.

23. This performance was not an isolated incident; indeed, the Bengals routinely stage this in their halftime shows. The use of the GoPro camera was novel. The resulting video is disturbing, though *USAToday* does not frame it as anything other than entertaining: see Alysha Tsuji, "The Bengals Gave a Monkey a GoPro and Set Him on the Back of a Shepherd Dog," *USAToday*, November 19, 2015, http://ftw.usatoday.com/2015/11/the-bengals-gave-a-monkey-a-gopro-and-set-him-on-the-back-of-a-shepherd-dog.

24. Rupert Goold, dir., *The Hollow Crown: Richard II* (BBC Two, 2012).

25. The International Primate Protection League, "Primate Import Statistics for 2013," January 17, 2014, https://www.ippl.org/gibbon/blog/u-s-primate-import-statistics-2013/.

26. Ibid.

27. Kaitlyn-Elizabeth Foley and Chris R. Shepherd, "Trade in Long-Tailed Macaques," in *Monkeys on the Edge: Ecology and Management of Long-Tailed Macaques and Their Interface with Humans*, ed. Michael D. Gumert, Agustín Fuentes, and Lisa Jones-Engel (Cambridge: Cambridge University Press, 2011), 20.

28. Suchinda Malaivijitnond, Yolanda Vazquez, and Yuzuru Hamada, "Human Impact on Long-Tailed Macaques in Thailand," in Gumert, Fuentes, and Jones-Engel, *Monkeys on the Edge*, 118–59, 141.

29. Una Chaudhuri, "Animal Acts for Changing Times, 2.0: A Field Guide to Interspecies Performance," in *Animal Acts: Performing Species Today*, ed. Una Chaudhuri and Holly Hughes (Ann Arbor: University of Michigan Press, 2014), 1–12, 5.

30. For these comments and more, see "Phra Prang Sam Yot," http://www.tripadvisor.com/Attraction_Review-g303912-d1010091-Reviews-Phra_Prang_Sam_Yot-Lop_Buri_Lopburi_Province.html.

31. Pierre Boulle, *La Planète des Singes* (Paris: Julliard, 1963).

32. See "Planet of the Apes (1968)," IMDb, http://www.imdb.com/media/rm3327645440/tt0063442?ref_=tt_ov_i.

33. Martin Robbins, "Planet of the Apes: Not Monkeys, Apes," *Guardian*, April 19, 2011, https://www.theguardian.com/science/the-lay-scientist/2011/apr/19/1.

34. Ibid.

35. Christopher D. Shea, "'King Lear with Sheep,' Yes, Sheep," *New York Times*, August 11, 2015.

36. Missouri Williams, "How My Production of King Lear with Sheep Ended Up Generating More Publicity Than I Ever Imagined," *Independent*, September 8, 2015, http://www.independent.co.uk/voices/how-my-production-of-king-lear-with-sheep-ended-up-generating-more-publicity-than-i-ever-imagined-10491410.html.

37. Donna Haraway, *When Species Meet* (Minneapolis: University of Minnesota Press, 2008), 33.

38. Ibid, 34.

39. Chris Bennion, "Review: King Lear with Sheep: Odd and Wondrous," *Telegraph*, August 14, 2015, http://www.telegraph.co.uk/theatre/what-to-see/king-lear-with-sheep-the-courtyard-review/; Ari Shapiro, "Review: King Lear with Sheep," *Wait, Wait, Don't Tell Me*, NPR, August 19, 2015, http://waitwait.npr.libsynfusion.com/bonus-ari-shapiro-reviews-king-lear-with-sheep.

40. See also Karl Steel, *How to Make a Human: Animals and Violence in the Middle Ages* (Columbus: Ohio State University Press, 2011).

41. Kiernan Ryan, *Shakespeare's Universality: Here's Fine Revolution* (London: Bloomsbury, 2015), 5.

42. Ibid, 10.

43. Ibid, 68–69.

44. Ibid.

45. Harlow, "The Nature of Love," 21, emphasis added.

46. Lauren Slater, "Monkey Love: Harry Harlow's Classic Primate Experiments Suggest That to Understand the Human Heart You Must Be Willing to Break It," *Boston Globe*, March 21, 2004, http://archive.boston.com/news/globe/ideas/articles/2004/03/21/monkey_love/.

47. Kathy Rudy, *Loving Animals: Toward a New Animal Advocacy* (Minneapolis: University of Minnesota Press, 2011), 173.

48. Ibid.

49. See William Shakespeare, *Macbeth*, in McDonald and Orlin, *The Bedford Shakespeare*, 1407–59 (1445).

Love

REBECCA R. SCOTT

Gravity is proof that the earth is in love with you and when you die and get buried she finally gets to go on your first date.

—MALLORY ORTBERG (@mallelis), Twitter

I am advocating the understanding that earthly heterogeneous beings are in this web together for all time, and no one gets to be Man.

—DONNA HARAWAY, *When Species Meet*

Unloveable

There's a public trail near my house that runs through a field of high grass and wild flowers, by some woods, and along a creek. I enjoy the trail but occasionally regard the creek as a poor substitute for the one I love and left behind in West Virginia. The creek by the trail where I live now is foamy when full, and on drier days is mostly notable for the dull opacity of its surface. The minnows and tadpoles observable in the flowing places are uneasy signs of health. It's a creek that's hard to love, flowing with runoff of fertilizers, pesticides, and oil. No children hunt crawdads in this creek; there are no makeshift dams, no explorers dreaming in the green. So you see, my trouble here is not just the challenge issued by William Cronon, of embracing the gamut of nature, from the most domesticated to the sublime, but of actually loving the damaged places, the scarred, the mutated, the unloveable.[1]

Modern Lovers

The first episode of the BBC sci-fi series *Black Mirror* involves a near-future prime minister who is forced by kidnappers to perform a sex act with a pig on national television. This interspecies "lovemaking" shocks and entertains the public but destroys the PM's marriage, because the act has undermined his dignity, his humanity, his ability to love. The pig's feelings are not part of the story. The episode highlights through absurdity an underlying incalculability in what it means *to love*. This slippery verb slides from the most intimate sexual contact to the most impersonal agape, from breastfeeding to star gazing. What and whom we are allowed to love, what or whom we can imagine loving, and what forms our loving may take describe the outlines of our world and of our conception of humanity.

According to the sixteenth-century commenter Elyot, "mutual concorde and love" are essential to the quality of humanity.[2] Tautologically, to be human is to be loved by a human. But loving is also the medium of imperial domination.[3] In this sense loving can "become . . . law, . . . become narrative."[4] Nature loving is not so different as it operates through points of intense intimacy and instrumental relations.[5] To love is an intimate, personal act that nonetheless requires public recognition and, more importantly, limitation, by institutionalization in marriage and the family, miscegenation laws, wilderness areas, nature-loving clubs, and so on.[6] This delimitation is a characteristic of Western modernity; we love categories and categorize love. Ask a Freudian about how desire, which in this case means that apparently inborn capacity for melding with the other, is truncated and restricted through the implementation of the law. Step outside these bounds and we are in the realm of the so-called primitive; of fetish, animism, the uncanny; the place where Sabine Spielrein addresses the wolf in the mirror, or where shape-shifters make a mockery of the law.[7] Loving is a weapon, in the colonization of Indigenous people, when "the absolute psychical distinction between man and beast" becomes the foundation for reifying settler hegemony.[8] Properly framed and limited, then, loving determines "civilized" humanity.

Western liberal humanism hinges on a definition of the human subject as choice maker, a will-bearing rational actor. Agents and objects are carefully segregated by the will to power.[9] But loving breaks this rationality down. Romances speak of the *coup de foudre* that changes two lives forever. Janice Radway notes that falling in love is traumatic for the alpha male heroes of certain genre romances, men whose icy stoicism is impenetrable by everyone but the heroine.[10] The romance reveals a contradiction; to love is to be impressed, to accept restriction, but liberalism depends on independence and disavows the fetters of interpenetration.[11] The hero's ability to love is quite circumscribed, yet, as the story goes, he needs it like he needs to breathe. The heroine is his safe harbor.

She's safe because heteronormative loving is settled by kinds, not individuals. Gender and race are, among other things, symbolic projects for organizing whom we love and how.[12] Heteronormativity depends on hyperseparation, on the assumption of radical disconnection that adds a spark to the romance.[13] The phrase "to make love" shines a rosy light but some of us are subjects of this making while others are objects. At its extreme, the patriarchal heteronormative model of "lovemaking" is a form of masturbation.[14] "The lover is a narcissist with an object."[15] Loving seems hopelessly determined by patriarchy. Centuries of dependence, discipline, and intimacy have infused hierarchy with wide-ranging affective intensities. Spare the rod and spoil the child. Domination and submission are woven into the affective economy of modern life.[16]

"I Love Coal"

The West Virginia coalfields are plastered with declarations of love: "WV Coal," "Friends of Coal," "I ♥ Coal."[17] Why do folks love coal? Maybe they love the rough edges of the coal camps where they were raised. They love the genealogy of coal in their blood. Maybe they love the hard manliness and extreme vulnerability of their husbands, fathers, uncles, brothers, and sons. Some love the bravery of the few women miners. Maybe they love the sacrifice coal marks on their bodies and the prayers and endless patience of their mothers and wives. They love powering the nation, through the dusty materiality of coal. They love running coal in the deep and dark where the blades of the underground miner (machine) cut through the coalface like butter. They love the terrific explosions and the gigantic trucks on a mountaintop removal (MTR) mine.

A romance novelist might say this is an all-consuming love. We love coal through the masterful edifice of its owners, who love us by consuming our labor. If we don't obey, they will leave us. And who loves coal more than the operators themselves? They must have it; they can't live without it and their jouissance of profit. Coal operators mark the landscape and people with the signs of their ownership.[18] Human Centipede-like, the subject loves the object and consumes it through the deadly gift of itself. The coal industry loves West Virginia with the passion of hundreds of thousands of tons of explosives. The dedicated project of dismantling the mountain to get to the coal can hardly fail to impress an onlooker with its determination, ardor, and power. To love this way costs the self and is expressed through fire.

After the explosions come the tears. As the Magnetic Fields tell it, if you don't cry, it isn't love. "Treehuggers" would rather define this as rape. The flattops, remains of former mountains, evoke painful mourning, but like politicians removing a rape exception from an abortion ban, coal boosters prefer to frame the loss as a productive transformation into usable flat land.[19] Either way, gendering furnishes a relationship with nature; a romance where "coal miners do it in the dark." Nature, like Lady Chatterley, remains a passive commodity for the consumption of her lover, Mellors: "It's thee down theer, and what I get when I'm i'side yee, . . . that's the beauty of ye, lass."[20] There is no question of leaving the coal in the ground; to love the coal is to burn it. Like a helpmeet designed by God, the coal was put here for us, coal boosters reason. Nature *for us* represents the limiting and reductive perspective of objectification, of classification into kinds—coal is to burn.[21] Loving and consuming blur together because loving becomes ownership; "You are mine." You are mine because of the kind you are, hasn't this always been the way of things? Some discipline and others obey.

"Explore, Enjoy, Protect"

Even in the era of MTR, there are plenty of "friends of coal" who deeply love the mountains.[22] One thing that they resent about environmentalists is the propagation of the stereotype that coal miners don't understand nature, or that they don't love it properly. Maybe they love it more than the environmentalists, as they claim, because they relate so intimately with it. People work in mining in order to be able to remain at home, after all. It seems just about everyone born in West Virginia eventually comes to love the place to tears. Friends of coal accuse environmentalists of idealizing nature abstracted from human needs: *Do you love a tree more than a child?* The question hinges on their fraught commitment to the place; they love West Virginia, but what will happen when the coal is gone? Who will love the postmining landscape?

The question of the tree or the child expresses perfectly both the job blackmail perpetrated by the coal industry in the mono economy of the coalfields and the "modern constitution" as diagnosed by Bruno Latour.[23] This constitution is based on a separation of powers, wherein political questions of human society are conceptually and discursively severed from the apparently objective study of nature. This separation of powers increasingly confounds us modern folk, as its often-monstrous offspring, nature–culture hybrids like mountaintop removal, continue to proliferate. Loving a tree, loving a child, on these terms, only impoverishes them both.

Nature loving reflects the perspective of a modern cosmopolitan citizen, educated in detached pleasure, able to appreciate Yosemite's natural cathedral.[24] Nature lovers elevate Nature, encapsulate it into a beautiful image of serenity, charismatic megafauna, and endangered Mother. But the human still stalks his lover, this time with a promise of salvation. The Hall of Biodiversity in the American Natural History Museum, paid for by generous endowments from Monsanto, Mobile, and Bristol-Meyers Squibb, directs the lover's eye to a crowded street in the Global South for an understanding of who is endangering our love object, while Disney's ecotourist replica ride educates tourists on the dangers of poaching and illegal mining.[25] In West Virginia the finger points to the miners.

Every romance requires a foil.[26] We moderns divide humans into worthy and unworthy Nature lovers, in a global reenactment of the romance of the white man saving the brown woman from the brown man, dooming ourselves in the process.[27] John Muir loved Alexander von Humboldt, and in turn we love Muir, who thought the Indians of Yosemite had "no right place in the landscape."[28] The appropriation of sacred landscapes as national parks enacts a collector's love, loving through appropriation and exclusion.[29] These relationships hinge on the politics of access to animals, resource extraction, and conservation, or a dizzying mix of all, in a colonizing move of territorial control.[30] The reductive and polarized way we moderns love Nature necessitates this rivalry; both the moral imperative of industrialization and the detached admiration of the Nature lover thrive on this hierarchy. The abstraction of Nature loving frees moderns from the identification with the concrete effects of modernity. Like the hero in a romance novel, green consumerism rises to the challenge of saving Nature from this exploitative relation, bringing her to safety through the warm embrace of the environmental Kuznets curve. Please, won't somebody deliver me from the deadly irony of loving Nature through the growth of modern disposable consumer capitalism?

Don't Fear the Reaper

Bill McKibben's classic treatise encourages us not to give up on our lover, Nature, despite her ruination.[31] Once free and wild, alas, now she is mired here with us in the mud. This story of paradise lost retells the legend of human emergence into a world that loves its favorite child, the prodigal.[32] It is a world-making story, resolutely focusing on the narrative of (modern) human development and damnation or salvation. In dreaming of "Earth-That-Was," "Old Earth," or (notably) "Terra," we express nostalgia for the world we lost.[33] Do we attempt to prepare our children for the loss through these fantastical iterations of Earth-That-Was? No matter, we are stuck in this place, living through this disordering slow-motion apocalypse. We can tell children all the stories we wish about the shiny future Federation in space, but in reality we are stuck here. We must slog through these plastic bag–laden seas, these parched deserts inundated by irrational monsoons; these infestations of pine beetles, emerald ash borers, or hemlock wooly adelgids; these scum-rimmed creeks; and these rampaging fires.

The Near Term Human Extinction (NTHE) Evidence Group on Facebook is part of a community of "doomers" who have concluded that there is no escaping the fate of catastrophic climate change in the immediate future, which *will* bring about a new Earth—one uninhabitable for humans. Refusing what they call "hopium," this group holds climate activists like McKibben and 350.org in something like contempt for their continued professed belief in the human ability to mitigate the disaster, countering that this is unrealistic even assuming the recommended measures, which so far have achieved little consensus, are taken. They worry about the methane time bomb and other interrelated feedback loops totally outside human control. They point out the obvious, that humans can't survive without the other species that are widely recognized by scientists to be undergoing the sixth extinction. Members seem to be largely from the Global North.[34] Some are survivalists, while many others express a feeling of being at the end of their agentic rope, feeling the frayed ends of possibility falling out of their hands. The main question is, how to live, knowing what they know? "Only love remains" is one answer.[35]

The compulsion to wash one's hands of the failed project of Western civilization suggests a certain platonic escapism, a reversal of the hubris fostered by the belief in human ontological isolation, the technological fix. This death wish may reflect a desire to avoid that slog through the mud that characterizes life on Earth despite our best efforts at manufacturing ease.[36] Glancing affectionately back at Freud again, we may notice that the death wish and pleasure principle are not unrelated. The generic conventions of a love story include a narrative arc toward a "whatever point" that settles things, a moment of idealized stasis.[37] In genre romance as much as in history this narrative arc trends toward completion. The Reverend Martin Luther King Jr. once expressed the hope that "the arc of the moral universe is long but it bends toward justice." This reflects the effort to construct a world where we love more widely, effectively, and better, in public, expanding justice.[38] But while aesthetic worlds may have trajectories and boundaries, and stories sometimes end in jouissance, we live in the nonrepresentational space of the open, distracted briefly from the gutter by looking at the stars, or meeting the tiger's teeth with the taste of a strawberry.[39]

Love the One You're With

In *Goodbye Gauley Mountain: An Ecological Love Story,* ecosexual performance artists Beth Stephens and Annie Sprinkle bring their camera to bear on the suffering heart of MTR in West Virginia. The film follows Stephens as she revisits her homeplace, moving from a reenactment of her conception in the space formerly occupied by her parent's bedroom to a nice long visit with her childhood best friend. The film presents Stephens and Sprinkle's conversation with this friend's husband, a MTR supporter who explains his position painfully for the camera. The scene ends with a kind of speechlessness, a lacuna that holds the agony of the coalfields front and center in the film. This speechlessness is echoed in the scene where Stephens loudly sings the state song, "The West Virginia Hills," in front of several uncomfortably silent state troopers trying to break up an anti-MTR demonstration at the capitol.

The film focuses on people loving the mountains; it features people mourning and crying their grief. Ecosexuality is defined in the film as sexuality that includes nonhuman nature in the panoply of pleasure, from the roundness of a ripe tomato to the caress of water rushing by one's body in a creek. The filmmakers embody a queer sexual agency, as pleasure seekers and lovers, that has nothing to do with the way sexuality is ordinarily objectified in media or in life. Ecosexuality queers the distinction between love and sex in a way that opens both categories to intersubjective complexity. Loving becomes less about ownership, and more about response, while sexuality becomes less a project of subjectification and more of an interaction in the open.[40]

In the novel *Woman on the Edge of Time*, Marge Piercy imagines a future communal ecological utopia as a place of genderless, open-ended polyamory, in opposition to an alternate dystopic techno-future where sex is reduced to a deadly total power exchange between hyper-hetero commodified machine people.[41] My students have had a lot more trouble understanding the former than the latter. In *The Handmaid's Tale*, Margaret Atwood describes the biblical theocracy of Gilead, where men classify women as either sexless wives or "handmaidens," who bear the brunt of men's sexual desire in lieu of the wives.[42] This scenario is revised in *Only Ever Yours*, a really depressing dystopic cli-fi where sex is strictly limited to male domination and female suffering.[43] Notably, each of these feminist fictions imagine male-centric societies ruining love and sex in ways that seemingly benefit no one, offering a backdoor view into the stunted alpha male romantic hero diagnosed by Radway, the impenetrable subject of heteronormativity.

Giovanna Di Chiro notes that it was the shock and horror that greeted sexually transformed amphibians that mobilized public outrage against endocrine disruptors.[44] In advocating a queer ecofeminism, Greta Gaard echoes Stephens and Sprinkle, noticing how erotophobia and heteronormativity structure the hierarchies and stories we take as natural. She traces the lineage from the imposition of heteronormative productive sexuality in Europe and the Americas to the objectification and commodification of the biophysical world, as the value hierarchies of gender and race intersect with speciesism and colonialism.[45] The reductive commodification of the biophysical, be that sex or other natural resources and processes, is so engrained in modern thinking that it is difficult to imagine nonobjectifying ways to love. To love without ownership, to inhabit without mastery, to have sex without obligation, these ideas are hard to get one's head around (as the fight against positive consent laws suggests). Loving differently may require "[slipping] between" ownership and renunciation to say, "I have no hope but all the same . . ."[46]

Goodbye Gauley Mountain is part of a series of ecosexual performance art pieces in which Sprinkle and Stephens marry aspects of nature such as the soil, the sun, and so on, as part of a radical queering of love, sex, and the human relation with nature. In the film they marry the mountains. The film is a political intervention in MTR in West Virginia, telling a regional story of love gone wrong, and perhaps for this reason "the mountains" must remain a category. The result is that *mountains* persist in being read as a generic feature of a human world, evoking a limiting condition in how we moderns love Nature. "I love mountains" is forced to respond to "I love coal," reflecting the powerful charisma of loving organized by kinds. Thus despite its expansion of love and sexuality, the ecosexual marriage to the mountains does not completely capture that elusive quality Ann Pancake evokes in her MTR novel *Strange as This Weather Has Been:* "What makes us feel for our hills the way we do?"[47] In this novel, two particular mountains are practically characters: Yellowroot, being mined, and Cherryboy, the next to go. "The memory picture of Yellowroot faded fast. And the feeling it left behind scared me worse than the mine site did."[48] Pancake enables us to feel their specific loss with keening intensity. What she captures is the agony of watching the mountain you love, that you grew up on, blown up and turned to a toxic poisoning mess.

> I did know you'd have to come up in these hills to understand what I meant. Grow up shouldered in them, them forever around your ribs, your hips, how they hold you, still astraddle, giving you always, for good or for bad, the sense of being held. It had something to do with that hold.[49]

The animism of Cherryboy and Yellowroot in the novel is less a domestication of these mountains that shelter the hollow where Pancake's narrator dwells than a recognition of the ultimate unknowability of these entities, their agency in her life, and her responsibility to them.[50] "I still stood with Cherryboy at my back. I couldn't see Cherryboy, but I could feel it behind me the way you can feel an animal hiding close by in the woods."[51] Response-ability is the ability to affect and be affected in an intersubjective–objective relation in which "no one gets to be Man." Pancake's book weaves a spell of embodied worldliness, kinship, and *becoming with* the mountains that forces a heartbreaking reckoning with the absolute irreducibility of each mountain that is razed.[52]

The one-upmanship of modern environmentalism, that claim to be the one who knows how to truly love Nature, demonstrates a reflex to critique that characterizes the search for the dead clarity of absolute morality. Ironic, of course, to critique this reflex, but moving toward real identification means moving away from idealization (a romance that impedes real relationship), even of ourselves.[53] Truly loving ourselves, loving where we live, requires more uncertainty, more pain, and a lot more work and an acceptance of the risk that we might be getting things wrong. It requires veering from critique to passionate immersion in a world of interconnections.[54] In her extended love poem to her dog, Donna Haraway describes the becoming together of multiple companion species—not only beloved pets, but the hunted boar as well, and the factory-bred chickens that currently help make humans what they are. "Human nature is an interspecies relationship," as she puts it.[55] I'm loving the bacteria in my gut and on my eyelashes.

A fixation with species distinction reflects an orientation to reproduction that sketches neat lines of descent in a romantic and comforting teleology, misrepresenting the openness of evolution.[56] Again the problem of humanity is that of being human enough to feel that "mutual concorde and love" because the human is always already nonhuman as well. Good enough humanity is conferred through "satisficing" or fulfilling the demands of DNA.[57] But DNA's demands don't satisfy human affective politics of race or nation. "Species reeks of race and sex," encoding rules for loving into a story about teleological reproduction.[58] Loving nature can't end with simply shifting "humans" from the center, if technically speaking, no such category can be said to exist. The category forbids us to love more often than it enables it.

This summer, the U.S. Supreme Court ruled same-sex marriage bans to be unconstitutional, a concrete, if imperfect, intervention in the struggle over whom and how to love. But the same summer saw other events. After the white supremacist shooting at Emanuel A.M.E. Church in Charleston,

Confederate battle flags were brought down from state capitals from Alabama to South Carolina. Witnessing the pain of their Black community members, a few former defenders of the Confederate flag admitted in public that their beloved flag must come down in the interest of human concord. At the same time, despite increasingly apparent racist police brutality, some insist that Black Lives Matter as a statement devalues white lives, demonstrating that loving Black people is a threat to white supremacy.

As opposed to what he calls the childish environmental stance of loving Earth as a child loves a mother, Charles Eisenstein suggests treating Earth as lover, as a way for industrial civilization to responsibly reenter community.[59] Yet I fear we moderns are overconfident in what it means to love Nature, too convinced of the sincerity of our romance, and too ready to speak for "humanity." Liberal sentimentality, acting in the name of love, often works to force an ideal on the other (of respectability, or purity) that forms a limit of acceptance. That story too easily collapses into the sublimated narcissism of the lover who idealizes his beloved in claiming her as his own.[60] Perhaps we must veer away from categorical loving in order to more effectively love our biophysical materiality and ourselves. Ahmed calls for a queer kind of loving based on identification, moving toward being like another, or in other words, veering off the straight and narrow to cross paths with each other.[61] Nature loving can only be launched from within idealized human communities, but identification with the biophysical world requires recognition of a web of similar needs, dependencies, and experiences. We can love Nature or Humanity in the abstract, but we can only actively love particular people and entities. Indigenous studies scholars insist further that community cannot be limited to the human, or to the living, without the logic of colonization sneaking back in. Kinship and belonging are nothing but concrete interactions with specific others, beyond disciplining categories of species, cognition, or even life as we define it, as the biological itself initiates the state of exception.[62] Far past enjoyment and far short of ownership, loving happens in innumerable dependencies and reciprocities with creatures, entities, and landforms beyond our categories, control, or understanding. If "nothing makes itself in the . . . world," we must indeed learn to love our monsters.[63]

Unloveable Redux

Besides the fish and tadpoles sometimes visible in the creek by my house, I see limestone rocks and boulders, thick, bushy green and flowering grasses, scrubby shrubs and tall trees, and an abundance of squirrels, rabbits, and deer. Despite the appearance of bounty, sometimes I grieve for children born into a time and place where water is so utterly degraded. Did you know people used to drink water from streams? I don't even consider touching the creek. I turn my eyes away, dream of home. But then a rabbit hops through the bushes to the water and it occurs to me— that rabbit loves the creek. I imagine it may not be the tastiest water, but the rabbit persists in kinship. Yes, what keeps her alive may also kill her; that's life.[64] The creek bears the heavy marks of abuse, but exists in itself, in a trickle or torrential force after a storm, and for the rabbit. She is thirsty; she has no real choice but to love the creek. Identifying with her thirst, I realize, neither do I. And I start to love the creek a little bit too.

Notes

1. William Cronon, "The Trouble with Wilderness; or, Getting Back to the Wrong Nature," in *Uncommon Ground: Rethinking the Human Place in Nature*, ed. William Cronon (New York: W. W. Norton, 1996).

2. Raymond Williams, *Keywords: A Vocabulary of Culture and Society* (New York: Oxford University Press, 1977), 149.

3. Laura Ann Stoler, *Carnal Knowledge and Imperial Power: Race and the Intimate in Colonial Rule* (Berkeley: University of California Press, 2001).

4. Chela Sandoval, *Methodology of the Oppressed* (Minneapolis: University of Minnesota Press, 2001), 142.

5. Cronon, "Trouble"; Carolyn Merchant, "Reinventing Eden: Western Culture as a Recovery Narrative," in Cronon, *Uncommon Ground*.

6. Sara Ahmed, *The Cultural Politics of Emotion* (New York: Routledge, 2004).

7. Avery Gordon, *Ghostly Matters: Haunting and the Sociological Imagination* (Minneapolis: University of Minnesota Press, 2008), 49; Jonathan Goldberg-Hiller and Noenoe K. Silva, "Sharks and Pigs: Animating Hawaiian Sovereignty against the Anthropological Machine," *South Atlantic Quarterly* 110, no. 2 (2011): 429–46.

8. Deborah Bird Rose, "Val Plumwood's Philosophical Animism: Attentive Interactions in the Sentient World," *Environmental Humanities* 3 (2013): 96; Billy-Ray Belcourt, "Animal Bodies, Colonial Subjects: (Re)locating Animality in Decolonial Thought," *Societies* 5 (2015), http://www.mdpi.com/2075-4698/5/1/1.

9. Anna Tsing, "Arts of Inclusion; or, How to Love a Mushroom," *Australian Humanities Review* 50 (2011): 19, http://www.australianhumanitiesreview.org/archive/Issue-May-2011/tsing.html.

10. Janice Radway, *Reading the Romance: Women, Patriarchy, and Popular Literature* (Chapel Hill: University of North Carolina Press, 1991).

11. Ahmed, *Cultural Politics*, 6.

12. Greta Gaard, "Toward a Queer Ecofeminism," in *New Perspectives on Environmental Justice*, ed. Rachel Stein (New Brunswick, N.J.: Rutgers University Press, 2004), 37.

13. Valerie Plumwood, *Feminism and the Mastery of Nature* (New York: Routledge, 1993).

14. Luce Irigaray, *This Sex Which Is Not One* (Ithaca: Cornell University Press, 1985), 25.

15. Julia Kristeva, *Tales of Love* (New York: Columbia University Press, 1987), 33.

16. Eva Illouz, *Hardcore Romance: "Fifty Shades of Grey," Best Sellers, and Society* (Chicago: University of Chicago Press, 2014).

17. Friends of Coal is the name of a coal industry lobbyist group.

18. Michel Serres, *Malfeasance: Appropriation through Pollution* (Stanford: Stanford University Press, 2011).

19. Rebecca Scott, *Removing Mountains: Extracting Nature and Identity in the Appalachian Coalfields* (Minneapolis: University of Minnesota Press, 2010).

20. D. H. Lawrence, quoted in Kate Millet, *Sexual Politics* (New York: Ballantine, 1970), 239.

21. Timothy Morton, *Hyperobjects: Philosophy and Ecology after the End of the World* (Minneapolis: University of Minnesota Press, 2013), 119.

22. This phrase comes from the website of the Sierra Club, http://www.sierraclub.org/.

23. Bruno Latour, *We Have Never Been Modern* (Cambridge, Mass.: Harvard University Press, 1993).

24. Anna Tsing, *Friction: An Ethnography of Global Connection* (Princeton: Princeton University Press, 2004).

25. Stephanie Rutherford, *Governing the Wild: Ecotours of Power* (Minneapolis: University of Minnesota Press, 2011), 14, 66.

26. Radway, *Reading the Romance*.

27. Ruth Frankenberg, "Introduction: Local Whitenesses, Localizing Whiteness," in *Displacing Whiteness: Essays in Social and Cultural Criticism*, ed. Ruth Frankenberg (Durham: Duke University Press, 1997).

28. John Muir, *Nature Writings: The Story of My Boyhood and Youth; My First Summer in the Sierra; The Mountains of California; Stickeen; Essays*, ed. William Cronon (New York: Library of America, 1997), 373.

29. Carolyn Finney, *Black Faces, White Spaces: Reimagining the Relationship of African Americans to the Great Outdoors* (Chapel Hill: University of North Carolina Press, 2014).

30. Tsing, *Friction*, 283; Belcourt, "Animal Bodies."

31. Bill McKibben, *The End of Nature* (New York: Anchor, 1989).

32. Merchant, "Reinventing Eden."

33. These terms for the planetary origin of human beings come from the TV series *Firefly* and *Battlestar Galactica* and the novel *Stranger in a Strange Land* by Robert Heinlein, respectively.

34. According to a map produced by members of the Near Term Human Extinction SUPPORT Group, a related closed group on Facebook.

35. Guy MacPherson, "Abrupt Climate Change: Only Love Remains," Nature Bats Last, May 24, 2015, http://guymcpherson.com/2015/05/abrupt-climate-cha nge-how-will-you-show-up-during-humanitys-final-chapter/.

36. Greg Kennedy, *An Ontology of Trash: The Disposable and Its Problematic Nature* (Albany: State University of New York Press, 2007).

37. Lauren Berlant, *The Female Complaint: The Unfinished Business of Sentimentality in American Culture* (Durham: Duke University Press, 2007), 266.

38. Michael Eric Dyson, *I May Not Get There with You: The True Martin Luther King, Jr.* (New York: Free Press, 2001).

39. Morton, *Hyperobjects*. Donna Haraway, *When Species Meet* (Minneapolis, University of Minnesota Press, 2008).

40. Haraway, *When Species Meet*, 81.

41. Marge Piercy, *Woman on the Edge of Time* (New York: Fawcett, 1985).

42. Margaret Atwood, *The Handmaid's Tale* (New York: Anchor, 1998).

43. Louise O'Neill, *Only Ever Yours* (New York: Quercus, 2015).

44. Giovanna DiChiro, "Polluted Politics? Confronting Toxic Discourse, Sex Panic, and Econormativity," in *Queer Ecologies: Sex, Nature, Politics, Desire*, ed. Catriona Mortimer-Sandilands and Bruce Erickson (Bloomington: Indiana University Press, 2010), 205.

45. Gaard, "Queer Ecofeminism," 2004.

46. Roland Barthes, quoted in Chela Sandoval, *Methodology of the Oppressed* (Minneapolis: University of Minnesota Press, 2000), 142.

47. Ann Pancake, *Strange as This Weather Has Been* (New York: Shoemaker and Hoard, 2007), 99.

48. Ibid., 167.

49. Ibid., 99.

50. Kim Tallbear, "Why Interspecies Thinking Needs Indigenous Standpoints," *Fieldsights—Theorizing the Contemporary, Cultural Anthropology Online*, April 24, 2011, http://culanth.org/fieldsights/260-why-interspecies-thinking-needs-in digenous-standpoints; Bird Rose, "Philosophical Animism."

51. Pancake, *Strange*, 167.

52. Haraway, *When Species Meet*, 82.

53. Ahmed, *Cultural Politics*, 140–41.

54. Tsing, "Arts of Inclusion."

55. Haraway, *When Species Meet*, 19.

56. Elizabeth Grosz, *Time Travels: Feminism, Nature, Power* (Durham: Duke University Press, 2005).

57. Morton, *Hyperobjects*, 175.

58. Haraway, *When Species Meet*, 18.

59. Charles Eisenstein, "The Ecosexual Awakening," Uplift, August 20, 2015, http://upliftconnect.com/the-ecosexual-awakening/.

60. Ahmed, *Cultural Politics of Emotion*, 128.

61. Ibid., 139.

62. Tallbear, "Interspecies Thinking"; Goldberg-Hiller and Silva, "Sharks and Pigs"; Belcourt, "Animal Bodies."

63. Haraway, *When Species Meet*, 32; Bruno Latour, "Love Your Monsters: Why We Must Care for Our Technologies as We Do Our Children," *The Breakthrough* (Winter 2012), http://thebreakthrough.org/index.php/journal/past-issues/issue-2/love-your-monsters.

64. James Hatley, "Blood Intimacies and Biodicy: Keeping Faith with Ticks," *Australian Humanities Review* 50 (2011): 63–75, http://www.australianhumanitiesreview.org/archive/Issue-May-2011/hatley.html.

Tend

———————➤   ◄——

ANNE F. HARRIS

Tend has a veering volatility. It bends around will and instinct, shaped by both, settling into neither: *tend* creates an oscillating ontological middle ground. That's where I seek to be with you for this essay. Three animal tales and their images will keep us there; stories from medieval, early modern, and contemporary worlds that have been captured in miracle story, woodcut print, and documentary film because the animals involved behaved beyond instinct, which made the humans question their own wills. We will be in complex company across time and scale: the Cistercian recorder of miracles, Caesarius of Heisterbach, and the bees that built a cathedral around a secreted Host; the Indian rhinoceros who was gifted to a king of Portugal then regifted to a pope and in between seized the artist Albrecht Dürer's imagination; and Timothy Treadwell, who is delicately understood by the filmmaker Werner Herzog through his camera, but whose killing by a grizzly bear is brutally explicable.

Temple Grandin writes of "Fast Fear, Slow Fear" in *Animals in Translation,* moving across the "hyper-fear systems" she argues are shared by animals and autistic people.[1] Grandin voices a reality that veers my own. Her work designing slaughterhouses is brutal; it is also merciful in making the design accountable to animal fear and assuaging it through her own understanding of fear systems. Donna Haraway asks about the effects of touch as it "ramifies and shapes accountability" between humans and animals, as they tend to one another.[2] Haraway's work in *When Species Meet* presents an intimacy based not only on tenderness but on the very lack of similitude, and often understanding, that keeps intimacy out of reach. Both thinkers push me beyond the telling of the story to wondering why and how *tend* can pull across the boundaries of human will and animal instinct not to collapse them into some utopic togetherness but rather to recognize that both will and instinct are open to tendencies and unexpected trajectories, through what Mel Y. Chen identifies as the "fibrillation and indeterminacy" of human–animal interactions.[3] The tendencies that both humans and animals display in these narratives will take us off the beaten path of the human–animal divide and into tragic territories that themselves may at times seek to argue for a reassertion of that divide. But what if we pre-tend that the divide isn't absolute, that it's permeable to tendencies instead? How can dwelling on (and within?) *tend* take us beyond the confines of will and instinct?

Language

First, the word itself. It is ancient: deep in our desire for language; roots in Sanskrit, Greek, Latin, and Old English send offshoots in intricate patterns that reach ever outward. *Tend* is grammatically unfixed. It can be both transitive and intransitive: I can tend my garden and I can tend to like gardens. A phrasal verb with a preposition (I can also tend *to* my garden), *tend* eludes a fixed syntactical framework. Tending *is* tender: slow, soft, searching, kind. Tending *can* tender: brisk exchange, material transaction, legal tender. *Tend* is linguistically contingent: it takes on prefixes and suffixes. The root word proliferates; it plays among adjectival forms and nouns. *Tend* portends, it asks you to pretend; it can be contentious, ostentatious, pretentious; it has tendencies, a tenderness, it can pay attention, have intentions. *Tend* tends to veer. It traverses the space of difference, reaching across in desire, transaction, or stewardship. To tend to someone (your love), to something (your garden); to look after it and worry about it troubles the anthropocentric georgic mode: *tend* is not control, it does not yield predictably. *Tend* strives, seeking to draw one thing closer to another, to bring itself nearer to not-itself. *Tend* unfurls to tendency and intention and pushes to rethink agency. Fire tends toward burning, the seed tends toward becoming a tree, humans tend to complicate things.

Herzog speaks of Treadwell having a "natural tendency for chaos."[4] I've spent time puzzling over this, grateful to Herzog for complicating things with the word *natural*. Is a tendency, like will and instinct, something that we have found useful to consider as a natural, unalienable, characteristic? Can one tend with artifice? This is a language and knowledge question worth answering, because if *tend* is tethered to our concept of the natural, we should wonder why. Is there such a thing as an *artificial* tendency? It seems oxymoronic. You can't fake a tendency; if you do, it becomes something else: a pragmatic move, a social grace. I tend to work better in the morning, to stay in the kitchen at parties, to prefer unreliable narrators—are those tendencies elements of individuation I should recognize as my discoverable character? Legal tender is cultural, tending to your affairs is cultural, but tenderness and tendency are deemed natural. *Tend* veers within itself; it is the root to offshoots that climb and spread in entirely different directions.

Middle Ground

If *tend* revels in language, if it is an etymological rhizome sending offshoots up throughout an expansive linguistic territory, how does it engage this allegorical idea of a participatory middle ground between will and instinct, perhaps even between humans and animals? There are many contested territories within which *tend* can spring up here. Will is shared by both human and animal (the willful ass, a child's will), and yet within human purview it takes on ethical and political dimensions (willpower, the will to power) generally denied to animals. Animals can kill but are incapable of murder—that show of will is reserved for humans. Will is not a fixed concept, but it is claimed as ethical and political by a specific subset of the species: adult humans of the age of reason. Instinct is insistently present throughout a human life, though it recedes at precisely that age of reason in which humanity has claimed a will for itself. Infants move instinctually to nurse and turn to familiar voices; bodies and minds at the end of life move with ever-greater mysteries of instinct. So when we tend, when we engage in the volatile act of reaching for something, we act somewhere between will and instinct; we act along the rhizome that *tend* stretches forth, that our desires keep constructing, as they stave off stability. *Tend* connects us with every other thing that reaches (plant, virus, time, political system) but promises no understanding—not even unto the thing itself that is tending. It offers pleasure, fear, satisfaction, love, confusion, endings, and, in its etymological versatility, play—and a fervent attempt to stay in a volatile, vorticular space.

Bees: Tend/Intention

Bees have come to be understood as exceptionally intentional. *Tend* around them has purpose as well as volatility. Their behaviors demonstrate complex and swiftly established hierarchies we name with queens, drones, and workers; their flights delineate extended trajectories of pollination, increasingly fragile filaments oscillating in climate change; and their movements in the hive reveal complex dances: a soft buzz of somatic semiotics, a dull roar of meaning. They are called to tend on multiple scales: the pollination of the planet, the care of larvae, the production of honey, the secretion of royal jelly. Bees don't appear to make mistakes, though they are vulnerable to human ones. Nature documentaries do not track miscalculations resulting in failed hunts or separations from the hive; instead, they revel in the smooth hum of an intricate society persistently making itself within a thrumming structure that simultaneously avails itself to the pleasure of eating honey and to allegories of political structures. Bees are so busy that they cannot rest ontologically. *Tend* projects them from the operations of the beehive into those of allegory, pleasure, decoration, and symbolism.

Bees have built this reputation for themselves, meticulously layering wonders upon miracles. Honey never spoils: its high acidity and low moisture content resist bacteria and decay and open up possibilities both medical and magical.[5] People, after all, have dreamed of entire lands of milk and honey. Hittite laws provide punishments for the theft of beehives;[6] and Egyptian papyri, Aristotle's *History of Animals,* and Virgil's *Georgics* all weave the symbolisms and partnerships of bees and humans. The intricacy and intentionality of bees have pollinated the planet and beguiled human interest. They have produced apiary expertise and humans who tend to bees; who keeps whom can be debated.[7] I am fascinated by a *New Yorker* profile of a preparator at the American Museum of Natural History who restores bees captured in Sardinia and put into ethanol in 1909 by bathing them and fluffing their fuzz: moves of microscopic tenderness across minute matted bodies otherwise forgotten.[8]

The usefulness, many argue the ecological necessity, of bees is continuously codified. So, too, their aesthetics. In the 1220s the Cistercian monk Caesarius of Heisterbach presented a miracle tale featuring bees as miraculous makers of exquisite architecture.[9] A woman, fearing for the health of her bees, secreted a consecrated host out of her church by not swallowing it. Once at home with her bees, she freed the host from her mouth and placed it into the beehive, where the bees immediately busied themselves in building an intricate chapel all around the host, complete with an altar upon which to place it. The chapel was not only intricate, it was complete: walls, windows, roof, towers, and doors surged all around and a central altar proudly held the host. For Caesarius, the bees' response was a matter of pious intention and recognition of the divine in the animal world that existed in sharp contrast with the irreverence and ignorance in the human one. Unlike the woman, whose kidnapping-by-mouth of the Eucharist signaled a singular shortcoming of respect for the sacrality of the body of Christ, the bees recognized and appropriately acknowledged the presence of the divine. Human intention is pragmatic and grotesque; the bees' is principled and beautiful, their *mirae structurae* (marvelous structures) simultaneously a miracle of their loftier intention and their better character. It wasn't just their tending to the Eucharist that earned the bees Caesarius's praise but also the beauty and wholeness with which they did it. As they tended to the ritual body of Christ, they performed their own religious act, their own opus Dei (work of God/prayer).

Miracles are momentarily part of a veer ecology in their departure from what is expected: their completion's assertion of a dominant Christian paradigm a-verts their remaining in such a volatile space for too long. In an unremarkable and predictable world, human beings would understand and revere a religious ritual designed for their salvation, and animals would live in expected ignorance of holy liturgy. But in those volatile moments of Caesarius's miraculous and veering world, humans extract saliva-slicked holy hosts from their mouths, and bees adorn the Eucharist with golden honeycombs. Using the trope of "even those who are lesser are greater," Caesarius finds shame in human tendencies and praise in the bees' attending to the host. *Tend* asks that you follow its action beyond its immediacy and into its future, especially on the contested concept of (God's) intention. *Tend* can change with the volatility of moral and ethical decisions, with reversals in intention. The miracle's conclusion seeks to revert the veer: the wonder of the bees' chapel moved the woman to confess all to her priest who came, with his parishioners, to marvel at the sight, drive the bees away, and collect the Eucharist so as to restore it with pomp and celebration to its rightful place on the church's altar. Human intention is corrected, normalized, and re-ritualized, and the bees' is relegated back to its routine.

But it is the bees' veer, and not the miracle's, that will prevail into the modern day. *Any* object positioned within a beehive will provoke bees to build around it in all the complexity of a honeycomb whose geometric shapes adapt to curvature and disruption. The contemporary artist Aganetha Dyck has collaborated with bees, placing found objects within their hives and displaying the results of multiple veering intentions.[10] Porcelain figurines of lyrical delicacy become poignant bodies lashed together by honeycombs. Gazes are now veiled, and movements hold a new

Figure 1. Aganetha Dyck, *An Inconvenient Proposal*, 2007. Beework and honeycomb on porcelain figurine, 7¼ × 7 × 3 inches. Photograph by Peter Dyck. Courtesy Michael Gibson Gallery.

resistance within their elegance. But movements also now have a visual amplification in the honeycomb's architectural echo and frame of form.

The bees take the shape of the object and tend to it: they pay attention, take it into account, and are intentional in their cohabitation with a new object. Dyck's figurines do not cause harm but they shift how the bees tend their hives—the bees tend differently, their intention is redirected, and the result is a surplus onto art—one that in this example conflates the woman's spurning of the man with the futility of both of their extractions from the bees' hold. The figurines could stand alone as crafts of fine porcelain, but with their honeycomb adornments they become collaborative works of art: "tend art"—art that addresses its multiple makers at their moment of intention. I have a hard time fixing the aesthetic as beautiful or grotesque—the geometric shapes of the honeycombs are pristine and stretch in measured evenness between the figures, but porcelain limbs are trapped and gestured gummed up in a messy morass.

It's the tension between the structure of the honeycomb and the structure of the human body that makes it difficult. Is the bees' intent to adorn or smother the intruding object? The question becomes more pressing in the revelation that, after over twenty years of working with bees, Dyck has developed a debilitating allergy to them.[11] What are the tendencies at play here? And how do they veer? For Dyck, despite her new affliction and the sadness it holds for her in the necessity to withdraw from her work with bees, the enduring pursuit has been power, specifically "the power of the small, instead of the power of something gigantic, like an elephant."[12] Lest we think *tend*, in its flexibility and adaptability, is inconsequential, lest we think *intention* in its search for action is weak, there are bees and miracles and art. *Tend* here encompasses both the bees' instinct to envelop a foreign object in a honeycomb and the artist's will to create a work of art. It is both compulsion and choice, or rather comes into being by participating in the two. *Tend* veers freely between miracle and fact.

Rhinoceros: Tend To / Attention

Our next veering will take place between reality and representation. In contrast to the usual first statements about the woodcut print of a rhinoceros by Dürer from 1515 (that it is not a realistic representation of a rhinoceros, that it is fraught with fantasy and the lack of eyewitness), I will argue that it tends to both realism and fantasy—that it emerges from both and continues to speak to both, without ever joining the two. The woodcut print is a widely reproduced and, to this day, broadly shared testament of an early modern fascination with the arrival in Portugal of a creature whose ilk had not been on European soil since the time of Pompey the Great.[13] It is also a highly personalized image, one to which Dürer laid claim prominently by placing his monogram directly onto the flat surface of the visual field, rather than, as he so often did, within the perspectival logic of the representational space.[14] This woodcut print is neither right nor wrong, neither reality nor fantasy; nor is it both. In looking at what this woodcut print tends *to* and at how the print is itself part of an extended series of human actions that sought to tend to the rhinoceros, to bring yet more attention to it, we can see beyond the binaries of fact and fiction and see the simultaneity that *tend* allows along the ontological curve that Dürer's engraving will travel.

News of the rhinoceros's presence in Lisbon reached Dürer via a letter written by the dignitary Valentin Ferdinand of Moravia to a merchant in Nuremberg identified simply as Brother, whom Dürer counted among his friends.[15] The letter itself is an ode to the early modern tendency to oscillate between realism and fantasy in writing about lands being colonized or lands whose colonization was richly desired. For the Portuguese court, India was such a place, and Valentin was deeply influenced by the courtly expectations fueled by reports coming from Goa, the Portuguese stronghold in the region. In the seamless interweaving of contemporary observation and ancient text, Valentin's letter is at the tipping point that Umberto Eco, when writing about Dürer's image, describes as that "certain point [when] the iconic representation, however stylized it may be, appears to be more true than the real experience."[16] Valentin saw the rhinoceros from India through the framework set for it by the Portuguese court. And Dürer manipulates that framework with his own interests in the verisimilitude of the exotic.

The letter begins with Valentin's eyewitness account of seeing the rhinoceros at the court of the king of Portugal, Manuel I, the Fortunate. It slips quickly to antiquity and Pliny's description of the enduring enmity between the rhinoceros and the elephant, and back again to Valentin's own experience attending a staged fight between the two animals (the elephant ran away and neither was hurt). Having finished his eyewitness account, Valentin launches into a description of India pulled from a "Wonders of the East" literature itself hundreds of years old by the mid-sixteenth century. Brahmans and "forests of splendid cinnamon" abound, along with "rubis, hyacinths, agates, sapphires and daisies" that emerge from the sea, and the rhinoceros is not heard about again.[17]

In an unusual move for a market-savvy artist such as Dürer, the wood-cut print he decided to make upon reading this letter was uncommissioned and relatively noncommercial.[18] What compelled Dürer to move the exotic descriptions he had learned about in words into visual form, accompanied by descriptive text pulled from Valentin's letter? What tendency did he respond to in making the rhinoceros image and framing it in language? He began his work with a pen and brown ink drawing (today in the Sloane Collection of the British Museum) and then worked within the format of a woodcut, notably cheaper than that of a metal engraving. He only printed one edition, in 1515, of the woodcut print. He would never know the popularity that the print would enjoy, nor could

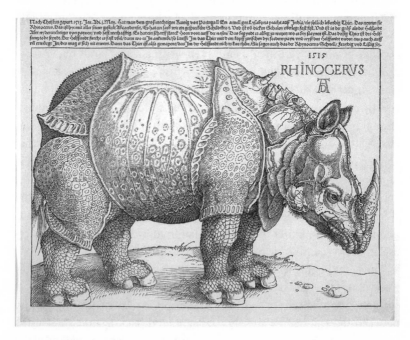

Figure 2. Albrecht Dürer, *The Rhinoceros,* 1515. Woodcut, 24.8 × 31.7 cm. Copyright by the Trustees of the British Museum / Art Resource, New York.

he have fathomed the multiplicity of media on which his rhinoceros image would be printed, including, for example, on the tea towel that hangs in my kitchen today. Dürer's rhinoceros (as its own image) has a lasting hold on the modern imagination, perhaps precisely because it represents the world-weariness of a pachyderm's stare and the fantasy of an armored skin and an extra horn.

The print emerges as the enduring last act in a long list of human actions that tended to the rhinoceros. The animal itself was an object of multiple exchanges, gifts deferred, and suspended negotiations. In February 1514, Alfonso d'Albuquerque, the governor of Portuguese India, gave Sultan Muzafar II, the ruler of Gujarat, a host of precious goods, including a silver goblet, a piece of Chinese brocade, a gold dagger with rubies on the handle, and nine measures of black velvet. He did so with the ambition of gaining the right to build a fortress at the port of Diu, which belonged to the territory of Gujarat.[19] The sultan was grateful to d'Albuquerque but demurred from granting him the right to build his fort, and chose to give him a rhinoceros in thanks instead. From then on, the rhinoceros was tended to by Europeans.

It is because of Dürer's print—I think because of the cast of the eyes, the weight of the lips, and the weariness of the head—that I wonder about the rhinoceros's experience, about his or her being tended to: being led onto a boat, sailed down the coast of India, docked in Marseilles, paraded before the French king François I and Queen Claude, and reboarded, to be welcomed into the menagerie of the king of Portugal.[20] It is because of Temple Grandin, and her work in articulating animal emotion and her own tending to animals, that I wonder about the relationship that developed between individuals who tended to this rhinoceros and the animal itself.

Grandin's own work with how animals think, feel, and learn (and how unacknowledged they are in doing so) draws my attention to the rhinoceros and his keeper in the hold of the ship as it traveled between India and Marseilles, and again between Marseilles and Lisbon. "True cognition," writes Grandin, quoting Dr. Marion Stamp Dawkins, an animal behavior researcher, "happens when an animal solves a problem under novel conditions."[21] The smells, the touches, the cries, the food, all the things to pay attention to—it must all have been so new, so deeply strange, to tend to. How does a creature enter expectations set for it by an ancestor's appearance before Pompey the Great? How did *this partic-ular* rhinoceros adapt to its novel conditions? Dürer records its presence, its novelty, and its impact on the human imagination. He becomes its keeper and tends to its image.

The keepers in Portugal became the mediators, were the ones to lead the rhinoceros into the arena holding the elephant, were the ones to reshackle it after the elephant had run away, crashing through the arena's walls. They were the ones to move the rhinoceros from that bewildering confrontation, that staging of ancient heroics, back to safety. "The single most important thing to remember," writes Grandin, "is that animals are afraid of tiny details in their environments. I like to use the term *hyper-specific* to describe animal fears. It comes from autism research, because *autistic people are extremely hyper-specific.* It's one of the main things that separate them from typical people."[22] In a poignancy I have tried to shake but have been unable to, I wonder about the role of the keepers when the rhinoceros was once again put on a ship, this time bound for Rome and the court of Pope Leo X, a gift once again deferred, this time by Manuel looking for a diplomatic advantage with the papal curia. But this ship never made land, wrecking off the Gulf of Genoa and sinking. Rhinoceroses can swim, but this one was shackled. Did it feel fear? Of course it did. Grandin leads me to think of what it noticed as the boat was sinking, how focused its fear was, how utterly it bracketed out the fantasies of a wondrous East that had been visited on it. The body of the rhinoceros washed up on shore and was stuffed and sent on to the Pope nonetheless—a representation of itself. It had been adorned with a green velvet collar decorated with gilt roses and carnations and a gilt iron chain.[23] We tend to our fantasies and try to bring them into the real with a surprising violence sometimes.

Grizzly Man: Tend for / Pretense

Watching *Grizzly Man* becomes more difficult each time. Not because of wonder or poignancy but because of my impatience with the pretense of both Timothy Treadwell, its fated subject, and Werner Herzog, its film-maker. I no longer want to follow where *tend* took Treadwell. His reaching for this heroic intimacy with grizzly bears seems so futile—because its justification was so quixotic (in the most original sense of the word), because the resulting deaths were so useless and because his familiarity with the bears, in violation of seven thousand years of ritualized distance between the local Aliqui Nation and the bears, smacked of white privilege.[24] Must *tend* be useful, wise, or wondrous then? Must Timothy "Tread Well," as Lowell Duckert puns to his students when teaching the film?[25] Is there an ethical obligation to "tend well"? What if the tendency itself is wayward? There is the idea, for example, a kinship one for Herzog as he puts it forward, that Treadwell had a "natural tendency for chaos." Herzog tends insistently toward the troublesome, too. In an essay reviewing Herzog's career-long fascination with animals, Paul Sheehan analyzes their importance as sites of the "violence and poetry of human incursions into the natural world."[26] Perhaps it is this insistence on a combination of "violence and poetry" that sits uncomfortably. I must recognize that *tend* sits in that uncomfortable incursion space, too; that it is wayward and volatile often without purpose. We have given "human will" frameworks from theology to psychology, and "animal instinct" the justifying arguments of biology and brain chemistry. But the many offshoots of *tend* (tendency, intention, attention, and pretense) are still elusive, not exactly framed but rather glimpsed by art.

And so to address the waywardness of *tend* here as it pulled from quixotic to tragic. Herzog's mastery in editing the over one hundred hours of videotape that Treadwell shot in his thirteen years in the Alaskan Katmai National Park and Preserve charts a path of waywardness: the needless death of his girlfriend Amie Huguenard; the bombastic speeches and declarations of love to the bears; the self-appointed role of protector of the bears; the revelation that he was born Dexter, not Treadwell; the facile fantasies of intimacy with the bears; the creepy worship of excrement because it had recently "been inside" a favorite bear; the horrific irony of being the only humans to be eaten by a bear in the park's history.

Herzog also provides an unwanted but effective intimacy: moments of vulnerability, of incertitude and self-deprecation; the recognition of the mundane within the self-imposed heroics; the discovery of Treadwell's gig as a medieval squire in a theme restaurant called Gulliver's; moments of beautiful filmmaking by Treadwell; his death; his fear. Of course these moments might seem like little more than sentimental interludes in the midst of the bleak tendency of the film as it shuffles doggedly to Treadwell's death. Herzog harnesses Treadwell's "natural tendency for chaos" to tragedy, amplifying his oeuvre's commitment to doomed inexorability.

Both the unnerving and the sentimental aspects of the film are met in Gilles Deleuze and Félix Guattari's scorn for the "individuated animals, family pets, sentimental Oedipal animals each with its own petty history" and Haraway's critique of that scorn.[27] In individuating the bears with names like Mr. Chocolate or Wendy or The Grinch, Treadwell entered into a familiarity with animals that Deleuze and Guattari roundly criticize as bourgeois. Haraway comes back to question the distance Deleuze and Guattari's contempt maintains and its own work in perpetuating human exceptionalism. Haraway's ideas of multispecies connectivity has been used to argue for Treadwell's queer ecology[28]—Treadwell's ecstatic love for bears both participates in privilege and does away with human exceptionalism. It tends. It veers.

Timothy Treadwell tended to bears, and the bears tended to themselves. The one that attacked and ate him was not known to him, having come out from farther inland to the edges of the preserve looking for food after most of the other bears had begun their hibernation. This bear's tendency, his waywardness, was then put on a collision course with Timothy's, whose presence on the preserve this late in the season was the result of a fight with an airline ticket agent. Surely it was a stupid fight. He and Amie would have never been in the preserve to satisfy the bear's excursion for food. Somehow, in the film, it becomes very important that mistakes were made, that Timothy's actions did not inevitably lead to his death, though in Herzog's worldview it was inevitable that he make them—that had he but respected the hibernation cycle as he had for the thirteen previous years, he would still be alive to foolishly claim his role of protector. Those who see a morality tale in Treadwell's death point to misguided human will inevitably being crushed by animal instinct. Herzog refuses to see the morality tale, and simultaneously refuses to sympathize with Treadwell. Sheehan cites Herzog's "quasi-medieval belief in the power and truth of images," and in this reverence, the last shots of the film showing Treadwell ambling away down a riverbed followed by two bears slowly lumbering through the water have all the doom and gentleness of *tend* before tragedy.[29]

Tend

In May 2013 I took students from my Ecology of Medieval Art class to a "Howl Night" at the Wolf Park research and rescue center in West

Lafayette, Indiana.[30] I will never know what our species were communicating to each other when we humans, coached by the wolves' caretakers, began howling at dusk and were answered by the wolves. But I will remember two things vividly: my own thrill, physical and tingling, at the wolves' response; and my laughter, nervous then full, when my then-eleven-year-old son and his friend began howling in the parking lot long after the official demonstration was over, and the wolves responded in kind. *Tend* calls out between unknowns: it makes the mundane miraculous in beehives; it holds a rhinoceros between fantasy and reality; it traces the waywardness of the ecstatic love of bears; it howls in the dark.

Notes

1. Temple Grandin, *Animals in Translation* (New York: Harcourt, 2005), 193.
2. Donna Haraway, *When Species Meet* (Minneapolis: University of Minnesota Press, 2008), 36.
3. Mel Y. Chen, *Animacies: Biopolitics, Racial Mattering, and Queer Affect* (Durham: Duke University Press, 2012), 89.
4. Werner Herzog, dir., *Grizzly Man* (Lions Gate Films, 2005).
5. Natasha Geiling, "The Science behind Honey's Eternal Shelf Life," *Smithsonian,* http://www.smithsonianmag.com/science-nature/the-science-behind-honeys-eternal-shelf-life-1218690/?no-ist.
6. Harry Hoffner, *Laws of the Hittites: A Critical Edition* (Leiden, Netherlands: Brill, 1997), 90–91.
7. On the early modern complexities of human fascinations with (and manipulations of) beehives, see Keith M. Botelho, "Thinking with Hives," in *Object Oriented Environs,* ed. Jeffrey Jerome Cohen and Julian Yates (Earth: Punctum Books, 2016), 17–24.
8. Laura Parker, "Bee's Knees," *New Yorker,* March 21, 2016, http://www.newyorker.com/magazine/2016/03/21/beautifying-century-old-bees.
9. Caesarii Heisterbacensis, *Dialogus miraculorum,* ed. J. Strange (Köln, Germany: Eberle, 1851), 172–73.
10. "Short Biography," Aganetha Dyck, http://www.aganethadyck.ca/ and Aganetha Dyck, *Aganetha Dyck* (Winnipeg: Winnipeg Art Gallery, 2012), 7.
11. Shannon Moore, "A Poignant Farewell: Agnetha Dyck at the Tom Thomson Gallery," *Magazine,* September 28, 2015, http://www.ngcmagazine.ca/corres pondents/a-poignant-farewell-aganetha-dyck-at-the-tom-thomson-gallery.
12. Skye Jordan, "Collectables Remixed with Real Honeycombs," *Visual News,* http://www.visualnews.com/2012/03/24/collectables-remixed-with-real-honey combs/.

13. Campbell Dodgson, "The Story of Dürer's Ganda," in *The Romance of Fine Prints,* ed. Alfred Fowler (Kansas City: Print Society, 1938), 45. *Ganda* is the Indian name for the rhinoceros that the early modern letter writers often use.

14. Giulia Bartrum, *Albrecht Dürer and His Legacy: The Graphic Work of a Renaissance Artist* (Princeton: Princeton University Press, 2002), 287.

15. A Fontoura Da Costa, *Deambulations of the Rhinoceros (Ganda) of Muzafar, King of Cambaia, from 1514 to 1516* (Lisbon: Portuguese Colonial Office, 1937), 5–6.

16. Umberto Eco, "Theory of Sign Production," in *A Theory of Semiotics* (Bloomington: Indiana University Press, 1976), 205.

17. Da Costa, *Deambulations,* 33–40.

18. Bartrum, *Albrecht Dürer,* 285.

19. Da Costa, *Deambulations,* 10–11.

20. Dodgson, "Story," 46.

21. Grandin, *Animals,* 243.

22. Ibid., 217.

23. Dodgson, "Story," 51.

24. A critique of race and class privilege in environmentalist movements, based on Peggy McIntosh's "backpack of white privilege" model, includes the ability to imagine and enjoy National Parks as "intact wildernesses because their establishment did not involve the forcible removal of [her] ancestors." Critiques of Treadwell's individualism here grids closely to those of Christopher McCandless, whose perishing was chronicled by John Krakauer in *Into the Wild* (New York: Villard, 1995). See also http://www.pachamama.org/news/race-and-class-privilege-in-the-environmental-movement.

25. Personal correspondence, March 4, 2016.

26. Paul Sheehan, "Against the Image: Herzog and the Troubling Politics of the Screen Animal," *SubStance* 37, no. 3 (2008): 126.

27. Haraway, *When Species Meet,* 29. Her critique focuses on Gilles Deleuze and Félix Guattari's "1730: Becoming-Intense, Becoming-Animal, Becoming-Imperceptible," in *A Thousand Plateaus; Capitalism and Schizophrenia,* trans. Brian Massumi (Minneapolis: University of Minnesota Press, 1993), 232–309.

28. Colin Carman, "Grizzly Love: The Queer Ecology of Timothy Treadwell," *GLQ: A Journal of Lesbian and Gay Studies* 18, no. 4 (2012): 507–28.

29. Sheehan, "Against the Image," 128.

30. http://wolfpark.org/planning/calendar/howlnight/.

Unmoor

STACY ALAIMO

Unmoor the fleet, and rush into the sea

—HOMER

Steve Mentz begins "Toward a Blue Cultural Studies" by declaring that "the new millennium is bringing humanities scholarship back to the sea." He suggests that looking "closely at the sea, rather than the just the land, challenges established habits of thought."[1] Indeed, he contends that "the scholarly benefits of the sea for many fields hinge on its unfamiliarity, and on the shock of novelty that comes from jolting one's habits and practices into a new structure."[2] In *Shipwreck Modernity,* Mentz discusses the marvelous 1627 woodcut of the Bookfish—a "codfish with a book in its belly"—a dramatic figuration of his contention that "real wisdom emerges from human encounters with the slimy deeps, if we are willing to go underwater to bring it up."[3] He surfaces, at the conclusion of this study, with "Seven Shipwrecked Ecological Truths,"[4] suggesting that marine encounters not only shift paradigms but disclose alternative ecological knowledges and matters of concern. I would like to complement Mentz's provocative trope of the shipwreck in this essay by veering back to the moment of setting sail.

Rushing into the sea, in a mad dash to conjure up concern for ocean ecologies that face accelerating threats, a blue ecocultural studies[5] must first unmoor. What ties must be loosened when environmentalism, with its terrestrial grounding, becomes submerged?[6] Two-dimensional maps flatten and distort as marine habitats must be comprehended through vertical dimensions. Many species traverse depths, crossing great distances in all directions. National boundaries do not extend to most waters, which dilutes conservation policy, legislation, and enforcement. The tragedy of the commons cannot be lamented as a mere historical relic, but is instead happening now in the largely unregulated open seas. The wilderness ideal, however problematic, which has infused much environmentalism in the United States, is often invoked when hailing the deep seas as the "last great wilderness,"[7] But the wilderness is a dreadfully anachronistic concept when the temperatures and acidity of seawaters have been humanly altered and so much has been taken out of and dumped into the oceans. Sustainability paradigms, which assume the ocean is a resource for human use, are readily adopted to promote business as usual, even though targets such as "sustainable seafood" seem desperate delusions when considered within burgeoning evidence of collapsing marine ecologies.[8] Marine biology—especially that of the deep seas—has lagged behind terrestrial biology in its understanding of ecological interrelations, in part because of the great difficulty and staggering cost of observation, sampling, and "field" work in the depths. Furthermore, ethical paradigms from animal studies are put under pressure in the sea. Salps, sea cucumbers, sponges, and corals, for example, can hardly be considered "companion species," even though they are "significantly other."[9] Humans may think with them but not live with them; we cannot trace embodied, co-constitutive relations or cultivate playful, respectful, interaction with say, the blobfish, despite its uncanny, humanoid face. Conversely, categorizing sea creatures as weirdly "alien," rather than as companions, expels or abjects these species from territories of human concern. Such figurations bolster the anachronistic narratives of discovery that seem to provide a refuge from human culpability—allowing us to imagine that deep-sea creatures exist beyond the reach of anthropogenic harms. Yet these marine species and even the as-of-yet undiscovered must live in Anthropocene waters that are warming, acidifying, and ravaged by industrialized fishing and mining. The fantasy that the ocean is so immense as to be untouched by human incursions and the temptation to enjoy oceanic feelings or other states of marine bliss may make unmooring a solipsistic affair, a structure of feeling lacking the ethical or political leverage that would propel ocean conservation projects. Similarly, to unmoor may evoke fantasies of wilderness and wildness—cutting ties and letting go—perhaps to encounter the newly discovered creatures from the deep whose already-iconic images stress their weirdness. Wilderness and weirdness seem to veer away from anthropocentrism, but they may end up reinstating the conventional parameters of the Western human.

Speculate

Although no one has a right to speculate without distinct facts ...

—CHARLES DARWIN, *The Voyage of the Beagle* (1839)

To unmoor, begin by imagining life in the depths, speculating with as many facts as can be assembled, even while disassembling and reassembling what "facts" entail. Vilém Flusser and Lois Bec in *Vampyroteuthis Infernalis,* itself a rather strange, even unclassifiable meditation on the vampire squid from hell, admit that the human and the octopod live far apart, and yet they insist nonetheless that the "vampyroteuthis is not entirely alien to us"; our common ancestor means that we "harbor some of the same deeply ingrained memories."[10] Notwithstanding evolutionary kinship, it is no small feat to conjure the Dasein of this mollusk, to "begin to see with its eyes and grasp with its tentacles":

This attempt to cross from our world into its is, admittedly, a "metaphorical" enterprise, but it is not "transcendental." We are not attempting to vault out of the world but to relocate into another's. Our concern is not with a "theory" but with a "fable," with leaving the real world for a fabulous one.[11]

Metonyms traverse trans-corporeal networks,[12] while metaphors demand some kind of imaginative leap. And yet, to unmoor would not be to rise and float in the air but to descend and hover in the depths—not to dissolve into the ocean as an immense imagined void or abyss but to grapple with the existence of a multitude of nonhuman lives in the seas. Melody Jue, in her astute analysis of *Vampyroteuthis Infernalis,* argues that "Flusser's ocean abyss serves as an epistemic medium for thought, bringing into relief . . . *the terrestrial bias of philosophy and critical theory.*" She argues that after reading *Vampyroteuthis Infernalis* we should ask ourselves "under what conditions have terrestrial knowledge structures evolved, and how would they appear radically different in an aqueous environment from the perspective of the vampire squid?"[13]

Departing from Flusser's quotation above, I would suggest that un-mooring does not entail "leaving the real world for a fabulous one," since the real could not be more fabulous and there is no solid epistemological divide between the two. Moreover, a marine science studies must seek to add reality, as Bruno Latour urges us to do, rather than subtract it.[14] As we descend in order to attempt to imaginatively inhabit the perspective of the vampire squid, this may entail, as Jue argues, a thoroughgoing de-familiarization of "terrestrial knowledge structures." This philosophical unmooring casts the terrestrial human out to sea. To return to Flusser's insistence that crossing from our world into that of the vampyroteuthis is not "transcendental," it is worth noting Donna J. Haraway's critique of the "god trick" of a limitless, "objective" disembodied vision, the "conquering gaze from nowhere."[15] To unmoor would mean shifting from the "con-quering gaze from nowhere," a place long occupied by the Western human knower, toward "situated knowledges"—thus veering away from veering away. Except, after grappling with one's own situated epistemological and political positioning, one would still need to engage with the scien-tific "captures" of marine life, the terribly mediated, compromised, yet sometimes passionate accounts of creaturely lives in the depths. Unmoor by practicing informed, intentioned speculation about other creature's perspectives, modes of being, and lifeworlds. Sit with these speculations long enough to shift the terrains of environmental concern to include even the as-yet-unknown species of the deep seas. But then cut away from any sense of certainty except that which insists other creatures can-not be contained within human conceptions. Even the familiar cat, next to Jacques Derrida, seeing him naked, darts away from conceptualiza-tion: "Nothing can ever rob me of the certainty that what we have here is an existence that refused to be conceptualized [rebelle á tout concept]."[16]

Despite Emily Dickinson's musings, the brain is not "deeper than the sea"; the brain cannot absorb it, because there is no "it" there, no "sea" as such (whatever that "as such" could mean) but instead multitudes of interacting species, ecologies, substances, and forces that make marine animal studies, like the marine sciences, a formidable venture. Drawing on phenomenology, rather than Haraway and other feminist epistemol-ogists who could be cited here, Michael Marder in *Plant Thinking: A*

Philosophy of Vegetal Life contends that the "spatiality of all living beings—unmoored from objective determinations and emancipated from a global, disincarnated perspective that disavows its own perspectivalism—will require that a different sense of what is above and below, etc., be laboriously worked out from the standpoint of each particular life-form in question."[17] Contemplating such a multispecies spatiality would be vertiginous enough on land, but to descend with these transmogrifying seascapes may be akin to the vertigo, giddiness, anxiety, and intoxication of nitrogen narcosis, as not only space, but time, may warp, twist, and ripple away.

Jakob von Uexküll considered the standpoint of different animals, proposing that even space and time are relative to each creature. For the deep-sea medusa, which moves in a constant rhythm, for example, "the same bell always tolls, and this controls the rhythm of life."[18] Dorian Sagan's introduction to Uexküll's *A Foray into the Worlds of Animals and Humans* notes that while literature, shamanic traditions, Tibetan Buddhism, and Vulcan mind melding promote the "art of putting oneself in another's shoes," "such explorations, such 'embodiments' remain rare in scientific literature."[19] Sagan's account of these practices does to some degree assume the very self–other binary it would subvert, as such categories persist in Western thought. Elizabeth DeLoughrey presents an alternative conception from Maori epistemology, where to know is to locate within a genealogy called a "whakapapa": "Because whakapapa incorporate the subject into planetary networks of kinship, including Tangaroa, the deity of the ocean, knowing and being are constitutive and interrelated."[20] It is doubtful that Western subjects would trace planetary kinship networks, even though the evolutionary origin story of all life beginning in the sea is frequently invoked by ocean conservationists. And yet DeLoughrey's argument and the Maori "whakapapa" are valuable provocations.

The question of the relation between knowing and being remains central, in my view, to a posthumanist, new materialist, conception of a blue science studies that veers into or begins with a marine animal studies so as not to end up in an empty ocean, an abyssal void. To unmoor, to set off for an oceanic animal studies, would require imagining the lives and worlds of myriad marine species. With a million or more species in the ocean, and many more that are not yet discovered, to unmoor by conjuring fabulous creaturely lives would be an immense praxis. While deep-sea creatures circulate in media and popular culture primarily as visual—often radiant—images, to imagine the life worlds of abyssal creatures may entail thinking with darkness, attempting to feel what it would mean to exist within waters that are only momentarily lit up by bioluminescence. Narrative is often hailed as a potent force for the environmental humanities, but so little is known about the lives, interactions, and habitats of deep-sea animals that narratives are difficult to construct. Moreover, the very shape of narratives—lines traced through eventful temporalities—could obscure the less-delineated lives of floating and drifting creatures in oceanic space. Instead, artists, theorists, and (post) humanists of all stripes may help us to unmoor through practices of informed speculation,[21] which draw on scientific captures but expand, stylize, and assemble them in such a way that they lure the public into conceiving and feeling the possibilities for life in aquatic zones.

Would such a practice be "compositionist," in Latour's terms, as it proceeds from the knowledge that the ocean is not "nature" in the Modernist sense, since it is not "always already assembled" but must be "composed" from "discontinuous pieces?"[22] Certainly this very sense of things is often echoed, though in different terms, by many marine scientists and ocean conservationists when they stress that new "discoveries" in ocean sciences—from new species to ocean acidification—happen within a context in which so little is known that it all must be "composed" from "discontinuous pieces" that are rather hard won. But Latour's quest for the "Common World," even though it is "slowly composed instead of being taken for granted and *imposed* on all," still suggests a unified transcendental perspective from which humans assemble and contemplate the arrangement of pieces.[23] While Latour rejects inanimism and welcomes the distributed agencies of nonhumans, the figuration of the Common World erases power dynamics as well as the differing perspectives of particular humans and nonhumans. Positing that "everything happens as if the human race were on the move again, expelled from one utopia, that of economics, and in search of another, that of ecology," Latour ignores much of the "human race" who would find that trajectory incomprehensible. His conception of the human not only evacuates histories, cultures, and other specific differences but it also banishes nonhuman animals from its territory. When he then asks, "How can a livable and breathable 'home' be built for those errant masses?" the reference to "eikos," the root of ecology, makes it possible to include nonhumans, and yet the emphasis on building, or worse, building *for,* suggests stationary structures, rather than multispecies agencies, habitats, and ecologies.[24] The "search for the Common," culminates with a domestic air, an enterprise that is too human and too terrestrial. Unmoor by thinking with the deep-sea medusa or thousands of other submerged, uncommon lives, that may move with currents and tides.

Keep Veering Away

> To unmoor upon the ebb, when it has made strong, veer away from the best bower, bring-to and heave in the small bower and keep veering away.
>
> — DAVID STEEL, *The Elements, and Practice of Rigging, Seamanship, and Naval Tactics* (1794)

Veering away from standard modes of scientific objectivity that reduce the investigations of marine animals to something akin to "an autopsy of a lifeless body," Flusser and Bec warn that fables should not "be woven of scientific texts"; and yet they admit they have "little choice but to rely on the contents of scientific literature."[25] Blue ecologies generally, as well as marine animal studies, must be modes of science studies, reflectively encountering mediated scientific "captures," whether these be images, data, hypotheses, results, arguments, or (meta)narratives. To contemplate or theorize about abyssal or even pelagic ecologies would be unthinkable without marine biology and its technologies, as ordinary people or citizen scientists cannot access the depths themselves. Nor would we expect to find much traditional ecological knowledge about the deep sea, given that humans cannot venture there without submersibles or special diving suits.[26] Tales of sea monsters abound, but such stories often sever connections—alienate—rather than mediate or transmit something of or about actual species. Flusser and Bec imagine sciences that "serve as luminescent organs, adorning fabulous tentacles with which the vampyroteuthis—one hopes—can be felt."[27] A tentacular science, with glowing organs, will have been shaped by its "object" of inquiry, as it bares itself to being touched by the creature it would know: the vampyroteuthis, would, ideally, "be felt." The ostensibly objective eye of science, which reduces lives, substances, and environments to objects, is reimagined as a biomorphic receptivity that is itself aesthetically lavish. The luminescent organs participate in the festivities rather than remotely recording them.

The marine biologist Edith A. Widder, in "Sly Eye for the Shy Guy," developed, with engineers, a luminescent organ called the Eye-in-the-Sea. The Eye-in-the-Sea camera system uses far-red light to remain unobtrusive but also features an optical lure that imitates bioluminescent displays.[28] The Eye-in-the-Sea, by the way, filmed the giant squid *(Architeuthis)* in its natural habitat for the first time in 2013. In a TED talk, referencing the video behind her on the screen, Widder describes what happened when the electronic jellyfish lit up at two thousand feet deep in the Bahamas, sparking more bioluminescent displays:

> We basically have a chat room going on here, because once it gets started, everybody's talking. And I think this is actually a shrimp that's releasing its bioluminescent chemicals into the water. But the cool thing is, we're talking to it. We don't know what we're saying. Personally, I think it's something sexy.[29]

Widder, an expert on bioluminescence, has developed other instruments that register light in the deep oceans and trace the distribution of bioluminescent displays. That may sound dry, but what if tentacular science could be a transspecies, somatechnic affair? Evidence that the creatures have been "felt" by the "luminescent organs adorning fabulous tentacles" would include Widder's own impassioned responses, the reactions of her audience, and the fact that she cofounded and leads ORCA, the Ocean Research and Conservation Association, advocating for stronger funding for marine science and ocean conservation projects.[30] Veer toward intimate science and impassioned politics. Widder has spent many hours thousands of feet down, in various submersibles, but it was her first experience in a Wasp suit at 880 feet below that sparked her "addiction" to bioluminescence, when siphonophores lit up even the inside of her suit and she was surrounded by "puffs and billows of what looked like luminous blue smoke," and it was "breathtaking."[31]

The marine scientist Ellen Prager, in *Chasing Science at Sea*, writes that even for researchers who study the deep ocean, the experience in a submersible can be astonishing, so much that it can be hard to stay focused on science. In these instances enthusiasm may turn objective scientific narrative into the layperson's words of excitement—ooh, ah, oh my god, and holy shit are a few of the descriptors that have been uttered under such circumstances, some even recorded for posterity."[32]

Few laypersons have the opportunity to experience the deep seas and find themselves exclaiming "holy shit" unburdened by professional protocol. There are a few outfits, however, that will take tourists down. The Roatan Institute of Deepsea Exploration in Honduras, led by Captain Karl Stanley, who "began building a submarine out of plumbing parts in his parents backyard at fifteen,"[33] takes people on descents as deep as two thousand feet, during a three-and-a-half-hour trip in a tiny, unregulated, uninsured, homemade submarine. (Confession: I've gazed at this very submarine with breathless longing and pressing terror. My desire to experience the ocean beyond the 120-foot-scuba-limit colliding with my claustrophobia. So tempting, still . . .) Text on the Roatan Institute's website encapsulates many familiar tropes of the deep sea. "We have finally made it to the land of perpetual darkness. This region of the ocean covers more than half the planet. The animals here have never known daylight or season. Many of the creatures date back before the dinosaurs. Fish encountered include jelly-nose eels, rough sharks, and chimera."[34] The deep oceans dwarf terrestrial environments and lure us with the illusion that time could be heavy and durable, rather than evanescent, preserving species instead of vanishing them. The animals seem to exist within their own time, without sun or season, not unlike the notion that the deep sea harbors "living fossils."[35] The descent promises encounters with exciting creatures: Veer from the obvious critique of touristic commodification, colonization, and spectacle to hear the call to recognize and think with nonhuman, nonterrestrial modes of life. Unmoor from academic modes of critique tethered to the dry detachment that insulates the human from less masterful affects. The artist Michelle Atherton, who, intrepidly, took the trip in Stanley's submersible, reflects on the experience in "Breathable Fabric. Submersion. And Parrotfish: Homotopia Three Unequal Speculations": "This is the total blackness of the Abyssopelagic zone in which we are descending, totally unmoored from the terrestrial."[36] When lights are turned on illuminating their surroundings, Atherton experiences an "aquatic immensity with no end": "Our orientation is scrambled; we have lost all sense of perspective." She explains that "in the deep, the day-to-day markers of space and time fall away."[37] She concludes, "We lose time, lose ourselves to a state of temporal drift."[38] Unmoor by sinking to the abyssal and benthic realms, where there is little to hold onto.

Veer out the Chain

To unmoor, veer out the chain to one anchor while getting the other anchor up.

—E. KNIGHT, *Small Boat Sailing on Sea and River* (1902)

Veering out the chain, in nautical parlance, means to slacken it in a controlled manner. This is no inebriated swerve or stumble but an intentional letting go in order to loosen the confines of the human. As the blue humanities call us to unmoor environmental thought from dry terrains, we may want to hold on to a certain skepticism regarding blissful oceanic feelings or even modes of philosophical critique that leave political activism without an edge, without leverage, without something to propel change. Floating too freely may be counterproductive. Mette Brylde and Nina Lykke warn, at the end of *Cosmodolphins*, that "self-abandoning, vis-à-vis the wild other," is a destructive mode of denial that ignores the fact that "we do have some opportunities for subjective and sociotechnical intervention."[39] Some of those interventions would entail the imposition of boundaries on what we relish as unbounded, in the form of the creation and enforcement of marine-protected areas, which would, ideally, protect even the sea vents, the deep sea mounts, and the seafloor. The deep seas are certainly not "alien" to industrial fishing operations or the companies planning to mine deep sea zones. Capitalist and neocolonialist mappings annex, rather than expel, these regions. Blue humanities and ocean conservation movements need to reckon with the particular challenges the deep seas face and relish the fabulous diversity of marine species, while avoiding conceptual maps that alienate. If there is to be an environmentalism, an activism, a philosophical, artistic, poetic reckoning with the scale of human alteration of the deep waters of the planet, it will require figurations of marine ecologies as they are continually being composed. Such figurations must circulate in culturally and politically potent modes. To unmoor means not only to sever what binds environmental thought to the shore but to cultivate alternative images, soundscapes, concepts, maps, stories, and highly mediated dispatches from the depths to mobilize enchanting cultural productions that will lure us toward an environmental ethics and politics that can traverse the scales of the seas.

Notes

Warm thanks to Jeffrey J. Cohen and Lowell Duckert for the invitation to contribute to this imaginative collection. More thanks for their generous and generative editing—they are the very best of collaborators!

1. Steve Mentz, "Toward a Blue Cultural Studies: The Sea, Maritime Culture, and Early Modern English Literature." *Literature Compass* 6, no. 5 (2009): 997.

2. Ibid., 998.

3. Steve Mentz, *Shipwreck Modernity: Ecologies of Globalization, 1550–1719* (Minneapolis: University of Minnesota Press, 2015), 178, 180.

4. Ibid.

5. The terminology is rather fluid here—*blue ecocultural studies, blue humanities, aquatic media studies, marine science studies, marine animal studies,* and *a new thalassology* are some of the terms currently in play. These burgeoning areas suggest the need to develop new concepts, approaches, frames, and arguments within environmental studies that can reckon with oceanic and perhaps other aquatic ecosystems without distorting them with a green, terrestrial lens. Along with Mentz's work, a partial, initial set of readings in these fields might include Dan Brayton, *Shakespeare's Ocean: An Ecocritical Exploration* (Charlottesville: University of Virginia Press, 2012); Elizabeth De Loughrey, "Heavy Waters: Waste and Atlantic Modernity," *PMLA* 125, no. 3 (May 2010) and other essays; Patricia Yaeger, "Sea Trash, Dark Pools, and the Tragedy of the Commons," *PMLA* 125, no. 3 (May 2010); Cecilia Chen, Janine MacLeod, and Astrida Neimanis, eds., *Thinking with Water* (Montreal: McGill-Queens University Press, 2013); Stefan Helmreich, *Alien Ocean: Anthropological Voyages in Microbial Seas* (Berkeley: University of California Press, 2009); Melody Jue, "Proteus and the Digital: Scalar Transformations in Seawater's Materiality in Ocean Animations," *Animation* 9, no. 2 (July 2014); Nicole Starosielski, *The Undersea Network* (Durham: Duke University Press, 2015); Stacy Alaimo, various essays and *Exposed: Environmental Politics and Pleasures in Posthuman Times* (Minneapolis: University of Minnesota Press, 2016).

6. Although nothing would seem less fluid than a stone, Jeffrey Jerome Cohen's eloquent musings on the lithic and its resistance to human knowledge may inspire us to unmoor: "Stone is a strangely sympathetic companion, a source of

knowledge and narrative, an invitation to an ethics of scale, the catalyst for humanist-scientist alliance, a disruption to everything we thought we knew." *Stone: An Ecology of the Inhuman* (Minneapolis: University of Minnesota Press, 2015), 65.

7. See, for example, Eva Ramirez-Llodra et al., "Man and the Last Great Wilderness: Human Impact on the Deep Seas," *PloS One* 6, no. 7 (July 2010): 1–25. The grand title harkens back to eras in which the majestic wilderness beckoned "Man," and yet the subtitle collapses the elevated sense of both "Man" and "Nature" into a prosaic account of anthropogenic effects.

8. A widely cited study in *Science* predicts that global fisheries will collapse by 2050. Boris Worm et al., "Impacts of Biodiversity Loss on Ocean Ecosystem Services," *Science*, November 3, 2006, 787–90. For introductions to the imperiled state of marine ecologies, see Sylvia Earle, *The World Is Blue: How Our Fate and the Oceans Are One* (Washington, D.C.: National Geographic, 2010); Callum Roberts, *The Ocean of Life: The Fate of Man and the Sea* (New York: Penguin, 2013).

9. Donna J. Haraway, *Companion Species Manifesto: Dogs, People, and Significant Otherness* (Chicago: Prickly Paradigm Press, 2003).

10. Vilém Flusser and Lois Bec, *Vampyroteuthis Infernalis,* trans. Valentine A. Pakis (Minneapolis: University of Minnesota Press, 2012), 5–6.

11. Ibid., 38.

12. For more on trans-corporeality, see Stacy Alaimo, *Bodily Natures: Science, Environment, and the Material Self* (Bloomington: Indiana University Press, 2010); *Exposed: Environmental Politics and Pleasures in Posthuman Times* (Minneapolis: University of Minnesota Press, 2016).

13. Melody Jue, "Vampire Squid Media," *Grey Room* 57 (Fall 2014): 85.

14. This refers to Bruno Latour's call for critical theory to *add* "reality to matters of fact and not *subtract* reality." "Why Has Critique Run out of Steam? From Matters of Fact to Matters of Concern," *Critical Inquiry* 30 (Winter 2004): 8.

15. Donna J. Haraway, "Situated Knowledges: The Science Question in Feminism and the Privilege of Partial Perspective," in *Simians, Cyborgs, Women: The Reinvention of Nature* (New York: Routledge, 1991), 188, 189.

16. Jacques Derrida, *The Animal That Therefore I Am,* ed. Marie-Louise Mallet, trans. David Wills (New York: Fordham University Press, 2008), 9.

17. Michael Marder, *Plant Thinking: A Philosophy of Vegetal Life* (New York: Columbia University Press, 2013), 62.

18. Jakob von Uexküll, *A Foray into the Worlds of Animals and Humans,* trans. Joseph D. O'Neill (Minneapolis: University of Minnesota Press, 2010).

19. Dorian Sagan, "Introduction: Umwelt after Uexküll," in von Uexküll, *A Foray,* 21.

20. Elizabeth De Loughrey, "Ordinary Futures: Interspecies Worldings in the Anthropocene," in *Global Ecologies and the Environmental Humanities: Postcolonial Approaches,* ed. Elizabeth DeLoughrey, Jill Didur, and Anthony Carigan

(New York: Routledge, 2015), 354. DeLoughrey cites M. Roberts and Peter Wills, "Understanding Maori Epistemology: A Scientific Perspective," in *Tribal Epistemologies: Essays in the Philosophy of Anthropology*, ed. Helmut Wautischer (Sydney: Ashgate, 1998), 43–77.

21. Jue reads Flusser's text in a similar manner: "Thus, as much as Vampyroteuthis Infernalis is a fable about photography, it is simultaneously (and nonexclusively) speculative fiction that takes seriously the conditions of the ocean as a novel and cognitively estranged starting point for philosophy." "Vampire Squid Media," 93.

22. Bruno Latour, "An Attempt at a 'Compositionist Manifesto,'" *New Literary History* 41, no. 3 (2010): 484.

23. Ibid., 488.

24. Ibid.

25. Flusser and Bec, *Vampyroteuthis Infernalis*, 72–73.

26. It is important to keep in mind, however, that, as DeLoughrey points out, due to "both history and geography, there is a large body of literature from the Pacific Islands that engages a complex oceanic imaginary." Elizabeth DeLoughrey, "Ordinary Futures," 354.

27. Flusser and Bec, *Vampyroteuthis Infernalis*, 73.

28. Edith A. Widder, "Sly Eye for the Shy Guy: Peering into the Depths with New Sensors," *Oceanography* 20, no. 4 (December 2007): 47.

29. Edith Widder, "The Weird, Wonderful World of Bioluminescence," TED talk, March 2011, https://www.ted.com/talks/edith_widder_the_weird_and_wonderful_world_of_bioluminescence/transcript?language=en. For a longer discussion of bioluminescence, cross-species communication, and posthumanism, see Stacy Alaimo, "Violet Black," in *Prismatic Ecology: Ecotheory beyond Green*, ed. Jeffrey Jerome Cohen (Minneapolis: University of Minnesota Press, 2013): 233–52.

30. In one interview, Widder states, "Twenty-five years ago, 7 percent of the national scientific research budget went to marine science; now it's only 3.5 percent, at a time when the oceans have never needed more help. Less than 1 percent of not-for-profit funds spent for conservation studies go to marine conservation." Lisa Capone, "Q&A: Edith Widder, MacArthur Fellow," *The Gulf of Maine Times*, Winter/Spring 2007, http://www.gulfofmaine.org/times/winterspring2007/qanda.html.

31. Edith Widder, "Glowing Life in an Underwater World," TED talk, 2010, https://www.ted.com/talks/edith_widder_glowing_life_in_an_underwater_world/transcript?language=en.

32. Ellen Prager, *Chasing Science at Sea* (Chicago: University of Chicago Press, 2008), 130.

33. James Nestor, *Deep: Freediving, Renegade Science, and What the Ocean Tells Us about Ourselves* (Boston: Mariner Books, 2015), 140.

34. Roatan Institute of Deep Sea Exploration. http://www.stanleysubmarines
.com/expeditions/2000ft-expedition/.

35. "Living fossils," "species from the evolutionary past," were once thought to
be abundant in the deep seas due to Charles Darwin's theory that "evolution
proceeded rapidly in fast-changing environments and slowly amid stasis." David
Dobbs, *Reef Madness: Charles Darwin, Alexander Agassiz, and the Meaning of
Coral* (New York: Random House, 2005), 159. The desire to discover living fossils
was a strong motivating force for the ambitious British *Challenger*'s explorations
of the deep seas from 1872–76. The scientists were disappointed, however, when
the expedition "failed to open a door into the early evolutionary history of life
on Earth." Tony Koslow, *The Silent Deep: The Discovery, Ecology, and Conservation of the Deep Seas* (Chicago: University of Chicago Press, 2007), 34. While the
concept of the living fossil has been discredited (the lines of evolution are not
so linear), nonetheless even today popular media report that scientists have discovered living fossils in the sea, such as a "remarkably primitive eel," a "living
fossil fish with lungs," and a "demon shark hauled from the ocean." See Mary
Beth Griggs, "'Living Fossil' Fish with Lungs," *Popular Science*, September 15,
2015, http://www.popsci.com/living-fossil-fish-has-lungs; "Scientists Call New
Eel Species a Living Fossil," Ocean Portal, Smithsonian Museum of Natural History, August 17, 2011, http://ocean.si.edu/blog/scientists-call-new-eel-species-liv
ing-fossil; "Living Fossil: Demon Shark Hauled from Ocean," *New York Post*,
January 21, 2015, http://nypost.com/2015/01/21/is-that-a-sea-monster-or-a-shark
-with-300-teeth/.

36. Michelle Atherton, "Breathable Fabric. Submersion. And Parrotfish: Homotopia Three Unequal Speculations," *E.R.O.S. Homotopia* 6 (2015): 218.

37. Ibid., 221.

38. Ibid.

39. Mette Bryld and Nina Lykke, *Cosmodolphins: Feminist Cultural Studies of
Technology, Animals, and the Sacred* (London: Zed Books, 1999), 225.

Whirl

TIM INGOLD

Herein wonder not
how 'tis that, while the seeds of things are all
Moving forever, the sum yet seems to stand
Supremely still.

—TITUS LUCRETIUS CARUS, *On the Nature of Things*

The nature of infinity is this: That every *thing* has its
Own Vortex; and when once a traveller thro' Eternity
Has passed that Vortex, he perceives it roll backward behind
His path, into a globe itself infolding; like the sun:
Or like a moon, or like a universe of starry majesty,
While he keeps onwards in his wondrous journey on the earth.

—WILLIAM BLAKE, *Milton*

Life in general is mobility itself; particular manifestations of life accept
this mobility reluctantly, and constantly lag behind. It is always going
ahead; they want to mark time. Evolution in general would fain go in a
straight line; each special evolution is a kind of circle. Like eddies of dust
raised by the wind as it passes, the living turn upon themselves, borne up
by the great blast of life. They are therefore relatively stable, and
counterfeit immobility so well that we treat each of them as a thing rather
than as a progress, forgetting that the very permanence of their form is
only the outline of a movement.

—HENRI BERGSON, *Creative Evolution*

Turning Life?

Life begins in whirling. By "life" I do not mean an interior property that distinguishes some things we call "animate" from everything else we call "inanimate." It is not to be found in the behavior of some magic molecule such as DNA that, in the right circumstances, can turn out replicas of itself. I mean, rather, the potential of a world given in movement to generate the forms of things, to hold them fast, and in turn to portend their dissolution. In a world of life, things are formed as eddies in the flow, that is as centers of stillness or dynamic stability that, far from having been fortified against the currents that would otherwise sweep them asunder, are constituted in these very currents. In this sense the whirling nebulae are phenomena of life; so too are the atmospheric storms or the oceanic maelstroms of our own earth, along with creatures of every kind that inhabit its lands and waters. It was in this vein that the Roman author Lucretius, in his essay-poem *De Rerum Natura,* described all things as having been formed in the never-ceasing flows of atomic particles that, falling rectilinearly through the void, swerve ever so slightly from their downward path so as to cause a cascade of collisions and combinations. Out of this commotion a world is formed, of things that, in our eyes, stand still. We see the forms and not the flow, and think of them not as movements but as objects in themselves. It is no wonder, then, that we look to some inner principle, some harbinger of agency or vitality, that would bring them back to life.[1]

In *Milton,* composed in 1804, the visionary poet William Blake was onto much the same theme. In the infinitude of space and the eternity of time, everything is formed as its own vortex in the currents of the cosmos. Blake imagines himself riding with the flow, a cosmic traveler adrift on its tides, sailing into the vortices of things. But only once he has passed through each, looking back, does he see them closing in or imploding, so as to give the appearance of bounded bodies that fall ever behind as he carries relentlessly on.[2] So too the philosopher Henri Bergson, in his equally visionary *Creative Evolution* of 1911, compares every living thing to an eddy or whirlwind in the "great blast of life." He is with Lucretius in thinking that life in general is given in movement,[3] and moreover in his conviction that for there to be particular living things it is necessary for this movement to veer from a course that would otherwise be absolutely straight—that is, for it to *turn aside*. It is as if, says Bergson, the living were to accept the movement of life with a degree of reluctance: as if they were to turn in on themselves, lagging behind—or marking time— while life itself moves on. But whereas Blake, riding the cosmos, looks back to see the things that once had sucked him in closing up and receding into the distance, for Bergson this closure is an illusion of the intellect in whose eyes, cast ever rearward, the "whirl of organism" reappears as an immobile figure set off by an outline against the ground of a ready-made world.[4]

Whirling, of course, is a movement, but as Bergson realized, it is not just *any* movement. It is, specifically, a movement that moves: one that, at every moment, veers off course. That is to say, it perpetually inflects or *turns*.[5] But nor is it just *any* turning. Of the mechanical clock, we may observe that its hands turn, or rotate, about a fixed and predetermined center. The turning of the whirl, however, does not so much presuppose a center as give rise to it, as a place of relative stillness, as a form. In the mechanism of the clock, the working parts—cogs, ratchets, hands—have all been manufactured in advance, and their movements are prescribed in their axial displacement from point to point. But in the whirl of organism, movement and form co-constitute one another: movement turns into form, and form into movement. Thus where Bergson distinguishes between the evolution of life in general and the special evolution of a particular life—the one straight, the other describing a "kind of circle"—we would do better, I think, to distinguish the *evolution* of the former from the *revolution* of the latter. By "revolution" I mean a turning that is also a turning *into*; revolving as becoming, becoming as revolving.

As such, the *revolution* of the whirl is quite different from the *rotation* of the clock. The whirl's revolution is formative, it is a movement of growth; the clock's rotation is mechanical, amounting to the spatial relocation of preexisting parts. Yet on second thoughts, this may not do justice to the clock. Suppose that it is of the predigital, preelectronic spring-loaded variety. In order for it to go we have had to wind it up. The very act of winding—performed, of course, by a living being—charges the material of the spring with incipient movement; as it is wound, the spring itself becomes a whirl. The cogs and hands of the clock might rotate, but at its heart is the revolution of the spring. And as Bergson said of the living organism, the spring—in turning in on itself—contrives to hold out against the passage of time. Yet eventually it marks time in its unwinding: in a cumulative sequence of escapements that punctate the evolution of life in its onward progress. As we shall see shortly, this has its precise parallel in the winding and reeling of thread.

On Describing a Circle

To better understand what we mean by the circularity of life, let me set an exercise that will no doubt be familiar to anyone who has undergone lessons in school geometry. The task is to draw circles, using only a pencil and a compass.[6] Try as you might, you will never be able to produce a perfect circle. This is because the human body, like the bodies of most living creatures, is not designed for rotational motion. In the history of technology, the introduction of the crank marked a key step in converting the reciprocal back-and-forth movement of the limbs to a rotary one.[7] With the compass, we attempt to achieve the same effect by gripping a shaft mounted at the apex between thumb and forefinger, and rubbing back and forth in a gesture known as "twirling." As I show below, twirling is whirling on a point, and the line that appears under your hand is the trace of this gesture, a form generated in its revolutionary movement. As such, it inevitably betrays the conditions of its production. For the velocity of the turn is never constant; moreover the uneven pressure of the pencil tip on paper inevitably leads to a line that varies in width and density. You have to start the line somewhere, and it is virtually impossible to conceal this starting point as it is always a little darker where the pencil point first makes contact with the paper, or where—in your effort to close the circle—the ending of your line overlaps its beginning.

Now we are taught, in school, to disregard these imperfections. Unavoidable they may be, but they are considered irrelevant to understanding the circle as a pure geometrical form. The pure form of the circle, however, is not a form of life. Neither does it turn; nor is it produced by turning. In its closure and finality—in masquerading as the outline of a figure rather than the trace of a gesture—it is the very opposite of the whirl.[8] Today, of course, the compass has all but disappeared from school classrooms, as more or less perfect circles can be printed to order from our computers. But I do wonder whether this facility has actually enhanced mathematical thinking. Many mathematicians argue to the contrary and rail against the insistence, in conventional teaching, on logical closure and final proof.[9] They would say—and I agree—that imperfection is the

key to the development of mathematical understanding, and that in drawing circles with compass and pencil, and never quite succeeding, the student can achieve a deep understanding of the phenomenon of circularity
that repudiates any distinction between the operations of the intellect
and skilled bodily practice. Such understanding is both open-ended and
in touch with life, as intuitive as it is cognitive. In this sense, geometry
is akin to music. With music, the more you practice a piece—the more
you inhabit it—the more your understanding grows. This growth is inexhaustible, not convergent on a limit. So too, by inhabiting the circle
and entering into the whirl of its inscription, you can begin to understand it from the inside, not as some timeless geometrical abstraction
but as the temporal revolution of the ever-turning wheel of life.

Questioning Words?

Etymologically, *whirl* is derived from the Old Norse *hvirfla*, "to turn" or
"to spin," and is one of a host of similarly sounding words, including
twirl, swirl, and *hurl.*[10] The precise derivation of these words is anyone's
guess. Did *twirl*, perhaps, come from a compound of *twist* and *whirl*?
Or did it come from the Old English *þwirl* (meaning "pot-stirrer")? Is
swirl just a variant of *whirl*, perhaps with its roots in the Dutch *zwirrelen*
rather than the Norse *hvirfla*? Did *hurl* have a quite different origin, in
the Low German *hurreln* ("to throw" or "to dash"), from which we also
get the word *hurry*? Perhaps then the resemblance of *hurl* to *whirl* is
fortuitous. But if it is, why does *hurl* have, as one of its meanings, "to
drive a wheeled vehicle," a meaning it has in common with *whirl*, whence
comes the Scots term *hurlbarrow* for what we more commonly call the
wheelbarrow? Is there a connection after all, between *whirl* and *wheel*?
No one knows the answer to any of these questions. Yet what the words
I have listed all share is a strong component of phonological iconicity.[11]
We can learn more of their meaning from simply pronouncing them and
from the feeling this induces. With twirl, as we have already seen in the
exercise of drawing a circle with a compass and pencil, the revolution
is centered on a point, vividly expressed by the hard, consonantal "t-"
before the "-wirl." It could be the compass point, or the pointed shoe of
the ballerina performing pirouettes on stage, or the finger-point of the
gentleman twirling his moustache. With swirl, by contrast, what counts
is the fluidity of liquid motion, with the hissing of the fricative "s" conveying a sense of escape rather than punctuality.

But perhaps the comparison with *hurl* is the most interesting. Say the word out loud and you feel the air being let out from the chest without obstruction. Say *whirl* and it feels quite different. It is as though the air were partially bottled prior to release: an incipient rather than an actual exhalation, a preparation for letting go. Indeed the difference between mouthing the two words, *whirl* and *hurl*, is a bit like that what happens in athletics, when the discus-thrower first whirls with his body, round and round, gaining angular momentum while the discus remains in his hand, only to release it, hurling the object as far as he can where it hits the ground, all energy spent. For the athlete the whirl is a gathering up, an acceleration: in the idiom of philosopher Gilles Deleuze and psychoanalyst Félix Guattari, it is "where things pick up speed".[12] The hurl, to the contrary, is a spin-off. So it is too with breathing in and breathing out. Breathing in is a gathering, a rewinding; breathing out a propulsive release. One sweeps around, the other launches forth through an opening at the center thus formed.[13] Of course one cannot literally speak on the inhalation; nevertheless one can deflect the flow of air to create an eddy, so as momentarily to hold it back, like a bubble that has still to burst. Why should the onomatopoeic variant of *whirl,* by which we imitate what we commonly think of as its sound, end with a rolled "r" that can be continued for as long as one has breath left to pronounce it? *Whirrrrr . . .* It could be the sound of a flying insect, or the helicopter's rotor as it lands or takes to the air, or the ship's propeller in water, or the bullroarer. Unlike the continuous hum or murmur, the roll of *whirr* suggests a movement that goes against the grain, a ratcheting, an infolding. This, in turn, sets up the aerial or aquatic vibrations that make it such a noisy affair. In the rolled "r," you can almost feel the revolutions of the whirl, right there in your mouth.

Most remarkable, however, is the fact that *whirl* begins with the very same sound that, in the English language, prefixes all of its interrogatives. As noted above, the whirl is not a resultant but a becoming, a turning *into* that which remains unknown, a ceaseless questioning from which any answer continually recedes. With the whirl, the thing is not yet settled, neither in its present nor in its future form. It is a problem, for which the steps toward a solution are not already given. Whither is it going? When will it arrive? What will it be? Why? We don't know. Whirl is a question, *where*? Hurl delivers the answer, *here*! The scribes of medieval

Europe must have known this when they converted what used to be pro-
sodic markers to help the orator in the declamation of a text into what we
now know as punctuation. The form of the question mark, with which
we are so familiar today, is no less than a miniature whirl.

¿

Moreover since every question solicits a response, it is addressed to the
ear of the listener. While the iconic resemblance of the question mark
to the involute form of the human ear may not be entirely fortuitous,
what the early architects of punctuation probably did not know is that
as sonic vibrations penetrate deep within the ear, they are funneled into
the spiraling tube of the cochlea. It turns out that every one of us carries
a miniature whirl in each ear! And when you place one cochlear form
over another, by covering the ear with a spiral shell, the combination
spontaneously generates its own sound, reminding us at once of the
maelstrom of the sea.

Spinning Threads

Tracked along another path, the chain of word associations leads into the
language of spinning, in which *whirl,* taken as a noun rather than as a
verb, is simply a variant of *whorl.* The whorl is a small, doughnut-shaped
disc, of some three or four centimeters in diameter, commonly carved
from stone, bone, or hardwood.[14] Fitted to the lower shaft of the handheld drop spindle, its function is to give weight and angular momentum
to the spindle when spinning a fleece from a distaff. Due to its relatively
imperishable material, the whorl is often the best evidence we have of the
practice of spinning in deep prehistory, where nothing remains of the
threads or the fabrics woven from them. Nowadays, when most thread is
industrially produced, much of it from synthetic material, the practice
has all but disappeared and is sustained—at least in Western societies—
only by a select band of hobbyists and craftspeople. But for the greater
part of human history, among peoples from around the world, spinning
was a ubiquitous activity carried on day after day, for hours on end, in
some societies only by women, in others only by men, and in yet others
by both men and women. Considering its ubiquity, however, spinning
has attracted extraordinarily little attention from historians and ethnologists. For most historians, it only comes into the picture with the onset

of industrialization; for ethnologists the focus has always been more on weaving than on spinning. It is all too easy to forget that there can be no weaving without thread and no thread that has not first been spun.[15]

I am as guilty of this neglect as anyone. In a work on the history and anthropology of the line I had distinguished two major classes of line, threads and traces, and had shown how threads transform into traces in the formation of surfaces, as in weaving, and traces into threads in their dissolution, as when a woven textile is unraveled.[16] It had seemed to me that weaving could be taken as a model for both making and thinking. To treat making as a modality weaving is to emphasize process over product, to see form and pattern as emergent within the process rather than preconceived and imposed on raw material, and to recognize too how material is held together or made to cohere through contrary forces of tension and friction. To treat thinking as a modality of weaving is to adumbrate an alternative mathematics, rooted in the routines of everyday life, that permits an open-ended exploration of the multiple possibilities of permutation and recombination, and of the patterns and symmetries that result. Making and thinking, thus conceived, could both be understood as ways of working with lines. Yet in all this, I had given no thought to how these lines are generated in the first place. Whatever happened to spinning? What if we were to regard making and thinking as modalities of spinning rather than weaving? Does not the turner spin his wood on the lathe, and the potter his clay on the wheel? Do we not turn over ideas in our minds, and get our heads into a spin? And do we not spin our narratives before weaving them into text? Might the whirl be as generative as the weave? Or yet more so?

I have scarcely begun to address these questions and can offer no more than a few speculations on the topic, partly provoked by my first lessons in how to spin from a distaff, using a drop spindle, under the guidance of one of the few specialists in the comparative ethnography of spinning, Tracy Hudson.[17] What took me most by surprise, in my initial attempts, was just how discontinuous an operation it is. Naively, I had imagined that with the revolution of the spindle, the thread would just come spooling out. Of course it did not, for one very interesting reason. This is that the twist travels upstream, and not downstream, from the revolving spindle up toward the distaff. Periodically, then, you have to interrupt the spin, reel the thread you have spun onto the shaft of the spindle (starting from above the whorl), and commence the spin again. The mistake I often found myself making was to reel on in a way that simply reversed the spin, thus undoing what I had just done. Though I have yet to think through the implications of this discovery for our understanding of thinking and making, it does lead me to question what I have certainly assumed up to now, namely that making and thinking entail an ongoing forward movement. What if every act of making and thinking took you *back* to the source from which your materials were derived, rather than further from them, such that the actual growth of the work would be an accumulation or ratcheting up of these successive backward movements? Is it only by the accumulation of successive gatherings or recollections, of memories, that we can advance? Is the future a succession of vortices in each of which we must necessarily find ourselves spinning into the past, only to reel on to the next?

Evidently the whorl winds up, not down; drawing tension into the thread, not releasing it. It is here that we can return to the parallel with the clock. For winding the thread, just like winding the clock, is a return toward the source, a revolution that goes against the grain, in which the materials—whether the metals of the spring or the fibers of the fleece—turn in on themselves, straining against the inexorable march of time. And just as with the rotations of the clock, or indeed with the rolled "r" of "whir," it is through the accumulation of escapements, reeled on the spindle, that the line of thread is advanced. What, then, is the relation between the whirl of the spindle and the line of thread? This is what I had missed, in my earlier comparison of the thread and the trace. With the trace, as when I draw circles with pencil and compass, the line issues directly from the movement of the tool and from the gesture of the hand that holds it. But with the thread, the line does not record the revolutions of the whorl, even though it is formed by them. Earlier, I observed that in the whirl, the turning movement continually gives rise to a center of relative rest—an "eye," if you will. And the thread-line, far from moving *around* the eye, issues *from* the eye itself as it moves. This observation brings me to my final theme, which concerns the meanings of *wind*.

The Whirling Wind?

Bergson, it will be recalled, compared living things to eddies of dust raised by the passing wind. It seems that the atmospheric wind is itself inclined to *wind,* turning on itself into a whirl. Wind and *wind* are of course the same word, nowadays distinguished only in pronunciation, which is why I have had to resort here to italics in order to distinguish the turning of the whirl from the draft of air. But if the gyre of the *wind* gives rise to things, that of the wind can rip them asunder. The very same vortex that grows the bodies of the living, as philosopher Michel Serres notes, can also destroy ships at sea: "It is order and disorder at once."[18] When the gyre is of moderate acceleration we call it a whirlwind, but if the acceleration grows to be of destructive force, it becomes a tornado, or on a greatly expanded scale, a cyclone. At its heart is the eye, and as the whirlwind moves we can track its path in the movement of the eye, and in the trail of dust or destruction that it leaves in its wake. That is, we can track its evolution. That the whirlwind evolves is not in doubt, since its movement is not like the transport of a solid body from one location to another across the sky. The whirlwind does not move like an airplane! On the contrary, it moves by perpetually winding up on its advancing front while unwinding at the rear.[19] Indeed, we might say of the whirlwind or the storm that it is continually losing the thread, much as happened to me in my first attempts at spinning, when every time I would reel my thread onto the spindle I would undo what I had just spun. That is why the storm leaves no trace of its evolution in the air, but only on the ground. Nevertheless, we must distinguish—as we have done before with regard to life in general and its specific cycles—between the evolution of the wind and its revolution.

Let's get back to Bergson. Here he is, once again insisting on the same distinction:

> The act by which life goes forward to the creation of a new form, and the act by which this form is shaped, are two different and often antagonistic movements. The first is continuous with the second, but cannot continue in it without being drawn aside from its direction.[20]

So it is, too, with the storm. Yesterday, a powerful storm struck the coast of eastern Scotland, where I live. I went walking along the shore, and had to struggle against the wind coming from the south. Later, the strength dropped, only to be replaced by a northerly. The storm itself, however, was tracking from west to east! Walking below as the storm passed overhead, my experience was of an airflow that had been "drawn aside," as Bergson would say, from its prevailing westerly direction and put into an anticlockwise spin. First the leading edge arrived, as the wind direction swung from west to south, then came the eye, and finally the trailing edge as the wind got up again from the north. And now, in the aftermath, only a gentle westerly remains. Could this image of the passing storm, I wonder, provide the key to solving a problem that has long perplexed me? Expressed in its most general terms, the problem is about how to understand the environment we inhabit.

I had come up with two possible answers. One was to think of the environment as a *meshwork*, woven by the myriad lines of living beings as they thread their ways through the world. The other was to think of it as an *atmosphere*, an aerial domain suffused with light, sound, and feeling. My questions, then, were: Is the environment a meshwork, or an atmosphere, or both? And if both, then what is the relation between them?[21] Perhaps it is like the relation between breathing in and breathing out. As we have seen, the wind *winds*. We breathe it in and, breathing it out as we walk along, we lay our linear trails through the world. But inhalation and exhalation are not the precise reverse of one another. It is instructive to compare breathing with swimming. In the breaststroke, the backward sweep of the arms and infolding of the legs is followed by a forward thrust that propels the swimmer through the water. Likewise when we breathe the air, the inhalation—a drawing in, a circulatory movement that deviates from or even reverses our direction of travel—is followed by a propulsive exhalation. Recall the discus thrower: the whirl prepares, the hurl delivers. *Whirl—hurl; whirl—hurl.* With every whirl, we draw in the atmosphere; with every hurl we weave a path in the meshwork. The same principle is at work, whether with the whirl of organism, the clockwork spring, the spindle whorl, or the whirlwind. In every case, linearity issues from circularity, and circularity from linearity, in an alternation that is foundational to all life.

Notes

1. What Lucretius teaches us, as political theorist Jane Bennett shows, is that we need not add vitality to inanimate things to bring them to life, since it is already immanent in the "primordial swerve" from which everything arises in the first place. The swerve, Bennett writes, "affirms that so-called inanimate

things have a life, that deep within is an inexplicable vitality or energy . . . a kind of thing-power." *Vibrant Matter: A Political Ecology of Things* (Durham: Duke University Press, 2010), 18. For an excellent account of the relevance of Lucretius's cosmology for contemporary anthropology and philosophy, see Stuart McLean, "Stories and Cosmogonies: Imagining Creativity beyond 'Nature' and 'Culture,'" *Cultural Anthropology* 24 (2009): 213–45.

2. It is conceivable that as he wrote these lines, Blake had in mind the theory of Descartes, that the universe is filled with matter that—having been given some initial momentum, a cosmic spin or whirl—has settled into a system of interlocking vortices that carry the sun, stars, planets, and all other heavenly bodies in their paths. I am grateful to Jeffrey Cohen for this suggestion. See Jeffrey Jerome Cohen and Lowell Duckert, "Introduction: Eleven Principles of the Elements," in *Elemental Ecocriticism: Thinking with Earth, Air, Water and Fire,* ed. J. J. Cohen and L. Duckert (Minneapolis: University of Minnesota Press, 2015), 3.

3. Whereas for Lucretius, however, the movement of life was ever downward, Bergson's "great blast" erupted upward. It is perhaps no accident that Bergson was writing during the heyday of balloon flight!

4. The felicitous phrase "whirl of organism" comes from an essay by another philosopher, Stanley Cavell. See Stanley Cavell, *Must We Mean What We Say? A Book of Essays* (Cambridge: Cambridge University Press, 1969), 52. I am grateful to Hayder Al-Mohammad for drawing this reference to my attention. More recently the feminist cultural theorist Astrida Neimanis has compared human bodies to oceanic eddies and herself to "a singular, dynamic whorl dissolving in a complex, fluid circulation." Astrida Neimanis, "Hydrofeminism; or, On Becoming a Body of Water," in *Undutiful Daughters: Mobilizing Future Concepts, Bodies, and Subjectivities in Feminist Thought and Practice,* ed. H. Gunkel, C. Nigianni, and F. Söderbäck (New York: Palgrave Macmillan, 2012), 96. The philosopher Michel Serres would agree. "Who am I?" he asks. The answer: "A vortex. A dispersal that comes undone." *The Birth of Physics,* trans. J. Hawkes (Manchester: Clinamen Press, 2000), 37.

5. For this idea of inflection as movement-moving, I am indebted to Erin Manning. "Perceiving the inflection," Manning writes, "does not mean being aware of it as though you could be outside it. It means moving in its tending. It means attending, in the event, to how movement diverges from its flow, attending to how movement moves." Erin Manning, *The Minor Gesture* (Durham: Duke University Press, 2016), 118.

6. On the exercise of drawing a circle and its implications, see Tim Ingold, "Bindings against Boundaries: Entanglements in an Open World," *Environment and Planning A* 40 (2008): 1796–1810.

7. On the technological consequences of the crank, see Lynn White Jr., *Medieval Technology and Social Change* (Oxford: Clarendon, 1962), 103–17.

8. As Serres has pointed out, in nature there are no perfect circles, only vortices: "No exact rounding off, no pure circumference, spirals that shift, that erode. The circle winds down in a conical helix. The Pythagorean or Platonic circle becomes the Archimedean helix. In other words, nature is not endowed with perpetual motion." See Serres, *Physics*, 58.

9. See, for example, Paul D. Lockhart, *A Mathematician's Lament: How School Cheats Us Out of Our Most Fascinating and Imaginative Art Form* (New York: Bellevue Literary Press, 2009).

10. This and other etymological references in this paragraph are drawn from the *Shorter Oxford English Dictionary*, 6th ed. (Oxford: Oxford University Press, 2007), and from the *Online Etymological Dictionary*, http://www.etymonline.com/.

11. On the idea of phonological iconicity, see Alfred Gell, "The Language of the Forest: Landscape and Iconism in Umeda," in *The Anthropology of Landscape: Perspectives on Place and Space*, ed. E. Hirsch and M. O'Hanlon (Oxford: Clarendon, 1995); Eduardo Kohn, *How Forests Think: Toward an Anthropology beyond the Human* (Berkeley: University of California Press, 2013), 27–33.

12. Gilles Deleuze and Félix Guattari, *A Thousand Plateaus: Capitalism and Schizophrenia*, trans. B. Massumi (London: Continuum, 2004), 28.

13. See Tim Ingold, *The Life of Lines* (Abingdon: Routledge, 2015), 66–68.

14. Ibid., 56–57. Topologically, the whorl takes the form of the torus, the formation and properties of which are discussed by literary scholar Valerie Allen in her essay "Airy Something," in Cohen and Duckert, *Elemental Ecocriticism*, 77–104. The whorl also bears comparison with the spinning top, of which Serres has the following to say: "Is it stable? Yes. Is it unstable? Yes, again. Is it rotating, does it follow a circumference? Yes, ever again. The top is a *circum-stance*. Can it move forward, lightfootedly? Yes. Can it lean? Yes, in all directions". All in all, says Serres, the spinning top "may serve as a little model of the world." *Physics*, 29.

15. On the importance of spinning for our concept of the line, see Victoria Mitchell, "Drawing Threads from Sight to Site," *Textile* 4 (2006): 340–61.

16. Tim Ingold, *Lines: A Brief History* (Abingdon: Routledge, 2007), 39–71.

17. Tracy Hudson visited my research group at the University of Aberdeen in February 2015, and I am very grateful for her inspiration and teaching. See Tracy P. Hudson, "Variables and Assumptions in Modern Interpretation of Ancient Spinning Technique and Technology through Archaeological Experimentation," *EXARC Journal Digest* 1 (2014): 1–14.

18. Serres, *Physics*, 29.

19. Ingold, *The Life of Lines*, 54.

20. Henri Bergson, *Creative Evolution*, trans. A. Mitchell (New York: Henry Holt, 1911), 129.

21. Ingold, *The Life of Lines*, 87–88.

Curl

LARA FARINA

> [Vines] are flexible in nature, and whatever they embrace they bind tight
> as if with a kind of arms.
>
> —ISIDORE OF SEVILLE, *Etymologies*

> Intelligence makes humans superb acclimators. It depends on nerves,
> cells with highly ramified shapes that are deeply interconnected. There is
> nothing remotely like this in plants . . . any resemblance to our
> intelligence is superficial, like that of a mandrake root to a person.
>
> —TOBIAS BASKIN, letter to the *New Yorker Magazine*

Lacking voices, faces, and quick motility, plants have seldom been considered candidates for the category of thinking beings in the West. Following Aristotle, early natural philosophers saw plants as the most minimal expression of "life," possessing vitality but no feeling or awareness of their own matter.[1] As Tobias Baskin's above response to Michael Pollan's "The Intelligent Plant," makes clear, resistance to extending sentience to *species plantarum* remains strong among bioscientists (Baskin is a cellular biologist studying botanic morphogenesis).[2] His use of the medieval figure of the mandrake root is meant to place plants' possible "resemblance to our intelligence" firmly in the realm of history's pseudo-scientific dead ends. Yet despite meeting with ridicule, the question of whether plants have feelings—and how humans might come to understand these—has been asked with some urgency in times of ecological crisis, both ancient and modern.

This essay follows the curling action of floral growth to explore a botano-poetics of sensation, one that attempts to feel *with* plants—to feel, that is, alongside, through, and from within vegetal sense-worlds. Feeling botanic, I argue, places us in the realm of the uncanny, the queer, and the numinous, a powerful but often emotionally incoherent domain that might best be understood as an affective preconscious, a "mind" of skin folding inward and sprouting through the edges of the body. Pursuing and practicing these uncanny feelings is one way to answer Jane Bennett's call to experience ourselves "as not only human." It is also a way to expand the affective alliances sought by recent ecocriticism, because it attends to feelings that are generally associated with, on the one hand, sexual perversity and, on the other, spiritual experience, neither of which have an easy place in current environmentalist thought. Of environmentalism's blind spots, Catriona Mortimer-Sandilands has recently argued, for instance, that "mainstream framings of environmental issues tend to ignore the homophobic and heterosexist relations that provide [their] social context."[3] While the politics of exclusion are very different concerning faith-based traditions, current ecological theory is nonetheless as wary of Western religiosity as Baskin is of sentient plants.[4] But following the curl of the vine may take us away from a safely secular subjectivity into perversely ecstatic affect. As we loop around past and present, feeling with plants turns us away from heteronormative embodiments and toward familial clusterings that are anything but straight. A queer vitality curls through the feeling of botanic desire.

In what follows, I invite you to curl in sympathy with an (un)holy triad of modern and medieval fictions; each of the three texts responds to ecological crisis, and each imagines a botanic world poised to redirect our habits of feeling. "Curling" with these scripts requires retrograde but noncircular movement, like the voicing of the word itself, which starts in the throat (k), moves to the teeth (r), and ends at the back of the tongue (dark l). To curl is to move against telos, backward in time (though not to an origin), and in affectively contradictory ways. It is the movement ascribed to heretics and perverts, as Jonathan Dollimore, reading Augustine, writes in his history of sexual "dissidence."[5] And like the viney embrace described above by Isidore of Seville, it is both intimate and frightening. We instinctively brace ourselves before rounding a bend at speed, tightening our muscles and shifting our weight against the disorienting fluctuation of gravitational force. But as Dollimore perceived, it is exactly through perversion—veering *away* from God—that the sacred becomes manifest; when you curl, moving away is also moving toward.

"twined with cable-like creepers and festooned with epiphytes"

Ursula Le Guin's short story "Vaster Than Empires and More Slow" revisits a speculative domain evoked but then foreclosed in Andrew Marvell's "To His Coy Mistress" (1680), the poem from which Le Guin takes her title.[6] Marvell's speaker famously argues to his paramour that men do not have "world enough, and time" to pursue a "vegetable love"—to explore, that is, a different kind of erotic affection, one which might be possible if we didn't experience space and time on a human scale. However enjoyable these unbounded feelings may be to imagine and write about, Marvell suggests, the presence of "time's wingèd chariot" demands a straighter course; "vegetable" pleasures should be put aside for a more expedient, familiar form of heterocoupling. Veering away from the poem's enjoined shortcut to pleasure, Le Guin's piece of science fiction attempts to linger in the uncannily vast and slow sensorium Marvell rejects by taking vegetable sentience, vegetable feeling, and even vegetable love as hypotheses for earnest exploration.

Initially, the vegetation of "Vaster Than Empires" is what it usually is in literary works: mere backdrop for human drama. The story follows the members of an "Extreme Survey" crew sent by a future League of the Earth as they establish first contact with a "pure phystosphere," an exceedingly distant planet covered with plants but lacking any animal or even bacterial life. The scale of the voyage through space/time is so enormous that only the insane, our narrator tells us, would volunteer, and, indeed, the crew is composed of escapists and deviants, each with a defining mental disorder and at least one sexual peculiarity. Not only the characters but the whole mission—and thus the basis for the speculative flight of this fiction—is attributed to the dysfunction of the human species as a family, with the peevish "Terrans" wanting to veer away from their "parent" culture, the "Hainish."[7] Thus, the large scale of interplanetary encounter seems like it will play out on the small scale of the nuclear domicile. We might expect the familiar Oedipal narrative, set, as all such narratives are, in an Eden that is really no Eden after all.

But unlike Eden, world 4470, so new it has no name, emerges as—if not exactly a character—a feeling being-space. We never learn the true nature of the green planet from the narrator; we only witness its effects on the crazed human characters as they struggle to come up with words, scientific explanations, and metaphors to deal with "it." An interconnected mass of roots, vines, leaves, and epiphytes bathed in drifts of pollen, "it" is referred to by the surveyors as the "trees" (in ironizing quotation marks), the "forest" (a term dryly noted by the team's psychologist to have "archetypical connections" leading to certain inevitable metaphors), and "that damned stupid potato."[8] Le Guin is obviously interested in the way plant sensitivities escape human language, causing words to double back on themselves in ironic knots of naming and unnaming, but she's nonetheless reluctant to give up representing the inner space of the "damned potato" entirely. The crew hypothesize a giant nervous system for the planet, proposing that each seemingly distinct plant functions as a single neuron connected to others via electro-chemical signals transferred through roots, vines, and pollen, making the whole planet a big brain, alone unto itself in isolating space. The conceit tackles biologists' insistence that neurons are needed for intelligence (*pace* Baskin) by blowing up the cellular scale to planetary size.

But the planet/brain model provides only explanation. The device Le Guin employs for giving readers some emotional access to "it" is the character Osden, an autistic Terran who was "cured" of his extreme withdrawal with the terrible result of becoming overly empathic and entirely exposed to the feelings of fellow humans. Osden is the survey team's official Sensor, and, as a medium for manifesting the interiority of others, he is unbearable company for the human crew (who try to kill him), but the only possible means of letting the plant mind speak for itself. Like a piece of parchment, this human animal is primed to be inscribed by the green world.

Osden is bodied forth as Le Guin's imagined sacrifice to the Kingdom of Plants; his appearance reveals the empath as a penetrated man:

He looked flayed. His skin was unnaturally white and thin, showing the channels of his blood like a faded road map in red and blue. His Adam's apple, the muscles that circled his mouth, the bones and ligaments of his wrists and hands, all stood out distinctly as if displayed for an anatomy lesson.[9]

The language-making parts of a human body (throat, mouth, and hands) are peeled open, exposed to the emotions of Osden's crewmates and the scrutiny of the reader. But Le Guin twists the surveyors' reading of the flayed man back in the other direction as well. While Osden has no protection from trans-corporeally circulating feeling, the other survey members are wounded by whatever names he can sum up to label their perverse desires and pathological affects, revealed to him in all their twisted horror. The result of an empathy good enough to register the feeling of a botanic Other, Le Guin suggests, is a "flayed" sensitivity and overexposure to emotional noise/feedback. In this exposed state, normally unheeded or unconscious somatic engagements—what Steven Shaviro calls "discognition"—manifest as perceptible sensations.[10] Osden gets caught in a perpetual looping of barely tolerable affect (resulting in fear, anger, disgust, and sorrow) that is alleviated by neither self-control nor catharsis, the "normal" human tactics for coping.

This affective wounding renders Osden's first "feeling with" the plant world a chthonic nightmare. Already psychically broken into by the other crewmembers, and, through them, doubly burdened with the "archetypal connections" summoned by the concept of "forest," the Sensor, "like the blind man trying to describe the elephant," recalls his moment of connection with "it":

> My face was in the dirt, in that soft leaf mold. It was in my nostrils and eyes. I couldn't move. Couldn't see. As if I was in the ground. Sunk into it, part of it. . . . My hands were bloody, I could feel that, and the blood made the dirt around my face sticky. I felt the fear. It kept growing. As if they'd finally known I was there, lying on them, under them, among them, the thing they feared, and yet part of the fear itself. I couldn't stop sending the fear back, and it kept growing and I couldn't move, I couldn't get away.[11]

Scraped raw and stuffed with dirt, the more usually privileged sensory organs are disabled. At the same time, "it" has become "they," "the fear," and the self, while still remaining "it." Subject positions are multiplied and flipped: Osden becomes the rooted one while "it" divides and replicates. Le Guin strives for language that can match the curl of becoming plant with pronouns that do triple duty: "*It* [the fear, the plant world, the self] kept growing"; "As if *they'd* [the plants, the selves-in-the-plant, and the plant-through-the-crew] finally known I was there."

Yet the curl of affect is not a closed loop. When Osden reenters "it" at the story's end, he takes with him the fellow feeling of the mission's captain, Tomiko, who has broken through her own fear spiral with Osden to emerge in something like (but not) love—"ontá," or "polarized hate."[12] Tomiko's affective connection with Osden, an emotionally paradoxical state of both-at-once, opens the feedback loop into a curl, allowing Osden to move with her as he is moved by "it." Now able to move through formerly closed parts of bodies, botano-Osden communicates to Tomiko, "All well. . . . Listen. I will you well."[13]

Le Guin's hero is the queerest in a group of queers, the crippest in a group of cripples, and a challenge to readers' sympathies. While the story suggests that the internally exiled (perverts, the disabled, and those without family) may have more to gain from external exile than do those who are comfortably ensconced in social privilege, it also portrays even these as mostly unwilling to enter the uncanny domain of radically nonhuman communion. The members of the crew do feel, obliquely, the feelings of the plant world, but rather than submerge in "it," the surveyors retreat into sleep, triple up on tranquilizers, and half-consciously turn their backs to their ship so as to keep a watchful eye on "the trees." Despite the fact that they have no futures to return to (because this is science fiction, literally so), they flee the place that offers an expanded self in the present. With the exception of Tomiko, the crew cannot imagine that feeling with the trees, like Osden's condition of affective exposure, will be anything other than crushing. Osden, meanwhile, is both martyr and mystic. Giving himself wholly to the vegetable Other in an act of something like love (for "it," "them," and himself-in-it/them), he abandons any pretense of heteronormative futurity to remain as the sole "colonist."[14] The word raises the question of how Osden might become part of a colony. Not through sowing his human seed, surely, but by curling and splitting, lending some of his feeling to "it" and receiving botano-sensation in return. For me, a medievalist, Osden's final "words" are reminiscent of nothing so much as those of the fourteenth-century anchorite and mystic, Julian of Norwich, when she "heard" the refrain, "alle shalle be wele, and alle shalle be wele, and alle manner thing shalle be wele," as she lay paralyzed with illness and in thrall to God.[15] While Le Guin claimed that "Vaster

Than Empires" "is not a psychomyth, but a regular science-fiction story,"[16] her work summons the weird affects that have, historically, been cultivated by the writing of ecstatic religiosity.

"vanished into dulse and seaweed"

Originally published in 1970, "Vaster Than Empires" was written during a time of rising Western concern about pollution, deforestation, and damaged ecosystems—phenomena that Le Guin and other science-fiction authors of the period address explicitly.[17] Deep ecology was on the cusp of its articulation.[18] But the interest in considering whether plants can think, feel, or communicate accommodated a very mixed politics. Plants were both newly radical and newly domestic in late 1960s and early 70s. On the one hand, the counterculture was taking up environmental activism and exploring the transformative possibilities of botano-delic practices; on the other hand, mainstream culture was caught up in the golden era of the American houseplant. Plants were cause for political protest, yet they were also becoming pets.[19] It is especially fitting, then, that Le Guin's fiction examines the structuring of the domestic family as the ground for the imbrication of human and vegetal kingdoms.

Curling backward through history, we find that this was not the first time that ecological crisis inspired speculation about the sense-worlds of plants, the possibility of human knowledge of these, and the consequences of that knowledge for groupings based on kinship. At times when kinship and kingdom have been more obviously intertwined, those consequences surely threatened to be even more destabilizing. Let's turn here to fourteenth-century Britain, a place and age of climate change that also imagined "Green Men" of various kinds.[20] While the most studied of the Green Man figures is doubtless the titular Green Knight of *Sir Gawain and the Green Knight,* another literary text, the Welsh "Math son of Mathonwy," or the fourth "branch" of the *Mabinogi,* as it is customarily called, is even more germane to considering not only what plants are to people, but what people are to plants.[21] This fantastical fiction anticipates Le Guin's staging of plant/human feeling but can be also itself recuperated for a queer ecocriticism by a reading inspired by "Vaster Than Empires."

In the late fourteenth-century Red Book of Hergest, the story of "Math" follows that of "Manawydan Son of Llyr," a magic-laden narrative in which ecological relations and land-management problems feature promi-nently.[22] In "Manawydan" the survivors of a devastating war wander a depopulated Welsh landscape, one rich in animal sources of food (fish, game, and honey) but bereft of agriculture and agricultural workers. After trying to subsist in Wales as a hunter-gatherer, then working in England as a craftsman of leather goods, Manawydan returns to his native land, bringing wheat seeds with him and attempting to restore a harvest. His first crops not only fail, they do so dramatically, turning from fine wheat to "bare stalks" overnight.

I want to make two observations about "Manawydan" as a precedent for "Math son of Mathonwy." First, while the story of Manawydan gives the crops' unmanageability an animal cause (we're told they are being destroyed by a vengeful retinue magically transformed into mice), plants actually did refuse to grow in fourteenth-century Wales, as they did else-where in the cooling European weather. European oaks—which supply scientists with readable climate histories—had unusually low levels of growth in the early to mid-fourteenth century, punctuated by years of no growth at all.[23] Weather-induced famine, blight, and the exhaustion of arable land were persistent and profound problems in this period.[24] Second, Manawydan's ability to solve and rectify the problem of the dis-appearing harvest has earned him the label of "culture hero" and "right-ful ruler" in critical readings of the *Mabinogi,* most of which connect successful rulership with the fertility of the land.[25] Following this line of thought, we could see Manawydan as a story in which uncooperative plants (of English origin no less) are successfully tamed by a clever native king, who both restores and modernizes Welsh land and culture. Mana-wydan's talent for working animal skin into usable goods (shields, sad-dles, and shoes) reflects the celebrated reputation of Welsh leatherwork and suggests a tradition of native success in turning the nonhuman into commodities for export, one that will be applied to grain crops in the future.

Yet if Manawydan's wheat ultimately surrenders to human rule, the vegetation of "Math son of Mathonwy" exacts a kind of revenge on the workers of skins. "Math," while temporally unhinged from the preceding tale of Manawydan, curls back to the beginning of the *Mabinogi,* depict-ing the undoing of its promise of a future based on royal lineage. The first *Mabinogi* tale, "Pwyll, Prince of Dyfed," features the problematic but eventually successful production of an heir to Pwyll's kingdom in the form of his son, Pryderi. "Math" portrays not only Pryderi's death but also the loss of older breeding stocks and familiar means of reproduction. Along the way, categories of native vs. alien and plant vs. animal are obliterated by sensational merges and splits of matter.

While King Math's heir, Lleu, is himself gestated outside of the womb, the undoing of succession occurs most dramatically through the character of Blodeuwedd, Lleu's adulterous bride. Math and his magician nephew, Gwydion, create her out of a mixture of flowers as a wife for their protégé. Not only is she a plant turned human, her botanical genealogy is itself hybridized. As medieval readers could have known from either herbals or simple observation, the plants from which she is made—oak, broom, and meadowsweet—are not species typically found together in the wild. Their preferred habitats are very different from one another's, and none of their needs, significantly, are best met by tilled land: oaks like loamy forest; broom seeks out rocky, dry soil; and meadowsweet thrives in marshes. The progeny of this trio would seem to demand an impossible, contradictory habitat: wild (untilled) yet unprecedented and unnatural.[26]

Not surprisingly, Blodeuwedd, the hybridized plant maiden with two fathers, wrecks havoc with human designs for fertility and reproduction. Rather than provide Lleu with heirs, Blodeuwedd immediately takes up with Gronw, a nobleman passing by on a hunt, conspires with him to kill Lleu, and is then herself transformed by Gwydion into a "flower-faced" night bird, the owl.[27] While Manawydan has been dubbed a culture hero, Blodeuwedd is always seen as either a perpetrator or a product of culture's undoing. She has been likened to Eve with her apple, connected with artifice and superficiality (because made of flowers), discussed as an example of medieval misogyny or, alternately, read as a caution against that misogyny and its objectification of the feminine.[28] In all of these views, diverse as they are, Blodeuwedd is a symbol of a broken relationship with the land, sterility, and dynastic failure—a botanic destroyer of futurity. She is even a singularity in medieval literature itself, with no known mythological origins and appearing nowhere else in the medieval corpus.[29]

But there are other plants in "Math" from which we might discern a kind of queer botanical family for the woman who is not, as the text says, "from any race on earth."[30] Before he gets to charming Blodeuwedd from flowers, Gwydion works his magic on other vegetation. Early in the story, he acquires Pryderi's valuable and scarce breeding stock of pigs in return for horses, saddles, and shields that he has conjured from mushrooms. Gwydion later tricks Lleu's estranged mother into giving Lleu his name "skillful hand" by means of a ruse that involves making leather shoes out of dulse and kelp.[31] The mushrooms and seaweed, however, refuse to stay enchanted for very long. Gwydion barely has enough time to get out of Pryderi's way before the shields go back to being mushrooms, and as soon as Lleu becomes "skillful hand," the fine cordovan leather of the shoes reverts to dulse. This seems a familiar fairy tale motif, with Gwydion playing fairy godmother, but these plants are not quite Cinderella's pumpkin.[32] They are all edible but none cultivated, none of them providing seeds to be sown, and they anticipate Blodeuwedd's queer intervention in the hereditary governance of Welsh land. Planned breeding projects (the pigs) and later matrilineal acknowledgements (from Lleu's mother) are fatefully derailed by the undomesticated plants' insistence on their own forms.

Then there's Math himself—a king without an heir who needs to spend his days with his feet, planted like roots, in the "curl" of a virgin's lap.[33] Like Blodeuwedd, Math is often criticized for failing to ensure succession. Helen Fulton says of him, "Math rules by aura rather than action. . . . [He] is a parody of the ancient myth of Celtic kingship, surrounded by magic and taboos, rousing himself to lead an army but ultimately ineffectual and barren."[34] Immobile, asexual, and possessed of inscrutable auratic powers, Math is Blodeuwedd's vegetable predecessor, King Mushroom. Under Math's reign, reproductive heterosexuality is pushed toward the grotesque; the punishment he devises for his nephew Gilfaethwy's rape of the virgin Goewin is to transform Gilfaethwy and his accomplice Gwydion into a mating pair of animals (first deer, then boar, then wolves) and keep their animal/human offspring as his own prized stock.

The specter of uncooperative, unworkable flora in "Math, Son of Mathonwy" raises the question of what it is that plants "feel" and/or "want." Our recent thinking about this question relies almost invariably on the familiar thesis offered by evolutionary genetics: what any species wants is to reproduce itself as much as possible.[35] Even Pollan's *The Botany of Desire*, an innovative consideration of human history from plants' perspectives, takes this focus on reproductive futurity as axiomatic.[36] "Math son of Mathonwy," circulated in a time when the oaks did not grow, when crops failed, and when some familiar species may not have frequented their usual haunts, gives us a different answer to the question, and it is certainly not colonization by reproduction. As rulers, both Math and Blodeuwedd are in good position to conceive powerful heirs, but they don't.

If Blodeuwedd is any indication, what plants—or at least plants in human guise—want is not a reproductive future but a present saturated with feeling. They want experience unrestrained by skin. Medieval scholastic writers, while thinking of skin as sensitive, also understood skin's function to be containing and protecting a fleshly body from an "*excess* of sensibles.*" They thought that "without the covering of skin, the sensitivity of nerve-filled flesh would make life unbearable."[37] When Blodeuwedd first sees and desires the hunter Gronw, he is in the process of skinning a deer. By gazing at Gronw, she takes in the possibility of unprotected, nerve-filled flesh. Does she envy skinless sensitivity? Covet our nerves? The surface of verdant botanica was often described by medieval encyclopedists in familiarly human terms as filled with veins, and covered with hair or pustules.[38] But the plants in Math complicate the analogy. Flowers, mushrooms, dulse, and kelp, none of them green, are either all skin or no skin—their surface and depths undifferentiated. What they seem to embody is a continuity of feeling—thorough penetration. And, indeed, the men in Blodewedd's life do get ripped open. When Lleu is speared by Gronw, he transforms into an eagle whose rotting flesh—poetically described by Gwydion as "Lleu's flowers"—continually drops away from him.[39] Gronw, in turn, is not even protected by a skin of stone when the restored Lleu spears him in retaliation. Feeling alongside, through, or as plants requires, once again, curling through affective layers, flipping the position of inside and outside, into a flayed sensorium.

"love wrappeth us and wyndeth us,
halseth us and all becloseth us"

Like "Vaster Than Empires," the *Mabinogi* is a secular work, one with only the most minimal expressions of piety. But both its representation of radically transformed matter and its contemplation of futurity's undoing have profound intersections with medieval Christian practice and thought. The centrality of the Eucharistic rite—as well as the sacrament's vulnerability to heretical interpretation—was the constant subject of both learned and popular theology. The sacramental transmutation of body into bread opened a line of speculation ready to be extended by fictional explorations of metamorphosis and hybridity.[40] In the interpretation of scripture, the text of Revelation, portraying the end of future time, provided medieval artists and writers with more material to elaborate into scenes of multimorphic mergings and splittings.[41] Apocalyptic poetry drawing on Revelation, and often featuring climate-changing weather, was especially prominent in fourteenth-century Wales, cropping up in the Red Book itself.[42]

Rather than elaborate on the theology of transubstantiation or apocalypse, however, I turn this time to another iconographic staple of medieval religiosity: one that puts a flayed human/nonhuman body on view. More than any other scriptural scene, the Passion of Christ was instrumental to the inculcation of devotional affect in the Middle Ages. As such, it was a medium for often quite extreme experiments in emotional/ somatic feeling. Those meditating on the Crucifixion were exhorted to feel as though they themselves were on the Cross or at Christ's side, even to imagine themselves entering the wound in his body.[43] A unique Crucifix image, produced very near the time the *Mabinogi* was inscribed into the Red Book, suggests an intriguing connection between this affective extension and feeling floral.

The only known rendering of a "Lily Crucifixion" in manuscript illumination occupies one of the first folia of the Llanbeblig Book of Hours, a late fourteenth-century Welsh codex likely to have been made for a female patron.[44] The motif is a rare one, found mostly in architectural media (as wall painting, carving, or stained glass) in Britain, dating from c. 1375–1500. Rarer still is the Llanbelig Hours' enfolding of the Lily Crucifixion into an Annunciation scene, where the angel Gabriel visits the Virgin Mary, herself commonly symbolized by the lily flower.[45] The lily, Mary's vegetal avatar, was sometimes thought to reproduce without seed, providing the model for a divine conception that circumvented the inheritance of original sin.[46] In the Hours' double-page layout, the crucified body of Christ stretches out on what appears to be a potted houseplant, the flowering lily, between the angelic and human figures (Figure 1).

Figure 1. Annunciation with Lily Crucifixion. Llanbeblig Hours (Aberystwyth, NLW MS 17520A), fol. 1v and 2r c. 1390–1400. Reproduced by permission of Llyfrgell Genedlaethol Cymru / The National Library of Wales.

The Llanbeblig Annunciation/Passion merging enables not only a temporal loop but a striking conflation of matter as well. The dying Christ coexists with the just-conceived Christ, Mary is both human recipient of Annunciation and botanic witness to the Crucifixion, and Gabriel, holding a palm frond and covered with feathers/foliage, appears a dazzling amalgam of animal, vegetable, human, and angel. The coloration of the frond and the angel's wing visually merge the two into a foliate curve: greenery supports and runs through these divine bodies. Further, botano-morphic sprouts and curls begin inside the framed panels and rupture their way into the margins, a nonnarrative, atemporal space that is touched also by Gabriel's foot, Christ's hands, and the flower of the lily.

The imagistic merging of categories of matter reflects both the material and the purpose of the book. The manuscript is animal skin, stretched, scraped, and overlaid with vegetable and mineral pigments. Showing a scene of a woman reading, it—like other Annunciation illustrations in books of hours—encourages its reader to imaginatively occupy Mary's place, using the book as a tactile device for affective identification. Mary's book, the color of parchment, visually continues the unpainted surface of the page, reinforcing the merge. Since medieval viewers of the crucifix were encouraged to feel with Christ on the Cross, to share, that is, his simultaneous pain and joy, the combination of Annunciation and Passion suggests that handling the book is a route to feeling as, with, and through the nonhuman divine in extremis. In this context, the illustration may also be read as a map of sense-worlds. The human body of Mary is confined both within the frame and inside the even-more limiting space of her seat. Her small hands contrast with Gabriel's elongated, feather-like, petal-like fingers, his tactile extravagance. But, in her lily form, Mary can extend into alternative sensory realm in which God's hands and angel's feet operate—in a medium of bare skin shot through with curling vines.

Is this the kingdom of "vegetable love"? It is certainly unhinged from "time's wingèd chariot." Sacred time loops backward and forward in this scene and through this book-artifact; if we see the Llanbeblig lily/cross as also recalling the "Tree of Life," as one scholar has suggested, even further temporalities are enfolded—the time of Resurrection, the time of Revelation.[47] Medieval affective piety encouraged its practitioners to experience their bodily selves as present at the Crucifixion. Would the reader of this book of hours then feel herself split into different but simultaneous presents? If so, this way of feeling floral, like that in "Vaster" and "Math," entails the destruction of futurity as we know it.

The Kingdom of Plants is a queer place, at least in these three works. By placing them together, I do not mean to draw a line of textual descent in which an ancient discourse is the root and modernity the flower. Better they be held side-by-side-by-side or allowed to curl through one another, enabling uncanny identifications. This way, the feathered, floral Gabriel can touch the floral, feathered Blodeuwedd, the owl who remains a "blossom-faced" creature. Botano-Osden, the "peeled turnip,"[48] can find that his cross is the Lily reading beside him, and know that all is/will be/has been well. These are blasphemous mergings, but we should not forget that the holy families of the West are queer to begin with. Just look at the Llanbeblig illustration: a seedless mother and her Son, who is also her parent, share a domestic intimacy with his foliated familiar. By turning to this religious imagery, I am not advocating either the adoption or abandonment of Christian belief, and I should point out that Le Guin is a Daoist and was at the time she wrote "Vaster." My point is, rather, that the iconographic and discursive traditions of medieval Christianity harbor rich indexes of curling, queer, and uncanny subjectivities. Following these vines of thought—whether using trinitarian texts to theorize an ontology of "the three-in-one," as Jane Bennett (citing Felix Guattari) suggests,[49] or crossbreeding ancient and modern artifacts into a new/old bastardized story about ecology, as I've done here—seems germane to imagining the imbrication of botanic sense-worlds with our own. At the very least, the terror-love of union with the Other described by mystics like Julian can help us feel queer with plants in these apocalyptic times.

Notes

This inquiry into botanic sensation was inspired by Carolyn Dinshaw's work with "Green Man" iconography, for which, see her "Ecology" chapter in *A Handbook of Middle English Studies,* ed. Marion Turner (West Sussex: John Wiley, 2013), 347–62. I also owe thanks to this collection's editors and to my department's Faculty Research Group for their enthusiasm and critical engagement.

1. See, for example, John Trevisa, *On the Properties of Things: John Trevisa's Translation of Bartholomeus Anglicus De Proprietatibus Rerum,* vol. 1 (Oxford: Oxford Clarendon Press, 1975), 95–96.

2. Michael Pollan, "The Intelligent Plant," *New Yorker Magazine,* December 23, 2013, 92–105. Pollan's article profiles biologists' "tart, dismissive" responses to

any work that even touches the notion of plant sentience, though he also notes that some dissenting researchers hope that new work on plant sensation and communication might challenge the privileging of cellular biology in the study of botany. As if on cue, biologists wrote the *New Yorker* to have the last word on the subject.

3. Catriona Mortimer-Sandilands, "Introduction: A Genealogy of Queer Ecologies," in *Queer Ecologies: Sex, Nature, Politics, Desire,* ed. Catriona Mortimer-Sandilands and Bruce Erickson (Bloomington: Indiana University Press, 2010), 27.

4. Bennett's critique of the Christian distinction between spirit and matter is one example. Jane Bennett, *Vibrant Matter: A Political Ecology of Things* (Durham: Duke University Press, 2010), 82–93.

5. Jonathan Dollimore, *Sexual Dissidence: Augustine to Wilde, Freud to Foucault* (Oxford: Oxford University Press, 1991), 103–47.

6. The title to this section is from Ursula K. Le Guin, "Vaster Than Empires and More Slow," in *The Wind's Twelve Quarters* (New York: Harper Perennial, 2004), 197. "To His Coy Mistress" can be found on the American Academy of Poets' website: https://www.poets.org/poetsorg/poem/his-coy-mistress. The title of Le Guin's story is line 12 of the poem; the lines quoted below are 1, 11, and 22.

7. Le Guin, "Vaster Than Empires," 189, 182.

8. Ibid., 196, 205, 213.

9. Ibid., 185.

10. Steven Shaviro, *Discognition* (London: Repeater Books, 2015), 9–11, 16–19.

11. Le Guin, "Vaster Than Empires," 203.

12. Ibid., 201.

13. Ibid., 216.

14. Ibid., 217.

15. *The Showings of Julian of Norwich,* ed. Denise N. Baker (New York: W. W. Norton, 2005), 39. Julian attributes her understanding of the revelation to three interlocking processes: (1) "bodyly" sight, (2) language, and (3) a "goostely" (18) or "inward" (31) sensory process that differs from normal sight and hearing.

16. Le Guin, "Vaster Than Empires," 181.

17. The quote header for this section is from *The Mabinogi and Other Medieval Welsh Tales,* trans. Patrick K. Ford (Berkeley: University of California Press, 2008), 101. For an overview of ecologically oriented science-fiction authors, see Andrew M. Butler, *Solar Flares: Science Fiction in the 1970s* (Liverpool: Liverpool University Press, 2002), 120–35.

18. Norwegian philosopher Arne Naess introduced the concept in a series of talks beginning in 1972. See Alan Drengson, "Some Thought on the Deep Ecology Movement," Foundation for Deep Ecology, 2012, http://www.deepecology.org/deepecology.htm.

19. The 1970s discourse of plant as domestic companion found its way into everything from gardening manuals like Jerry Baker's *Plants Are Like People* (Los Angeles: Nash, 1971—with multiple print runs) to Peter Tompkins and Christopher Bird's best-selling *The Secret Life of Plants* (New York: Harper and Row, 1973) to *Doonesbury* strips featuring lifestyle hippie Zonker Harris and his poetry-reciting houseplants (some of these strips are reprinted in G. B. Trudeau, *Dude: The Big Book of Zonker* [Kansas City, Mo.: Andrews McMeel, 2005], 82–83, 100, 112, 132–34). The name of the surveyor's ship is a word meaning "Baby or Pet." Le Guin, "Vaster Than Empires," 182.

20. On climate in the fourteenth century, see Richard C. Hoffman, "*Homo et Natura, Homo in Natura*: Ecological Perspectives on the European Middle Ages," in *Engaging with Nature: Essays on the Natural World in Medieval and Early Modern Europe*, ed. Barbara J. Hanawalt and Lisa J. Kiser (Notre Dame: University of Notre Dame Press, 2008), 11–38. Examples of plant/human hybrids in the late medieval period include the flower maidens of Alexander romances, "foliate head" sculptures like the roof bosses in Norwich Cathedral, and the legendary "green children" of Woolpit, Suffolk. On these phenomena, see Peggy McCracken, "The Floral and the Human," in *Animal, Mineral, Vegetable: Ethics and Objects*, ed. Jeffrey Jerome Cohen (Washington, D.C.: Oliphaunt, 2012), 65–90; Carolyn Dinshaw, "Ecology," in *A Handbook of Middle English Studies*, ed. Marion Turner (West Sussex: John Wiley, 2013), 347–62; Jeffrey Jerome Cohen, "Green Children from Another World; or, The Archipelago in England," in *Cultural Diversity in the British Middle Ages: Archipelago, Island, England*, ed. Jeffrey Jerome Cohen (New York: Palgrave, 2008), 75–94.

21. On the Green Knight, see, for example, Gillian Rudd, *Greenery: Ecocritical Readings of Late Medieval English Literature* (Manchester: Manchester University Press, 2007), 91–132.

22. The *Mabinogi* tales survive in two fourteenth-century manuscripts, The White Book of Rhydderch (National Library of Wales: Peniarth 4), and the more complete Red Book of Hergest (Oxford: Jesus College 111). While composition of the works has sometimes been hypothesized as twelfth-century, Helen Fulton has made a compelling case for reading the tales in a thirteenth- and fourteenth-century context, when Wales was "politically included in the centralized English monarchy." "The Mabinogi and the Education of Princes in Medieval Wales," in *Medieval Celtic Literature and Society*, ed. Helen Fulton (Dublin: Four Courts Press, 2005), 230–47.

23. Hoffmann, "Homo et Natura," 18.

24. A. D. Carr, *Medieval Wales* (New York: St. Martin's Press, 1995), 99–103.

25. This equation of fertile, productive land with good kingship is voiced by critics who see the *Mabinogi* as preserving older mythologies of sacral kingship but is also upheld by scholars who read the texts in relation to thirteenth- and fourteenth-century Anglo-Welsh politics. See, for example, Fulton's article,

which responds in part to the dubbing of Manawydan as "culture hero" by Andrew Welsh in his article "*Manawydan fab Llyr*: Wales, England and the 'New Man,'" in *Celtic Languages and Celtic Peoples*, ed. Cyril J. Byrne et al. (Halifax: St. Mary's University, 1989), 349–62.

26. Since common broom *(planta genista)* provided the nickname Plantagenet for Henry II's father, Geoffrey of Anjou, and at some later point the surname for the Angevin Kings of England, Blodeuwedd could be read as either a royal bastard and/or a human of mixed ethnicity as well. However, the dating of the surname is uncertain: see John S. Plant, "The Tardy Adoption of the Plantagenet Surname," *Nomina* 30 (2007): 57–84.

27. *The Mabinogi*, 108.

28. See, for example, Susana Brower, "Magical Goods, Orphaned Exchanges, Punishment, and Power in the Fourth Branch of the *Mabinogi*," in *Welsh Mythology and Folklore in Popular Culture*, ed. Audrey L. Becker and Kristin Noone (Jefferson, N.C.: McFarland, 2011), 81–90; Joseph Falaky Nagy, "Are Myths Inside the Text or Outside the Box?" in *Writing Down the Myths*, ed. Joseph Falaky Nagy (Turnhout, Belgium: Brepols, 2013), 15–16; Alfred K. Siewers, *Strange Beauty: Eco-critical Approaches to Early Medieval Landscape* (New York: Palgrave, 2009), 58–65.

29. Jessica Hemming, "Ancient Tradition or Authorial Invention? The "Mythological" Names in the Four Branches," in *Myth in Celtic Literature: CSANA Yearbook 6*, ed. Joseph Falaky Nagy (Dublin: Four Courts, 2007), 83–104 (103).

30. *The Mabinogi*, 102.

31. Ibid., 101.

32. Mushrooms are generally placed in the category of "plants" in medieval reference works. While Trevisa's catalogue of plants has no entry for fungi, Pliny designates truffles and other mushrooms as plants that grow without root or seed. *Naturalis Historia*, ed. Karl F. T. Mayhoff, 19.11–14, available online at http://www.perseus.tufts.edu. *Musheron* is an entry in some versions of the glossary of botanic pharmaceuticals known as the *Alphita* (such as that in the British Library, MS Sloane 284 [c. 1400]).

33. Sarah Sheehan notes, "The noun *croth*, often translated as 'lap,' has the base meaning 'womb,' 'uterus,' or 'belly.' Also, the word that tends to be translated as 'in the fold of,' *ymlyc*, is a compound of *yn* (in) + *plyc* (fold, curve), which can mean either 'in a fold or curve' or simply 'within.'" "Matrilineal Subjects: Ambiguity, Bodies, and Metamorphosis in the Fourth Branch of the *Mabinogi*," *Signs* 34, no. 2 (2009): 319–42 (322).

34. Fulton, "The Mabinogi," 245.

35. A claim made most influential by the publication of Richard Dawkins's *The Selfish Gene* in 1976 (Oxford: Oxford University Press, 1976).

36. Michael Pollan, *The Botany of Desire: A Plant's-Eye View of the World* (New York: Random House, 2001).

37. Katie L. Walter, "The Form of the Formless: Medieval Taxonomies of Skin, Flesh, and the Human," in *Reading Skin in Medieval Literature and Culture*, ed. Katie L. Walter (New York: Palgrave MacMillan, 2013), 119–140 (122).

38. See, for example, Pliny the Elder's description of the surfaces of the oak tree in his *Natural History* 16.10.

39. *The Mabinogi*, 107.

40. The header to this section is from Julian of Norwich, 9. On metamorphosis in relation to Eucharistic piety, see Caroline Walker Bynum, *Metamorphosis and Identity* (New York: Zone, 2001).

41. A well-known, spectacular example is the depiction of the Locking of Hell in the Winchester Psalter (British Library Cotton MS Nero C iv), fol. 39r (c. 1220–29), viewable online: http://www.bl.uk/collection-items/illustration-of -the-damned-swallowed-by-a-hellmouth-from-the-winchester-psalter.

42. See Pierre-Yves Lambert, "Visions of the Other World and Afterlife in Welsh and Breton Tradition," in *Apocalyptic and Eschatological Heritage: The Middle East and Celtic Realms*, ed. Martin McNamara (Dublin: Four Courts Press, 2003), 98–120. One such Doomsday poem includes the line, "Wind will smelt trees (into a dew)" (110).

43. On the textual instruments of affective experimentation in medieval Britain, see Lara Farina, *Erotic Discourse and Early English Religious Writing* (New York: Palgrave MacMillan, 2006); Sarah McNamer, *Affective Meditation and the Invention of Medieval Compassion* (Philadelphia: University of Pennsylvania Press, 2009); Paul Megna, "Better Living through Dread: Medieval Ascetics, Modern Philosophers, and the Long History of Existential Anxiety," *PMLA* 130, no. 5 (2015): 1285–1301. For a recent discussion of the iconography of the wound in relation to an "ecology of the inhuman," see Anne Harris, "Hewn," in *Inhuman Nature*, ed. Jeffrey Jerome Cohen (Washington, D.C.: Oliphaunt Books, 2014), 17–38.

44. A calendar in the manuscript contains an obituary for Isabella Godynogh (d. 1413), a likely owner. National Library of Wales website: https://www.llgc.org .uk/en/discover/digital-gallery/manuscripts/the-middle-ages/the-llanbeblig -book-of-hours.

45. E. J. M. Duggan, "Notes Concerning the 'Lily Crucifixion' in the Llanbeblig Hours," *National Library of Wales Journal* 27, no. 1 (1991): 39–48 (42, 40). Duggan points to the common medieval belief that the Annunciation and Crucifixion both occurred on March 23 as the likely reason for their iconographic combination (40).

46. Trevisa notes the belief that the lily lacks the "vertu seminal" in its seed, though it is unclear whether he concurs (981). Mary's association with the lily has been variously explained: as a gloss on the "lily of the valleys" in The Song of Songs (2.1–2), as a result of the symbolic equation of white and gold with purity and grace, and as a result of its medicinal uses for healing. On Marian devotion

in Wales, see Jane Cartwright, *Feminine Sanctity and Spirituality in Medieval Wales* (Cardiff: University of Wales Press, 2008), 8–66.

47. Sarah Jane Boss, *Empress and Handmaid: On Nature and Gender in the Cult of the Virgin Mary* (London: Cassell, 2000). On trees and Christian mysticism, see Alfred Kentigern Siewers, "Trees," in Cohen, *Inhuman Nature*.

48. Le Guin, "Vaster Than Empires," 188.

49. Bennett, *Vibrant Matter*, 114, discussing Guattari's *The Three Ecologies*, trans. Ian Pindar and Paul Sutton (London: Continuum, 2008).

Hope

TERESA SHEWRY

A poet calls for rain, hail, and floodwater to have a future, to "laugh again."[1] The disturbing laughter of this work—Hone Tuwhare's "Haiku (1)," first published in 1970—affirms the potential power and exuberance of marginalized water, but it also signals tensions that are bound up in hoping for a future involving floodwater.[2] The life jettisoned in floods—silt-choked grass; a drowned sheep, bent against a fence post— bears the marks of the extractive economy that settler farmers interwove with water in Aotearoa New Zealand, a landmass inundated by some 560 billion cubic meters of rain and snow every year and that desires "rain's opium," in the words of poet David Eggleton.[3] Governmental discourses, from natural hazards to floodplain management, emphasize containment of these heavy waters. The image of rain, hail, and floodwater that laughs in "Haiku (1)," in contrast, affirms the liveliness of water. But it also casts such a future as unstable and ambivalent in its meanings, given that flooding, rain, and hail have many dimensions.

In this essay I argue that to hope is to engage in a complex communal life with multiple, contradictory implications at once. A future involving floodwater could include the endangered life of floodplains, such as lowland podocarp forest that relies on periodic inundation to persist but that has largely been eliminated from floodplains to make way for farms. But such a future implicitly devalues and could eventually shape the destruction of other life, subsumed in the potential floodwaters. Indeed, Tuwhare's call to rain, for certain readerships, may call for grieving, because rain and lament are linked in many Māori archives.[4] In calling for water to

455

laugh, Tuwhare creates imaginative space in which to grasp and navigate these tensions. If to hope is to veer, calling up countercurrents against a torrent of violence, by laughing, or by seeking to make others to laugh, we create an unstable context that can disrupt and spotlight habituated, naturalized social logics, turning toward other possibilities and shaping the conditions required to hope. Simultaneously, in laughing, we cast our own engagements as ambiguous in their agency, as beyond the grasp of full understanding and intentionality. More than navigating environmental violence that we cannot accept, in hoping and laughing we express creative relationships with water. We also give expression to stresses, tensions, and incomprehensions that mark the survival of these relationships in curtailing conditions of environmental upheaval.

Upside Down

To hope, I must turn toward troubled waters. So suggests Tuwhare in "Haiku (1)," addressing a "snivelling" creek and calling for rain, hail, and floodwater to "laugh."[5] A Ngā Puhi poet born in Kokewai, New Zealand, in 1922, Tuwhare published thirteen poetry collections during his lifetime and wrote many poetic dialogues with water, directly addressing rivers, rain, and the ocean and emphasizing their expressive capacities. Far from offering a means of drifting away from this world, his call for a future in which water laughs is activated and propelled by his interaction with a distressed creek. Remembering a past involving water's laughter and subsequent devastation also shapes Tuwhare's effort to call back water. To hope is to be unaccountable to linear temporalities in which the past has no importance or agency in relation to current injustices and potential futures. As Matthias Fritsch writes, drawing on Walter Benjamin, we might consider "the relationship between the promise of a liberated future and the memory of the violence that explains the need for such a promise."[6] There is no reason to hope if we take this world as complete or perfect. To hope expresses anger, disappointment, and loss.

Tuwhare's distressed creek evokes the collision of empire, rain, and floods in nineteenth-century Aotearoa New Zealand. Early explorers and settlers comment extensively on swamps, rivers, and, especially, rain. The archipelago's apparent excesses of water preoccupy colonial scientists. In 1869 the scientific journal *The Transactions and Proceedings of the Royal Society of New Zealand* runs several articles on the benefits and methods of swamp draining, with J. C. Crawford affirming, "I can speak from experience when I say that nothing pays better in Great Britain than judicious drainage of land."[7] In an 1871 article in the same journal, engineer A. D. Dobson writes of rivers that are exerting emotional and economic havoc by rapidly widening their banks, cutting new channels, and repeatedly flooding: "The Wairau . . . has always been a source of great anxiety, danger, and loss, to the inhabitants of the plains near Blenheim."[8] He describes settlers as agents in flooding in felling trees in catchment areas and establishing European-style farmland along rivers. After forest is clear-cut, the land can no longer soak up so much water.

Through the years of draining, stopbanking, felling, burning, sheep and dairy farming, and sewage dumping, these waters have been massively changed. David Young writes of the Whanganui River and its people, collectively Te Ātihaunui-a-Pāpārangi, over more than one hundred and fifty years, tracing creeping land alienation and land confiscations, British military buildup at the river and the devastating violence of nineteenth-century wars, and the destruction of catchment and riparian forest.[9] In the late 1800s, a government-supported riverboat industry sought to allow boats passage by removing Māori eel weirs; clearing and blasting rapids, driftwood, stones, and shoals; and altering banks and channels. After World War II, the Whanganui (Wanganui) city council transformed the river's overflow wetlands into a rubbish dump. Blood and offal from a freezing works (slaughterhouse) as well as raw sewage were offloaded in the river, which today remains heavily silted in the context of deforestation, farming, and a hydroelectricity scheme. By the 1950s, writes Young, people would describe Whanganui, not so long ago a city intersected by a green and stony river, as a place of disorienting upheaval "where the river flows upside down."[10]

How do imperatives to hope act in histories of dispossession and violence? It is helpful to understand hoping as encouraged or discouraged in social processes but also as shaping these processes.[11] Riverine upheavals, in particular, have been justified by wild and violent claims on futures, from fortunes to racial and national superiority. Colonial-era poetry suggests how promising futures are a point of orientation and silencing within landscapes of massive and contested upheaval. In Sir Charles Christopher Bowen's 1861 poem "Moonlight in New Zealand," our narrator camps beneath shadowy mountains and beside a torrent that fills the

night with a "sullen roar" as it moves through "ghastly white" shingle.[12]
The river is a rare point of life here, albeit one with an uncooperative and
angry disposition. At the end of the poem, Bowen situates this place in
aspirational terms. By night, the land is "one vast cemetery" inhabited by
"ghosts of Maori warriors."[13] But "to-morrow's dawn will glow / To gild
the cradle of a mighty race."[14] Bowen figures his present world—volatile
and haunting in its violence, loss, and displacement, as well as in heterogeneous presences—as taking readers toward a good future. In positing
this future as prescient, indeed as already scripted, Bowen makes invisible how conflicted his current world is. One would not know it here, but
"Moonlight in New Zealand" was published in the thick of nineteenth-
century wars between specific Māori groups and government forces,
wars that had been flaring since the 1840s and would continue into the
next century. Bowen positions his imagined future simply as reflecting the realities of his world, as making sense given what is unfolding.
We might rather understand his future as having a part in shaping the
materiality of his world, in speaking of the futility of struggle and justifying empire.

To hope, I believe, can involve a more volatile and threatening approach to both the present world and the future. In hope, we establish,
feel, and express a relationship between things of this world (and through
this world, the claims of the past) and the future in terms of openness
and potential, loosening the hold of imagined futures said to be inevitable already. This is an embodied, unassuming, wordless relationship that
is shaped by and at the same time nourishes ideas, or what one hopes
for: rain, hail, and floodwater, perhaps.[15] Establishing a relationship of
openness between present and future creates space not only for varied
imaginaries of what might come but also for what we do now to enrich
the future. In "Haiku (1)," Tuwhare's repeated use of imperative verbs
(*stop, come, rain, hail, flood,* and *laugh*) charges a present-world action
(the poet's call to water) with potential (how the water might respond).
The poem embeds a communal life with water by directly addressing
that water. In this action, the river can be experienced in terms of openness in where it might go. Tuwhare projects laughter into the unknowns
of that future.

Different relationships with water do not come easily in this world
and to imaginaries of what might come. To hope is not to assume a future

but to relate to it in terms of potential, even danger: "The category of danger is always within it," writes Ernst Bloch. "If it could not be disappointed, it would not be hope."[16] Danger emerges in hoping for a future that may fail to emerge and in refusing to acclimatize to the conditions of the present world. Expanded capitalist extractive industries form particularly powerful claims on water's current realities and futures. These futures are not inevitable. They need infrastructures and actions. They are backed up by military and police interventions against people who are in the way. In 2011, for example, the New Zealand government sent navy, police, and air force vessels to stop the iwi (tribe) Te Whānau-ā-Apanui from protecting its marine territory, the Raukūmara Basin, from *Orient Explorer,* a ship contracted by Brazilian energy company Petrobras to carry out a seismic survey for deep sea oil and natural gas. The following year Petrobras withdrew from the Raukūmara Basin, and in 2013, under urgency, the government amended the Crown Minerals Act to allow for "non-interference zones" that physically shield offshore mining ships and structures from people.[17]

In speaking directly to water in "Haiku (1)," Tuwhare undertakes a creative engagement that cannot be defined only in terms of the dominant trajectories of expanded extractive industries. Often outside the imagination of mainstream media and sometimes in tension with environmentalist discourses, many social movements and initiatives, too, struggle for different possibilities in the lives they share with water. They could be missed, in the Whanganui's stones and steep blue-grey papa (mudstone) cliffs: holes, drilled by river people across many generations using poles to propel canoes upstream. Canoeists still use them, although many are placed too high now that the river runs low, its water diverted for a hydroelectricity scheme. Ducks and kingfishers have taken them up as homes, but the remembrance of this customary relationship with the river persists, part in people's struggles over more than one hundred and fifty years to resist the grip of an oppressive social life on the river.[18] The people of Whanganui, writes David Young, are "as tenacious and robust as the kiekie, the kānuka, rātā, and pukatea spreading back over the abandoned river farms."[19] He traces their assertions of sovereignty, authority, and special ties to the river from the 1830s onward in land occupations, forcible evictions of settlers, new confederations and spiritual movements, river blockades, and legal claims, as well as boycotts of the legal system. In 2014 the Crown and Whanganui Iwi signed a deed of settlement that acknowledges Te Awa Tupua—"an indivisible and living whole" encompassing all physical and metaphysical dimensions of the river from mountains and tributaries to sea—as a legal person with all attendant rights, powers, duties, and liabilities.[20] The settlement recognizes the inseparable connection and responsibilities between Whanganui River iwi (tribes), hapū (sub tribes), and Te Awa Tupua.[21] But the struggles of the people of the Whanganui River have also involved refusals of the legal system, and remain open-ended and heterogeneous, currently flaring over water diversion for hydroelectricity and sediment runoff from farms.

Tuwhare's "Haiku (1)" turns me toward such different communal ties to water in currents of injury and upheaval. But the narrator's imperatives for water to laugh remind me that to hope involves embodiment. It involves extension across what others do and reflects a particularly intense engagement with certain things (rain, for example) but inseparably is physically, socially, and ecologically situated.[22] Expressing a connection of embodiment and communal life, in hoping we could never define all the experiences on which we rely. To hope for rain, for example, may address people in very different ways, given that rain can have disparate material implications shaped by deeply unequal social and economic structures, such as housing patterns and transportation infrastructures that curtail people's capacity to move out of floodwater's way. In calling for a deluge where floodwater, rain, and hail laugh in "Haiku (1)," Tuwhare's approach is ironic, evoking an alluringly exuberant but unstable, even disturbing, future.

Laugh

Laugh. To laugh can respond to humor, but it can also involve a visceral expression of shock, anger, and distress. It can, in turn, be shocking, disrupting a social hierarchy or a story otherwise habituated as solemn and natural. By laughing, we actively share life with others, perhaps strengthening intimacy with them in activating shared experiences but perhaps isolating and diminishing them. The politics and implications of laughing are elusive.

Angelique Haugerud connects laughing and hoping in an anthropology of satirical activism around inequality, wealth, and democracy in the

United States. She argues that activists make space for different conversations by using irony and satire to destabilize and reframe official, corporate, and news media framings of political debates. Their work "shines a spotlight on power's fault lines and contingencies."[23] By wielding placards such as "Corporations Are People Too" and "Widen the Income Gap," activists cast economic inequality as a site of interventions involving power and politics rather than as a natural development.[24] Emphasizing such contingency, satire invites a sense of openness and potential and so works toward "inspiring hope and laughter as well as anger."[25] But it does not involve a politics of certain plans and outcomes. Haugerud argues that satirical activism takes agency as unstable and even its own understandings as limited, breaking with comic narrative structures that offer comforting resolutions.

I read Tuwhare's call for water to laugh in "Haiku (1)" as destabilizing in these terms, first as disrupting established narratives about water and making space for a different, imagined future that includes rather than controls and denigrates water's exuberance, but second as casting that future as unstable and ambivalent in its meanings, given that rain, hail, and flooding have complex implications. Tuwhare's imaginary of water sees potential liveliness in rivers that are currently marginalized. It cuts against a colonial New Zealand literary imagination that takes floodwater as angry, as suggested in the "sullen" disposition of the torrent of Bowen's "Moonlight in New Zealand."[26] A direct address to others—a person, rain, a river, the ocean—is a recurring motif of Tuwhare's poetry. Reading the poem "Child Coming Home in the Rain from the Store," Cassie Ringland-Stewart suggests that Tuwhare links communication with other beings, here a stone, to "rejuvenation and vitality in the natural world."[27] Talking with other beings, she suggests, endows them with personal life and responsiveness, makes space for their voices, and affirms shared familial connections. Speaking with a stream encourages floodwater's laughter, and insofar as this laughter calls to mind well-being, it reminds us that floodwater and survival are particularly tightly interwoven for some beings. Matai and tōtara, for example, inhabit areas shaped by periodic flooding. Widespread in floodplain podocarp forest before European settlement, these endemic trees were heavily felled for timber and to open space for agriculture. Both trees can grow new root systems after being surrounded by silt, while tōtara also regenerates after being carried far down rivers: "The long-term survival of matai/totara forests clearly depends on the periodic deposition of fresh alluvium and the availability of these sites for seedling establishment. However, such sites are in high demand for pastoral farming."[28] Tuwhare calls for a future that can be inhabited by beings like matai and tōtara that are currently endangered by upheavals to floodplains.

But if Tuwhare's imaginary of rain, hail, and floodwater speaks to long-standing struggles for alternatives to river-related oppression, the ambivalent meanings of laughing suggest rain's implications are nevertheless unstable and complex. Storms and floods are currently devastatingly ensnared and active in their connections with cutover forests, infilled wetlands, and the retreat of glaciers, among other largely unchecked processes. Facing the contradictory tendencies of flooding can be overwhelming and could shape the abandonment of commitments to the overflow dynamics of rivers and the lifeforms that rely on these dynamics. But it also makes me think of radical, creative points of pressure where floodwaters and storms can move through terrain differently, in part due to social initiatives to claim vital relationships with rivers, revitalize catchment and riparian vegetation that modulates the movement of water, and build alternatives to capitalist agriculture on and around rivers. "To understand the importance of pressure," writes Nikhil Anand, "is to recognize that water is accessed by enabling both physical and social relations, and water supply can be curtailed as much by politics as by topography."[29] Water interacts with pressure, moving not only in relation to rain, snow-melt, or vegetation, but also politicians, capital, engineers, and the ordinary activities of people who tap into pipelines or turn on taps.

Exerting pressure on water to veer, or to move differently, can open up new possibilities while bringing, or being appropriated into, new forms of marginalization and dispossession. Such marginalization is explored in a poem published forty years after "Haiku (1)," "Warming" (2010), where Eggleton calls for the movement of water from ocean to atmosphere where it could fall onto drought-stricken lands and associates this call with extinction.[30] A New Zealand poet of European, Tongan, and Rotuman (Fijian) heritage, Eggleton published "Warming" in *Time of the Icebergs*, a collection that broadly reflects on cultures of climate change, especially in New Zealand. In "Warming," a narrator reflects from an atmospheric vantage point on a moment with little social traction around climate change, on cars and wharf cranes where "time is reluctant to turn the page." At sea, the sun hits the "foam-white, / dissolved wings of a billion butterflies," an image of both sea-foam and of massive loss of life, evoking the far-reaching implications of ocean and atmospheric change. "Pick up that foam," Eggleton writes, "pick it up and drape it / across the dry riverbeds of the skies."[31] He calls for someone, or something, to bring foam to the parched atmosphere and so, presumably, to land in the form of rain. The image of foam as "dissolved wings of a billion butterflies" evokes the fragility and iridescent colors of sea-foam hit by the sun. But it also links an intervention in parched conditions—an effort to pressure water to move to a drought-stricken context—to the dissolution of butterflies. The image turns us toward immense destruction that has already unfolded in connection with global climate change. It also brings to mind the social and environmental injustices interwoven with climate change mitigation, particularly in the forms of nuclear energy, hydroelectricity, and biofuels. Insofar as to hope is to feel potential in the relationship between certain things of this world and the future (a nuclear power plant's or oil palm plantation's potential to reduce reliance on coal and to shape a more livable world, at least for some), it may make it more difficult to acknowledge the multivalent implications, even the violence, that can emerge in these things. Eggleton's call for sea-foam, as dissolved butterfly wings, to be picked up and carried into a parched atmosphere, is a call for an intervention shaped by remembering the loss of human and nonhuman life interwoven with both climate change and efforts to address it. The poem calls for an engagement that is shaped by the dead, the dissolved butterflies. Remembering is a claim on the importance of taking responsibility for how efforts to address climate change not only combat but unfold in power relationships with disparate implications, including violence and loss. Some such memories have shaped struggles to delink climate justice from capitalist enterprise, in particular.[32]

But even given such critical and creative engagements, the fall of rain, the waters blasting through stopbanks and liquefying earth, the break of drought could never be good realities or futures from every perspective. To hope, I must struggle against the long-standing cultural heritage that takes this activity simply as *good*. Despite common critiques of the willfulness, passivity, and, sometimes, the sheer and utter incomprehensibility

and un-relatability of hoping, I still find it hard to shake off the idea that
to hope is to sparkle with potential. But to hope does not confirm the
moral and political rectitude of our actions and imaginaries. It is vital
to resist hoping as if that could define every meaning of reality and its
futures. Agency is uncertain, multivalent, interconnected, and subject to
limitations in understanding. To hope I must veer away from optimism,
if the latter implies confidence in achieving a certain goal, assumes the
rectitude of that goal, and insists on a future that we could control. To
draw from Bloch, "Hope is the opposite of security. It is the opposite of
naive optimism."[33] In calling for water to return to a drought-stricken
or drained environment, we may be doing more (and perhaps more
damage) than what we can—or are willing to—understand ourselves to
be doing. Such unpredictability, Sylvain Perdigon suggests, speaks also
to the precarious relevance of hoping because the failure of actions to
achieve planned ends may prove to be otherwise than simply a disaster.
He writes of a discrepancy that opened up between the understandings
that Palestinian refugee Abu Saeed invested in a car and the eventual
ends of that car. Abu Saeed operated the car as a taxi without a driver's
license and despite being unable to own a taxi license as a Palestinian in
Lebanon at that time. He persisted in hoping, in an intimate, embodied
experience of moving into an unknown future, despite also living with
deeply affecting images of the future as "an unfinished certainty," a future
where uncertainty lay only with when his car would be seized, he would
be arrested, and his wife and child would face the catastrophe of his possible imprisonment.[34] But life was veering on a different line of creativity, unbeknown to Abu Saeed, for the car eventually caught fire through
a fuel leak on the carburetor, and Abu Saeed later narrated the car's loss
as helping him to avoid prison. Perdigon describes Abu Saeed's narrative
as "an affirmation, of the unpredictable nature of things, developing
before our tired eyes, and of the relevance of an affect of hope."[35] He identifies abandonment in hoping in conditions where one is dispossessed of
agency in what is happening and what will happen, conditions that affect
us all but that are much exacerbated in certain social contexts.

   In establishing a relationship between hoping and laughing, Tuwhare's
"Haiku (1)" helps me think through the navigation of a commitment to
alternative possibilities for water while being alive to the tensions and
unknowns in this commitment. Tuwhare's imperative to laugh addresses

water but may also speak to the reader. As I have suggested, making someone (or something) laugh can create the conditions to hope by disrupting otherwise naturalized social dynamics. In laughing, Tuwhare's potential floodwater interrupts a habituated idea that such water should be controlled or even eliminated, making imaginable different possibilities for living with water. But laughing, or attempting to make others laugh, also destabilizes such commitments by turning humorous and wary attention onto their parameters and implications, as suggested in the response: Are you joking? In laughing, we raise questions about our own ability to understand, to bring the change we hope for, and to find closure in our struggles. As Haugerud writes, "Activists who embrace irony remain passionately committed to change—even as they implicitly acknowledge the limitations of human consciousness and of language itself in representing the world."[36] Tuwhare's suggestion that future floodwater might laugh is mischievous, hinting that the narrator may be undermined in seeking a "good" ending and, indeed, may achieve something different from what was anticipated in calling to water. Laughing can involve serious commitments, but it also bears unique potential to destabilize and make space for further imagination and creativity in these commitments.

By some accounts, such a concern to destabilize pertains to those who hope but not to scholars because we simply do not (or should not) hope. Michael Taussig comments that Western intellectual work perhaps "correlates lack of hope with being smart, or lack of hope with profundity."[37] A marginal collective of scholars, in contrast, consider their work as involved in hoping and not simply as reflecting a world beyond where others hope, whether as failing or strength. Their experimentation in traction on scholarly work includes Elizabeth A. Povinelli's recent critique of critical theory as hoping for new possibilities of life in relation to beings and objects that endure in material, embodied, and lethal spaces of intensified potentiality.[38]

To consider ecotheorists among those who hope can bring out of the shadows those moments, archives, and methods through which we work to foreclose or open up a sense of possibility, and the question of what we are hoping to exert or allow through such work. Hoping (and laughing) turns me toward my engagement with Tuwhare's poem, "Haiku (1)," in particular. This poem is inscribed in the pavement of Queen Street, a major commercial street in Auckland, above an ancient stream, Waiho-rotiu. The stream runs in a bricked sewer beneath the city toward the sea through an area that was once a swamp. While many people simply walk onward, others stop and orient downward, disrupting the steady flow to somewhere else. The poem reminds of the stream's presence in the city and of its marginalization in a colonial history. "Haiku (1)" has since been taken up in another public space in Auckland, carved on artist Selwyn Muru's *Te Waharoa o Aotea,* the gateway to Aotea Square, where it is more directly linked to Māori histories and presence in the city.

Hope. To read a poem, inscribed on pavement, as evoking the persistence of water in Auckland, but far beyond that, as an imaginative experiment in animating a relationship with this water based on personal address to change the reality and possibilities of this stream that first came to run in a sewer in the trajectories of colonialism. To explore a broader range of communal relationships with and conceptualizations of rivers than are allowed for in universalizing narratives that address only environmental degradation and abandonment. *Laugh.* To turn in unease on what I elide in hoping that the imagination of a river's presence could in any simple way support substantially different lives with it, in this city of deep inequality where living with rivers interacts with, for example, the practice in capitalism of working multiple jobs, of living without a home. To loosen the grip of ideas and feelings that such inequalities are natural and inevitable and to suggest the importance of linking rivers with social and economic justice struggles. To hope, laughing, throws into view disturbing tensions and shapes possibilities in our engagements with water amid the gale forces of injustice.

Notes

1. Hone Tuwhare, "Haiku (1)," in *Small Holes in the Silence: Collected Works* (Auckland: Random House, 2011), 97.

2. Concern to engage speaking beings such as rivers and floodwater also shapes hoping in science studies. For example, Bruno Latour argues for the creation of collectives in which nonhumans, in connection with scientists, participate in discussions. Bruno Latour, *Politics of Nature: How to Bring the Sciences into Democracy* (Cambridge, Mass.: Harvard University Press, 2004).

3. Dave Hansford, "Liquidation," *New Zealand Geographic,* January/February 2014, n.p.; David Eggleton, "Aotearoa Considered as a Scale Model (for Hone

Tuwhare, 1922–2008)," in *Time of the Icebergs: Poems by David Eggleton* (Dunedin, New Zealand: Otago University Press, 2010), 63.

4. Basil Keane, "Tāwhirimātea—the Weather," in *Māori and the Natural World—Te Taiao*, ed. Jennifer Garlick, Basil Keane, and Tracey Borgfeldt (Auckland: David Bateman, 2010): 33.

5. Tuwhare, "Haiku (1)," 97.

6. Matthias Fritsch, *The Promise of Memory: History and Politics in Marx, Benjamin, and Derrida* (Albany: State University of New York Press, 2005), 11.

7. J. C. Crawford, "On Thorough Drainage," in *Transactions and Proceedings of the Royal Society of New Zealand* 2 (1869): 212.

8. A. D. Dobson, "On the Destruction of Land by Shingle-Bearing Rivers, and Suggestions for Protection and Prevention," *Transactions and Proceedings of the Royal Society of New Zealand* 4 (1871): 153.

9. David Young, *Woven by Water: Histories from the Whanganui River* (Wellington, New Zealand: Huia, 1998).

10. Ibid., 2.

11. Focusing particularly on aspirations of national security and economic advancement in Australia, Mary Zournazi argues that "we must . . . explore hope through the societies we live in: the alienations that affect us in individual and collective ways; the grief, despair and loneliness; and the new social, ethnic and class relations that come out of these alienations" (16). Mary Zournazi, *Hope: New Philosophies for Change* (Annandale, Australia: Pluto Press, 2002).

12. Sir Charles Christopher Bowen, "Moonlight in New Zealand," in *The New Place: The Poetry of Settlement in New Zealand, 1852–1914*, ed. Harvey McQueen (Wellington, New Zealand: Victoria University Press, 1993), 38.

13. Ibid., 39.

14. Ibid.

15. I am drawing from Elizabeth A. Povinelli: "Affects may be ultimately determined by the given system of ideas that one has, but they are not 'reducible to the ideas one has'". "The Persistence of Hope: Critical Theory and Enduring in Late Liberalism," in *Theory after 'Theory,'* ed. Jane Elliot and Derek Attridge (New York: Routledge, 2011), 105.

16. Ernst Bloch and Theodor W. Adorno, "Something's Missing: A Discussion between Ernst Bloch and Theodor W. Adorno on the Contradictions of Utopian Longing," in *The Utopian Function of Art and Literature: Selected Essays,* trans. Jack Zipes and Frank Mecklenburg (Cambridge, Mass.: MIT Press, 1988), 16–17.

17. Crown Minerals Amendment Act 2013, section 55.

18. Waitangi Tribunal, *The Whanganui River Report* (Wellington, New Zealand: GP, 1999), 76; Young, *Woven,* 247.

19. Young, *Woven,* 13.

20. *Ruruku Whakatupua—Te Mana o Te Awa Tupua*, 2014, 6, https://www.govt.nz/dmsdocument/5947.pdf.

21. *Ruruku Whakatupua—Te Mana o Te Iwi o Whanganui,* 2014, 7, https://www.govt.nz/dmsdocument/5950.pdf.

22. See Sylvain Perdigon's discussion of a Palestinian refugee whose capacity to hope was extended across several family members as well as an old car: "Yet Another Lesson in Pessoptimism—A Short Ethnography of Hope and Despair with One Palestinian Refugee in Lebanon," *Asylon(s)* 5 (2008).

23. Angelique Haugerud, *No Billionaire Left Behind: Satirical Activism in America* (Stanford: Stanford University Press, 2013), 13.

24. Ibid., 14.

25. Ibid., 12.

26. Bowen, "Moonlight," 38.

27. Cassie Ringland-Stewart, "Talking with Hone Tuwhare," *Ka Mate Ka Ora: A New Zealand Journal of Poetry and Poetics* 6 (2008), http://www.nzepc.auckland.ac.nz/kmko/06/ka_mate06_ringland.asp.

28. G. D. McSweeney, "Matai/ Totara Flood Plain Forests in South Westland," *New Zealand Journal of Ecology* 5 (1982): 125.

29. Nikhil Anand, "The PoliTechnics of Water Supply in Mumbai," *Cultural Anthropology* 26, no. 4 (2011): 543.

30. David Eggleton, "Warming," in *Icebergs,* 8.

31. Ibid.

32. See Julianne A. Hazlewood's discussion of struggles for alternatives to engaging climate change through biofuels in "CO_2lonialism and the 'Unintended Consequences' of Commoditizing Climate Change: Geographies of Hope amid a Sea of Oil Palms in the Northwest Ecuadorian Pacific Region," *Journal of Sustainable Forestry* 31, nos. 1–2 (2012): 120–53.

33. Bloch and Adorno, "Something's Missing," 16.

34. Perdigon, "Yet Another Lesson."

35. Ibid.

36. Haugerud, *No Billionaire,* 11.

37. Michael Taussig and Mary Zournazi, "Carnival of the Senses," Zournazi, *Hope,* 44.

38. Povinelli, "Persistence of Hope."

Afterword

On the Veer

—————————▶   ◀—————————

NICHOLAS ROYLE

An afterword, as Jacques Derrida once remarked, "ought never to be a last word. It comes *after* the discourse, that's true, but detached enough from it or wandering away from it enough not to accomplish, finish off, close or conclude."[1] An afterword should, in a word, veer. Heterogeneous to the text or texts that precede it, the afterword entails "an interruption," Derrida suggests, something that "cannot be scarred over": "There's no suture once there's the after-the-event, the *Nachträglichkeit* of an *afterword*."[2] All of this may seem incontestable: the time of an afterword is different, out of joint with what it follows. The afterword is about deferred effect, delayed meaning, afterwardsness *(Nachträglichkeit)*. But what is happening when, as here in these pages, essay after essay speaks of these very effects?

What is named *veer ecology* is in many ways a terrifying enterprise. It has to do with what Timothy Clark calls "Anthropocene disorder," which he sees in terms of a crisis of scale and agency, as "a derangement of linguistic and intellectual proportion, a breakdown of 'decorum' in the strict sense."[3] *Decorum* means "appropriate behavior": it is a question of what is "suitable," "proper," and "becoming." It is a Latinate word, originally deriving from the noun *décor*, meaning *grace*. A breakdown of *decorum* "in the strict sense" could be said to entail not only the breakdown of appropriate behavior (actions, responses, discourse, thinking) in a local context but, more radically, the falling apart of *globalatinization—*

469

of everything that links globalization with Latin (and, above all perhaps, with what is called the "Anglo-American" language), and the very "world order" (capitalist, U.S.-centered, fundamentally Christian) that it, for a time, sustained.[4] Clark goes on to illustrate this breakdown in terms of what he calls "new kinds of affect":

> The environment is now increasingly experienced not just as an object of physical perception but in new kinds of affect: an unusual flood can lead to rage against the crass advertising of a cheap airline; the violence and noise of a newly built local road may induce a sense of nausea at the hypocrisies of systems of politics dominated by the short-term demands of international capitalism. The very glare of the sun takes on fearful qualities: the sight of the sea and the sky is no longer of entities totally untouched by the human. This is also the realm of feeling sick at the sight of a mountain gouged out by mining. It means finding TV programmes that celebrate the natural world increasingly unbearable, for they already feel like re-animations of the dead.[5]

The linguistic, intellectual, and affective dimensions of the "breakdown of 'decorum'" impinge profoundly on the project of veer ecology. There is a poignant (because so imperilled) adherence, in the essays gathered here, to notions of scholarly propriety and what is academically appropriate. But there is also a compelling impression, from one chapter to the next, that the norms and conventions of "academic writing," and even of the value and purpose of any kind of writing, are in rapid transformation. These texts resemble "objects undergoing alteration," they are *scholarly writing in transition,* no longer quite recognizable as examples of the genre of academic essay that predominated in the late twentieth- and into the early twenty-first century. They mix scholarly insight with poetic reflection, drawing on the extraordinary resources of literary texts from the medieval to the contemporary, as well as on science, philosophy, anthropology, film, music, the Internet, and everyday experiences and events. They make unexpected moves, they swerve, slow down, zoom in or out, often to surprising (and at moments, uncanny) effect. They seek to illuminate a single word *(unmoor, whirl, curl, sediment, rain, hope, try, decorate, attune),* while never losing sight of its embeddedness in the pressing urgency of ecological questions and challenges—climate change, global warming, the environment, the Anthropocene, human and nonhuman animals, plant life, and so on. And constantly, also, they articulate and respond to the uncanny phenomenon and feeling of *Nachträglichkeit*, in particular the sense that something irrevocable, for which we struggle to find words, has *already happened*. Or: something irrevocable is happening at a speed that eludes comprehension and even perception. This is the spectral sense of time that Clark evokes when he speaks of "finding TV programmes that celebrate the natural world increasingly unbearable, for they already feel like re-animations of the dead."

At once supplementing and disrupting the serene terms of science, veer ecology makes clear that the challenges, threats, and terrors of climate change or global warming are primarily physical (veering is what is happening, it is *out there*) but are also caught up in language. As a woman in a Margaret Atwood story observes, trying to recall and describe the experience of pregnancy: "I'm stuck with [words], stuck in them."[6] But veer ecology is not static. It reckons with the promise and danger of the unpredictable. It affirms movement. The foregoing essays explore not only how language is changing in response to what has happened and what is happening but also how language can innovate and invent, alter or start differently—to change how people think and feel, and what they do. In these pages there is a deep attention to the poetic character and active possibilities of language, as well as to the actual or potential violence of its forms—in law, in clichés, in blind stereotypes or unthinking idioms, in so many anthropocentric figures of speech and thought. There is also, throughout these pages, a gentleness and respect, an involvement with decorum and dignity, even or perhaps especially when it is not designated as such. Veering is at the heart of being, anxiety, and desire. Moreover, it is not peculiar to humans: other animals veer. Life is veering. And—at once in spite and because of the law, the clichés, and the stereotypes—language is also veering.

If, as *Veer Ecology* indicates, the academic essay is mutating in unprecedented ways, so too perhaps is the thing called *literature*. It is indeed tempting to suggest that the survival of literature will depend on its capacity to voice and elaborate on the kinds of concern evident in the foregoing pages. Don DeLillo's *Zero K* (2016) is a provoking case in point.[7] This novel is about cryonics and the desire to elude death. In other words, it is about what the philosopher John Gray refers to as "the Immortalization Commission."[8] Gray writes:

> During the late nineteenth century and early twentieth century science became the vehicle for an assault on death. The power of knowledge was summoned to free humans of their mortality. Science was used against science and became a channel for magic.
>
> Science had disclosed a world in which humans were no different from other animals in facing final oblivion when they died and eventual extinction as a species. That was the message of Darwinism, not fully accepted even by Darwin himself. For nearly everyone it was an intolerable vision, and since most had given up religion they turned to science for escape from the world that science had revealed.[9]

Gray's study explores this revolt against death, from late nineteenth-century spiritualism and psychical research to contemporary cryonics. His argument is that those who look for "a technical fix for death" disavow or fail to acknowledge that "human institutions are unalterably mortal." In particular, he notes, even in the coming few decades "climate change may alter the conditions in which humans live radically and irreversibly."[10] In corresponding fashion, DeLillo's *Zero K* offers a fictional elaboration of what Gray calls "techno-immortalism,"[11] as well as of the constitutive links between novelistic writing and magical thinking.

The desire to prolong life by whatever possible medico-technological prosthetic means can nonetheless give a different inflection to what is currently happening with the concept of anthropocerism. On the one hand, radical ecological thought is concerned with the ongoing deconstruction of anthropocentrism: veer ecology entails, perhaps first and foremost, what Jeffery Cohen and Lowell Duckert call the *disanthropocentric*, acknowledging that what the environment environs, what veers in it, never was, never is, and never will be merely or primarily the human. On the other hand, *Zero K* suggests that anthropocentrism is being, in a different sense, deconstructed from within: there is a kind of uncanny division emerging out of the obscenely inequitable distribution of wealth currently prevailing across the world, whereby a tiny proportion of the population can fund or buy access to forms of biomedical, technological enhancement (more powerful vision, faster brains, exponentially increasing longevity, etc.) that render them in decisive ways differently human. (Here adjectives such as *posthuman* or *allohuman* are of perhaps limited help or interest.) If such an internal splitting of anthropocentrism is not an explicit focus in the preceding pages, it is nevertheless a specter haunting the Anthropocene.

It would be facile to call *Zero K* a "dystopian" or "futuristic," let alone "sci-fi" novel. A more apposite way of figuring it might be in relation to the remarkable images published in *Nature* in the same month as DeLillo's novel appeared. Described in a research letter titled "Early Neanderthal constructions deep in Bruniquel Cave in southwestern France,"[12] these are photographs taken deep in a cave and far from any daylight, of annular constructions made by Neanderthals from snapped-off stalagmites approximately 175,000 years ago. There is something chilling and inconceivable about these images of subterranean building, and a sense that *Zero K* taps into something akin to this uncanny ancientness. "If the planet remains a self-sustaining environment, how nice for everyone and how bloody unlikely," a character remarks toward the end. And then: "Either way, the subterrane is where the advanced model realizes itself."[13] Up on the earth's surface, by contrast, there is simply the ongoing "loss of autonomy," "the sense of being virtualized."[14]

Like the essays in *Veer Ecology*, *Zero K* is deeply preoccupied with questions of language and environment, with the nature of the human, the eco- and *oikos,* with floods, fires, and destruction, with the strangeness of age and deep time, as well as with the performative and generative, that is to say the radically fictive or poetic possibilities of words, even (as the narrator suggests) "down into sub-atomic levels."[15] DeLillo's novel takes itself, and the form and discourse of literary fiction, into a new kind of underworld or underspace. At its troubled center is the communicating passage to where human bodies lie in suspended animation. There is a struggle for words to describe this. It is "a space that became an abstract thing, a theoretical occurrence. I don't know how else to put it. An idea of motion that was also a change of position or place." Finally the novel offers a name: "It's called the veer."[16] It is, the narrator explains, "an environment . . . suited to rigorous thinking."[17]

Notes

1. Jacques Derrida, "Afterw.rds: or, at Least, Less Than a Letter about a Letter Less," trans. Geoffrey Bennington, in *Afterwords*, ed. Nicholas Royle (Tampere, Finland: Outside Books, 1992), 197.

2. Ibid.

3. Timothy Clark, *Ecocriticism on the Edge: The Anthropocene as a Threshold Concept* (London: Bloomsbury, 2015), 139.

4. For more on globalatinization, see Jacques Derrida, "Faith and Knowledge: The Two Sources of 'Religion' at the Limits of Reason Alone," trans. Sam Weber, in *Religion*, ed. Jacques Derrida and Gianni Vattimo (Cambridge: Polity Press, 1998), esp. section 30, pp. 29–30; and for a rich commentary on the topic, see Michael Naas, *Miracle and Machine: Jacques Derrida and the Two Sources of Religion, Science, and the Media* (New York: Fordham University Press, 2012).

5. Clark, *Ecocriticism*, 139.

6. Margaret Atwood, "Giving Birth," in *Dancing Girls and Other Stories* (London: Virago, 1985), 226.

7. Don DeLillo, *Zero K* (New York: Scribner, 2016).

8. See John Gray, *The Immortalization Commission: Science and the Strange Quest to Cheat Death* (New York: Farrar, Straus, and Giroux, 2011).

9. Ibid., 1.

10. Ibid., 210.

11. See Ibid., 209.

12. See *Nature*, May 25, 2016, at http://www.nature.com/news/neanderthals -built-cave-structures-and-no-one-knows-why-1.19975.

13. DeLillo, *Zero K*, 238.

14. Ibid., 239.

15. Ibid., 141.

16. See Ibid., 138–39.

17. Ibid., 151.

Acknowledgments

Acknowledge is an impossible verb. Generous and generative, acknowledge is an imperative not to be satisfied: after all, we owe the coming into being of this ambitious volume to, well, *everything*. The writers gathered here have been true companions. From crowdsourcing the collection's subtitle to framing its convivial table of contents, our collaborators took up our invitation to *veer* with the world. Choosing their words carefully, they set an environmental story in motion. We thank them for their intellectual adventurousness and their openness to the unpredetermined. Douglas Armato ensured that *Veer Ecology* found a hospitable home at the University of Minnesota Press. We thank him for his vision and his recurring welcome. The external readers chosen by the press made this a better book. Erin Warholm-Wohlenhaus made everything work, down to the smallest detail, guiding the volume's restless capaciousness. Working with the University of Minnesota Press is always a pleasure. Scott Mueller copyedited this manuscript with admirable care, and Douglas Easton prepared its index. A hope for enduring curiosity is what we ultimately wish to extend to you, the reader. The Errata section is an open-ended catalog meant to suggest additional ecological companionships. May some of these words "turn" in unexpected ways.

Lowell Duckert thanks the students of his three (so far) Foundations of Literary Study sections who embraced "veering" as a viable theory of literature; the students of his first Environmental Criticism class, especially Andrew Munn, whose enthusiasm helped establish a permanent course offering at West Virginia University; Catherine Venable Moore;

and the lingering histories uncapped by Brucella and Norman Jordan at the African American Heritage Family Tree Museum. In their third collaborative endeavor, Jeffrey Jerome Cohen continues to exemplify companionship; without his buoyancy, family, and friendship there would be no "Try." And finally, all of his efforts are aided by Erin, the one messmate who will foreve(e)r move him.

Jeffrey Jerome Cohen thanks his students and colleagues at the George Washington University, and especially the Columbian College of Arts and Sciences for a research chair that enabled much of the work in this volume to be completed. He is grateful to Julian Yates for Noachic solidarity and for audiences around the world for their feedback on his climate change project. Lowell Duckert has been a steadfast, inspirational, and never predictable companion (the very best kind): his good heart and his friendship and his puns have been sustaining as well as a spur to impossible things. No book of this kind would have been possible without the love and wanderlust and inspiration of Wendy, Alex, and Katherine. Long may we veer, together.

Errata

(words for wandering not included here)

Imagine	Conserve	Walk
Essay	Pluck	Quake
Exhaust	Compose	Live
Freeze	Conserve	Better
Slide	Acclimate	Thrive
Grade	Commune	Time
Plant	Compete	Bury
Reinhabit	Evolve	Split
Entangle	Warm	Dance
Object	Predate	Mangle
Promise	Restore	Dream
Near	Rewild	House
Hybridize	Hold	Stroke
Crane	Revive	Curate
Impart	Extinguish	Caress
Cultivate	Protest	Winter
Embrace	Intensify	Shelter
Exterminate	Color	End
Storm	Toxify	
Multiply	Face	
Hazard	Free	
Renew	Map	

Contributors

STACY ALAIMO is professor of English and Distinguished Teaching Professor at the University of Texas at Arlington. She is author of *Undomesticated Ground: Recasting Nature as Feminist Space, Bodily Natures: Science, Environment, and the Material Self,* and *Exposed: Environmental Politics and Pleasures in Posthuman Times* (Minnesota, 2016) and editor of *Material Feminisms* and *Matter.*

JOSEPH CAMPANA is a poet, arts writer, and scholar of Renaissance literature. He is author of *The Pain of Reformation: Spenser, Vulnerability, and the Ethics of Masculinity* and three collections of poetry, *The Book of Faces, Natural Selections* (which won the Iowa Poetry Prize), and *The Book of Life.* He coordinates an arts and media cluster, presenting art concerned with energy, for the Center for Energy and Environmental Research in the Human Sciences at Rice University, where he is Alan Dugald McKillop Associate Professor of English.

JEFFREY JEROME COHEN is professor of English and director of the Medieval and Early Modern Studies Institute at George Washington University. His scholarship places posthumanism and ecotheory into conversation with medieval materials. He is author and editor of numerous books, including *Medieval Identity Machines, Of Giants: Sex, Monsters, and the Middle Ages, Monster Theory: Reading Culture, Prismatic Ecology: Ecotheory beyond Green, Elemental Ecocriticism,* and *Stone: An Ecology* *of the Inhuman* (which won the 2017 René Wellek Prize for best book in comparative literature), all from the University of Minnesota Press.

LOWELL DUCKERT is associate professor of English at West Virginia University, specializing in early modern literature and environmental criticism. With Jeffrey Jerome Cohen, he has edited "Ecomaterialism" (*postmedieval* 4, no. 1, 2013) and *Elemental Ecocriticism* (Minnesota, 2015). His work reconceives human and nonhuman relations today by plumbing water worlds of the past. He is author of *For All Waters: Finding Ourselves in Early Modern Wetscapes* (Minnesota, 2017).

HOLLY DUGAN is associate professor of English at George Washington University. Her research and teaching interests explore relationships among history, literature, and material culture. Her scholarship focuses on questions of gender, sexuality, the boundaries of the body, and the role of the senses in late medieval and early modern England. She is author of *The Ephemeral History of Perfume: Scent and Sense in Early Modern England*.

LARA FARINA is associate professor of English at West Virginia University. She is author of *Erotic Discourse and Early English Religious Writing* and has published on medieval and modern reading practices, queer theory, tactility, and "disabled" sensation. She has coedited with Holly Dugan a special issue of the journal *postmedieval* on "The Intimate Senses" and is writing a book about the sense of touch.

CHERYLL GLOTFELTY is professor of literature and environment at the University of Nevada, Reno, and cofounder and past president of the Association for the Study of Literature and Environment. She is coeditor of *The Ecocriticism Reader: Landmarks in Literary Ecology, The Bioregional Imagination: Literature, Ecology, and Place,* and *The Biosphere and the Bioregion: Essential Writings of Peter Berg*.

ANNE F. HARRIS is professor of art history and vice president for academic affairs at DePauw University. A fascination with the material production and experiential reception of multiple media in medieval art guides her work. Research and teaching on manuscripts, ivory, wood, stained glass, and alabaster explore the interactions of materiality, art, and audience, and the tendency of materials to shape human responses both medieval and modern.

TIM INGOLD is chair of social anthropology at the University of Aberdeen. Following ethnographic research in Finnish Lapland, he has worked on human–animal relations, human ecology and evolutionary theory, environmental perception and skilled practice, and, most recently, on the synthesis of anthropology, archaeology, art, and architecture. His recent books include *The Perception of the Environment, Lines, Being Alive, Making, The Life of Lines,* and *Anthropology and/as Education.*

SERENELLA IOVINO is professor of comparative literature at the University of Turin and past president of EASLCE. She has written on environmental ethics and ecocritical theory, bioregionalism and landscape studies, ecofeminism, and posthumanism. Her recent publications include *Material Ecocriticism, Environmental Humanities: Voices from the Anthropocene* (both coedited with Serpil Oppermann), and *Ecocriticism and Italy: Ecology, Resistance, and Liberation,* winner of the Book Prize of the American Association for Italian Studies.

STEPHANIE LEMENAGER is Barbara and Carlisle Moore Professor of English and professor of environmental studies at the University of Oregon. Her publications include *Living Oil: Petroleum Culture in the American Century, Manifest and Other Destinies,* and the forthcoming *Weathering: Toward a Sustainable Humanities,* which treats the role of the humanities in the era of global climate change.

SCOTT MAISANO is associate professor of English at the University of Massachusetts, Boston. His most recent publications include *The Tempest: A Critical Guide.* His research explores the works of Shakespeare, the English Renaissance, literature and science in the sixteenth and seventeenth centuries, and theories and theaters of artificial life.

TOBIAS MENELY is associate professor of English at the University of California, Davis. He is author of *The Animal Claim: Sensibility and the*

Creaturely Voice and coeditor (with Jesse Oak Taylor) of *Anthropocene Reading: Literary History in Geologic Times.*

STEVE MENTZ is professor of English at St. John's University in New York City. His latest book is *Shipwreck Modernity: Ecologies of Globalization, 1550–1719* (Minnesota, 2015). He is author of *At the Bottom of Shakespeare's Ocean* and *Romance for Sale in Early Modern England*, editor of *Oceanic New York*, and coeditor of *Rogues and Early Modern English Culture* and *The Age of Thomas Nashe.*

J. ALLAN MITCHELL is professor of English at the University of Victoria and author of *Ethics and Eventfulness in Middle English Literature* and *Becoming Human: The Matter of the Medieval Child* (Minnesota, 2014). He is coeditor of a special issue on the Critical/Liberal/Arts (with Myra Seaman and Julie Orlemanski, *postmedieval* 6, no. 4, 2015).

TIMOTHY MORTON is Rita Shea Guffey Chair in English at Rice University. He gave the Wellek Lectures in Theory in 2014 and has collaborated with Björk. He is author of *Dark Ecology: For a Logic of Future Coexistence* (forthcoming), *Nothing: Three Inquiries in Buddhism* (forthcoming), *Hyperobjects: Philosophy and Ecology after the End of the World* (Minnesota, 2013), and *Realist Magic: Objects, Ontology, Causality.*

VIN NARDIZZI is associate professor of English at the University of British Columbia. He has published *Wooden Os: Shakespeare's Theatres and England's Trees* and is working on a new book project called "Marvellous Vegetables in the Renaissance." With Stephen Guy-Bray and Will Stockton he coedited *Queer Renaissance Historiography: Backward Gaze* and with Jean E. Feerick *The Indistinct Human in Renaissance Literature.*

LAURA OGDEN is associate professor of anthropology at Dartmouth College. She has conducted ethnographic research in the Florida Everglades and with urban communities in the United States and is working on a long-term project in Tierra del Fuego, Chile. She is author of *Swamplife: People, Gators, and Mangroves Entangled in the Everglades* (Minnesota, 2011).

SERPIL OPPERMANN is professor of English at Hacettepe University, Ankara, and the current president of EASLCE. She is coeditor of *The Future of Ecocriticism: New Horizons, International Perspectives in Feminist Ecocriticism* (with Greta Gaard and Simon Estok), *Material Ecocriticism* (with Serenella Iovino), *Environmental Humanities: Voices from the Anthropocene* (with Serenella Iovino), and editor of *Ekoelestiri: Cevre ve Edebiyat (Ecocriticism: Environment and Literature)* and *New Voices in International Ecocriticism*. Her recent work is focused on material ecocriticism, posthumanism, and the Anthropocene.

DANIEL C. REMEIN is assistant professor of English at the University of Massachusetts, Boston. His work in poetry and criticism moves between medieval poetics (especially Old English) and the poetics of avant-gardes in English since the mid-twentieth century. He is author of *A Treatise on the Marvelous for Prestigious Museums* and is working on a book about *Beowulf* and the Berkeley Renaissance.

MARGARET RONDA is assistant professor of English at the University of California, Davis, where she teaches courses in American poetry and environmental literature and theory. Her articles have appeared in *PMLA, Post-45, ELN, Genre: Literature and Culture,* and *the minnesota review*. Her forthcoming book is titled *Remainders: American Poetry at Nature's End*.

NICHOLAS ROYLE is professor of English at the University of Sussex, England. His books include *Telepathy and Literature, After Derrida, The Uncanny, Jacques Derrida, How to Read Shakespeare, In Memory of Jacques Derrida,* and *Veering: A Theory of Literature*. He is coauthor (with Andrew Bennett) of *This Thing Called Literature* and *An Introduction to Literature, Criticism, and Theory* (5th edition). He has also published two novels, *Quilt* and *An English Guide to Birdwatching*.

CATRIONA (CATE) SANDILANDS is professor of environmental studies at York University, where she teaches and writes at the intersections of the environmental humanities, sexuality and gender studies, and materialist thought and politics (especially about vegetality and human/plant biopolitics). She is a 2016 Pierre Elliott Trudeau Foundation Fellow.

She is author of *The Good-Natured Feminist: Ecofeminism and Democracy* (Minnesota, 1999) and coeditor of *This Elusive Land: Canadian Women and the Environment* and *Queer Ecologies: Sex, Nature, Politics, Desire.* "Vegetate" is part of a larger, in-progress work, *Plantasmagoria: Thinking Vegetal Politics.*

CHRISTOPHER SCHABERG is associate professor of English and environmental studies at Loyola University New Orleans. He is author of *The Textual Life of Airports: Reading the Culture of Flight, The End of Airports,* and *Airportness: The Nature of Flight.*

REBECCA R. SCOTT is associate professor of sociology at the University of Missouri. She is author of *Removing Mountains: Extracting Nature and Identity in the Appalachian Coalfields* (Minnesota, 2010). Much of her work focuses on environmental affects, fossil fuels, and climate change.

TERESA SHEWRY is associate professor of English at the University of California, Santa Barbara. She is author of *Hope at Sea: Possible Ecologies in Oceanic Literature* (Minnesota, 2015) and coeditor of *Environmental Criticism for the Twenty-first Century.*

MICK SMITH is professor and Queen's National Scholar in the School of Environmental Studies at Queen's University. He is author of *An Ethics of Place: Radical Ecology, Postmodernity, and Social Theory* and *Against Ecological Sovereignty: Ethics, Biopolitics, and Saving the Natural World* (Minnesota, 2011).

JESSE OAK TAYLOR is associate professor of English at the University of Washington in Seattle. He is author of *The Sky of Our Manufacture: The London Fog in British Fiction from Dickens to Woolf,* which won the Association for the Study of Literature and Environment (ASLE) book award in ecocriticism, as well as coeditor (with Tobias Menely) of *Anthropocene Reading: Literary History in Geologic Times* and coauthor (with Daniel C. Taylor and Carl E. Taylor) of *Empowerment on an Unstable Planet: From Seeds of Human Energy to a Scale of Global Change.*

BRIAN THILL is professor of English at Golden West College, where he also serves as director of the Writing/Reading Center and the Civic Literacy Project. He is author of *Waste (Object Lessons)*. His writing has been published in the *Atlantic*, the *Guardian*, *Jacobin*, *Mediations*, *3:AM Magazine*, and *the Los Angeles Review of Books*.

COLL THRUSH is professor of history at the University of British Columbia, Vancouver, and affiliated with UBC's Institute for Critical Indigenous Studies. He is author of *Native Seattle: Histories from the Crossing-Over Place*, coeditor with Colleen Boyd of *Phantom Past, Indigenous Presence: Native Ghosts in North American Culture and History*, and author of *Indigenous London: Native Travellers at the Heart of Empire*.

CORD J. WHITAKER is assistant professor of English at Wellesley College. He works on the development of racial ideology and its intimate relationship with medieval rhetorical theory. His writing has been published in the *Journal of English and Germanic Philology*, the *Yearbook of Langland Studies*, and *postmedieval*, where his special issue "Making Race Matter in the Middle Ages" has achieved critical acclaim. He is completing the book *Black Metaphors: Race, Religion, and Rhetoric in the Literature of the Late Middle Ages*. He is the recipient of fellowships and grants from the Andrew W. Mellon Foundation, the Ford Foundation, and the National Endowment for the Humanities.

JULIAN YATES is professor of English and material culture studies at the University of Delaware. He is author of *Error, Misuse, Failure: Object Lessons from the English Renaissance* (Minnesota, 2003), a finalist for the Modern Language Association's Best First Book Prize in 2003; *What's the Worst Thing You Can Do to Shakespeare?* coauthored with Richard Burt; *Object-Oriented Environs in Early Modern England*, coedited with Jeffrey Jerome Cohen; and *Of Sheep, Oranges, and Yeast: A Multispecies Impression* (Minnesota, 2017).

Index

487